SEMICONDUCTORS

SEMICONDUCTORS
BASIC CONCEPTS AND APPLICATIONS

Edita Stefanic

Kruger Brentt
Publishers

2024

Kruger Brentt Publishers UK. LTD.
Company Number 9728962

Regd. Office: 68 St Margarets Road, Edgware, Middlesex HA8 9UU

ISBN: 978-1-78715-271-7

For information on all our publications visit our website at http://krugerbrentt.com/

PREFACE

Semiconductor Devices holds paramount importance in the realm of electronics and technology. This comprehensive guide serves as a cornerstone for individuals seeking to understand the intricate world of semiconductor devices. By delving into the fundamentals of semiconductors, exploring the operation of essential components like PN junctions, diodes, and transistors, and detailing their applications in various electronic circuits, the book bridges the gap between theory and practical implementation. Its significance lies in its ability to empower students, researchers, and professionals with the knowledge and skills needed to design, analyze, and optimize electronic systems. From amplifiers to power amplifiers, from traditional Bipolar Junction Transistors (BJTs) to modern Metal Oxide Semiconductor FETs (MOSFETs) and Insulated Gate Bipolar Transistors (IGBTs), the book covers a spectrum of topics crucial for anyone working in the semiconductor industry. As the electronics landscape continues to evolve, the understanding gleaned from this book becomes increasingly relevant, making it an indispensable resource for those navigating the complexities of semiconductor technology. Whether one is entering the field or seeking to deepen their expertise, "Semiconductor Devices: Theory and Application" stands as a foundational text, fostering a comprehensive understanding that fuels innovation and advancements in electronic systems.

The book is organized into chapters, each focusing on a specific aspect of semiconductor devices. We begin with the "Fundamentals of Semiconductor," establishing a solid foundation in the principles that underlie the behavior of semiconductor materials. The journey continues through the exploration of PN Junctions and Diodes, fundamental components in semiconductor physics, and their applications. Subsequent chapters delve into Bipolar Junction Transistors (BJTs), amplifiers, and power amplifiers. Junction Field Effect Transistors (JFETs) and Metal Oxide Semiconductor FETs (MOSFETs) are thoroughly examined, along with their applications in small signal amplifiers. Additionally, the book covers Class D power amplifiers and the operation of Insulated Gate Bipolar Transistors (IGBTs).The material is presented in a structured and accessible manner, with a focus on clarity and depth of understanding. Each chapter begins with an overview of the topic, followed by theoretical explanations and practical applications. Illustrative examples and diagrams are included to facilitate comprehension, and each chapter concludes with exercises to reinforce key concepts.This book is intended for students and professionals in electrical engineering, electronics, and related fields.

We are grateful to all those persons as well as various books, manuals, periodicals, magazines, journals etc. that helped in the preparation of this book. In spite of the best efforts, it is possible that some errors may have occurred into the compilation and editing of the book. Further queries, constructive suggestions and criticisms for the improvement of the book are always welcome and shall be thankfully acknowledged.

Edita Stefanic

CONTENTS

CHAPTER-1

FUNDAMENTALS OF SEMICONDUCTOR

1.1 INTRODUCTION: AN OVERVIEW

Semiconductors form the bedrock of modern electronics, serving as the fundamental building blocks for various electronic devices. These materials, often derived from silicon, exhibit a unique property of conductivity that lies between that of insulators and conductors. The intricate behavior of semiconductors allows for the controlled flow of electrical current, making them essential in the creation of transistors, diodes, and integrated circuits. The manipulation of semiconductor properties through processes like doping and fabrication techniques has enabled the miniaturization of electronic components, leading to the development of increasingly powerful and compact devices. Understanding the fundamentals of semiconductors involves delving into concepts such as electron-hole pairs, band theory, and semiconductor devices, laying the groundwork for advancements in electronics, telecommunications, and technology as a whole.

First, a Little History

Just as the late eighteenth through nineteenth centuries are known as the industrial age due to the rise of mechanization, the twentieth century can be referred to as the beginning of the electronic age. The first half of the century was dominated by electronic vacuum tubes that made possible devices such as radio, television, radar and long distance telephone. The technology of the vacuum tube was displaced mid-century by the introduction of *solid-state semiconductors*. The first working prototype transistor was invented at Bell Labs in 1947 by John Bardeen, Walter Brattain and William Shockley. This device, properly referred to as a *point contact transistor*, was quickly superseded by the *bipolar junction transistor*, a major topic of this text.

Commercial production of the transistor and related devices improved the performance of existing applications and made possible a range of new ones.

Semiconductors proved to be smaller, lighter, more reliable and less expensive to build than their vacuum tube counterparts. The last 30 or so years of the century saw the rapid expansion of the integrated circuit where numerous transistors are combined in a single device. Initially such a device may have contained the equivalent of a dozen or so individual semiconductor devices, but today that number has grown to the billions1. This extreme density has given rise to now common-place applications such as cell phones, GPS devices, laptop computers, tablets and our global communications infrastructure.

The science writer, Arthur C. Clarke, once observed that "Any sufficiently advanced technology is indistinguishable from magic". Indeed, although today the typical citizen living in an industrialized country makes use of numerous electronic devices each day (sometimes without even being aware of it), they typically have scant knowledge of how these devices "work their magic". Obviously there is no magic, only the application of scientific principles mixed with human ingenuity. Further, just as it is true that many more people can use a cell phone than design one, it is also true that there is a greater need for people who can design, manufacture and maintain devices based on semiconductors than for people who design the semiconductors themselves. The scope of this text, then, focuses on the operation and application of semiconductor devices rather than the design of the semiconductors themselves.

Variable Naming Convention

One item that often confuses beginning students of almost any subject is nomenclature. Before we begin our discussion of semiconductor devices it is important that we decide upon a consistent naming convention. Throughout this text we will be examining numerous circuits containing several passive and active components. We will be interested in a variety of parameters and signals. In order to keep confusion to a minimum we will use the following conventions in our equations for naming devices and signals.

R	Resistor (DC, or actual circuit component)
r	Resistor (AC equivalent, where phase is 0 or ignored)
C	Capacitor
L	Inductor
Q	Transistor (Bipolar or FET)
D	Diode
V	Voltage (DC)
v	Voltage (AC)
I	Current (DC)
i	Current (AC)

Resistors, capacitors and inductors are differentiated via a subscript that usually refers to the active device to which it is connected. For example, RE is a DC bias resistor connected to the emitter of a transistor while rC refers to the AC equivalent resistance seen at a transistor's collector. CE refers to a capacitor connected to a transistor's emitter lead. Note that the device related subscripts are always shown in upper case, with one exception: If the resistance or capacitance is part of the device model, the subscript will be shown in lower case to distinguish it from the external circuit components. For example, the AC dynamic resistance of a diode would be called rd. If no active devices are present or if several items exist in the circuit, a simple numbering scheme is used, such as R1. In very complex circuits a specific name will be given to particularly important components, as in Rsource.

Voltages are normally given a two-letter subscript indicating the nodes at which it is measured. VXY is the DC potential from node X to node Y while vXY indicates the AC signal appearing across node X to node Y. A single-letter subscript, as in VX, indicates a potential relative to ground (in this case from node X to ground). The exceptions to this rule are power supplies, that are given a double letter subscript indicating the connection point (VCC is the collector power supply), and particularly important potentials that are directly named, as in vin (AC input voltage) and VR2 (DC voltage appearing across R2). If an equation for a specific potential is valid for both the AC and DC equivalent circuits, the uppercase form is preferred (this makes things more consistent with circuits that are directly coupled, and thus can amplify both AC and DC signals). Currents are named in a similar way but generally use a single subscript referring to the measurement node (IX is the DC current flowing through a conductor into or out of node X). All other items are directly named. By using this scheme, you will always be able to determine whether the item expressed in an equation is a DC or AC equivalent, its approximate circuit location, and other factors about it.

1.2 ATOMIC STRUCTURE

Atomic structure is a foundational concept in the field of chemistry and physics, elucidating the arrangement of particles within an atom. At the core of an atom lies a nucleus containing protons and neutrons, surrounded by orbiting electrons. The number of protons in the nucleus determines an element's identity, defining its chemical properties on the periodic table. Electrons, which carry a negative charge, occupy distinct energy levels or orbitals around the nucleus.

In our effort to understand the operation of semiconductors, a fundamental question we might ask is "What is the internal structure of an atom?" Please understand that it is nonsensical to ask what an atom might "look like" because its components are all smaller than the shortest wavelengths of light that humans can see. Instead, we simply need a model to explain its observed behavior.

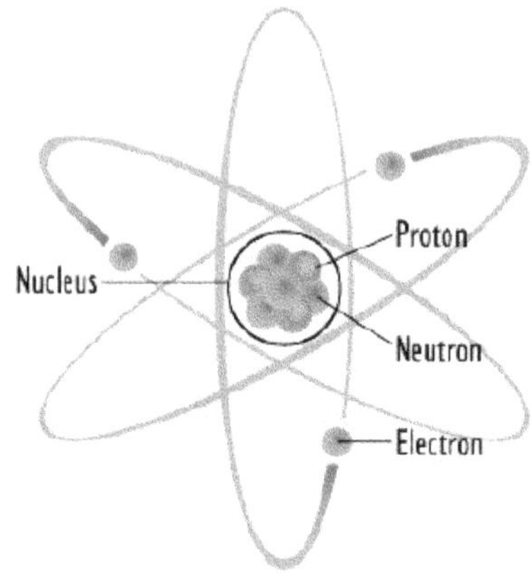

Figure 1.1 Planetary atomic model: Pretty, well-known and wrong. Image source (modified)

Perhaps the most prolific model in the popular imagination is the *planetary model* shown in Figure 1.1. In this model, the core, or *nucleus*, is drawn at the center and contains positively charged protons and non-charged neutrons. Revolving around this core are negatively charged electrons, each following a nice, regular, planar path much like a planet around the sun. Unfortunately for us, this model is starkly incorrect, although it has found use as a symbol for nuclear regulatory agencies and a DEVO album cover from the 1970s.

Before we come up with a more accurate and useful model, let's take a closer look at the sub-components; namely the proton, neutron and electron. First off, most of the mass of any given atom is from the protons and neutrons. Protons and neutrons have similar masses, about 1.67E–24 grams each. The mass of an electron is roughly 2000 times smaller. The radius of a proton is approximately 0.87E–15 meters and the mean distance to the nearest electron is about 5.3E–11 meters. This means that this electron is about 60,000 times farther away from the proton than the size of said proton. To put this into perspective, that's roughly the same as the ratio between a golf ball and a sphere with a radius of 3/4ths of a mile or 1200 meters. This would be the case for a hydrogen atom as it consists of a single proton and electron. The magnitude of this ratio is not much different for other substances, including things like crystalline carbon (diamond) and quartz (a molecule of silicon and oxygen) that are very hard and solid. If you think about that for a moment, you realize that the idea of "solidity" is in some ways an illusion because the vast majority of what we call "something" is really just empty space. For example, chances are that you are sitting down while reading this. You probably feel your buttocks pressed against the chair. Both of these things are considered solid yet at the atomic level the vast majority of both items is nothingness. In reality, the feeling of solidity is just the result of the interaction of atomic forces between the two. So if someone suggests that you might have a bit too much to spare in the *department of the posterior*, you can inform them that it's really nothing.

One of the major issues with the planetary model is the idea that electrons whirl around the nucleus in stable, planet-like orbits. That's simply not true. First, the electron inhabits a region of 3D space, it does not simply move through a plane. Second, due to the Heisenberg Uncertainty Principle, we can't precisely plot the position and trajectory of a given electron. The best we can do is make a plot of where the electron is *likely to be*. This is called a *probability contour*. Imagine that you could record the position of an electron relative to the nucleus. A moment later you record its new position, a moment after that you record the next position, and on and on for thousands of measurements. If you attempted to plot them all, you would wind up with a cloud of dots around the nucleus. This cloud is referred to as an *orbital*. You wouldn't know how the electron got from one position to the next but you would get a general idea of where it was likely to be. Do not confuse *orbital* with *orbit* (like a planetary orbit). They are two different beasties.

There are several potential orbitals. Due to quantum physics, only certain orbitals are allowed. The permissible electron energy levels are first grouped into shells, then subshells and finally orbitals. It is important to remember that orbitals indicate the electron energy level. That is, a higher orbital implies a higher energy level. Further, orbitals fill in first from lowest energy level to highest energy level. These are important ideas that we will leverage in future discussions.

Shells are denoted by their principal quantum number, n; 1, 2, 3, etc. The higher the number, the more subshells it can contain. Subshells are organized by their orbital shape and are designated by letters, the first four being s, p, d, and f. Shell 1 contains only subshell s while shell 2 contains subshell types s and p. Shell 3 contains subshell types s, p and d, and so on.

Thus, we see designations such as *1s*, *2s* and *2p*. These subshells may also have variations within them. There is one variation on *s*, three variations on *p*, five variations on *d*, etc. These variations are the orbitals and each orbital can hold a maximum of two electrons.

Putting this all together, we find that the first shell can contain a maximum of two electrons: two in the single *s* subshell orbital (*1s*). The second shell can contain a maximum of eight electrons: two in the *s* subshell (*2s*) plus two in each of the three *p* subshell orbitals (*2p*). In like manner the third shell can contain a maximum of 18 electrons: two in *3s*, six in *3p* and two in each of the five *d* subshell orbitals (*3d*). You can condense this into a simple formula, $2n^2$, where n is the shell number.

Figure1.2 Electron probability contour for innermost orbital, 1s. Image source

Figure 1.2 shows the electron probability contour of the innermost orbital, namely *1s* (i.e., principle quantum number *1*, subshell *s*). As you can see, it is spherical in shape. The nucleus is located at the center, obscured here. All *s* orbitals are similarly spherically shaped although the internals change. *1s* is the lowest energy orbital.

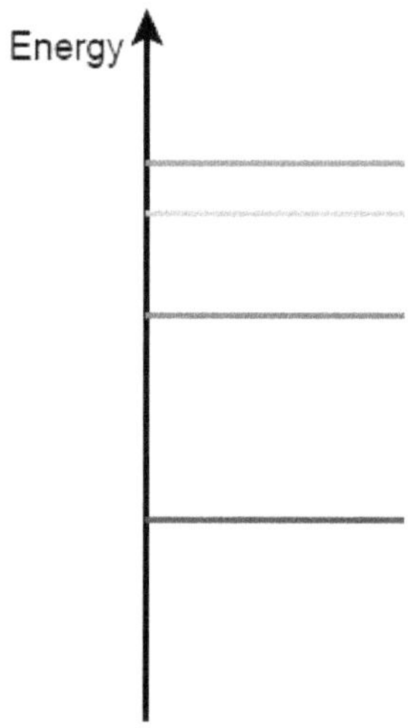

Figure 1.3 Electron probability contour for orbital 2p. Image source

Orbitals are not limited to simple spherical shapes. Higher order orbitals can take on a variety of forms. Figure 1.3 shows the electron probability contour for the *2p* orbitals (recall there are three *p* variations, one each oriented along the X, Y and Z axes). The nucleus is situated in the small void between the two lobes. Obviously, this is nothing like the well-behaved elliptical orbits of planets around the sun.

Probability contours can be very complex. For the highest orbitals, especially when combined with the lower orbitals, the contour combinations can become reminiscent of the sculptures of a deranged clown forming herds of imaginary balloon animals.

As interesting as these graphics are, they are cumbersome to work with. Consequently, a more functional graphic is called for. Such a device is the Bohr model, named after Danish physicist Niels Bohr. An example is shown in Figure 1.4.

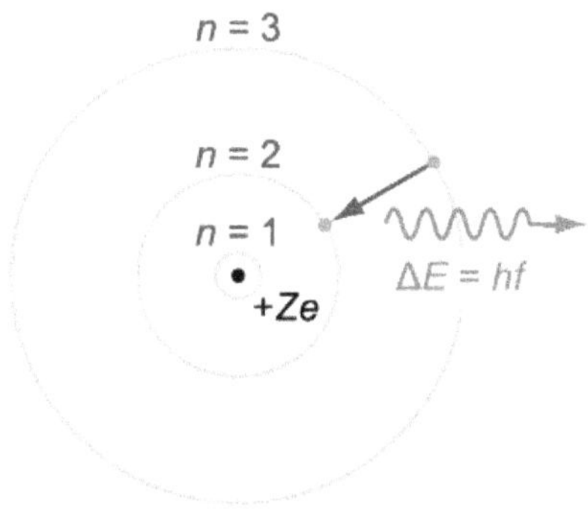

Figure 1.4 Generic Bohr model.

It is important to understand that the Bohr model is an energy description of the atom, not an attempt to mimic its physical appearance or structure. The nucleus is placed at the center. It is surrounded by concentric rings that represent the electron shells. The higher the number, the larger the ring and the greater the energy level. If an electron were to move from a higher level to a lower level, the energy difference is radiated out. This could be in the form of heat or light. This is a point worth remembering. For example, this transition is what makes light emitting diodes (LEDs) function. The inverse is also possible, namely that by absorbing energy, an electron can move into a higher orbital. This is an equally powerful concept, as we shall soon see.

Using the Bohr model we can create diagrams to represent individual elements. For example, copper has an atomic number of 29 meaning that it has 29 protons and 29 electrons. The electron shell configuration is 2-8-18-1. That is, the first three shells are completely filled and there is a single electron in the fourth shell. This single outer electron is only loosely bound and thus makes copper a very good conductor. The Bohr model for copper would simply show four rings, the first three being filled and with a single electron in the fourth ring.

Figure 1.5 shows the Bohr model of an atom of Silicon, atomic number 14, with an electron shell configuration of 2-8-4. In this version, the individual electrons are drawn in each shell and the atomic number is indicated at the nucleus. Again, please do not imagine this representing individual electrons orbiting the nucleus in lanes. This is an energy level depiction.

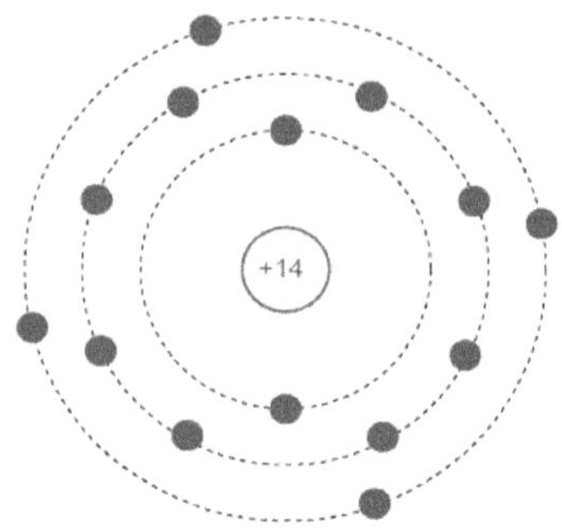

Figure 1.5 Bohr model of Silicon.

Often, it is useful to simplify this model further by omitting the filled inner shells. Also, the atomic number is replaced by the number of electrons in the outermost, or *valence*, shell. This is shown in Figure 1.6. The valence shell is particularly important as it gives insight into the general behavior of the material.

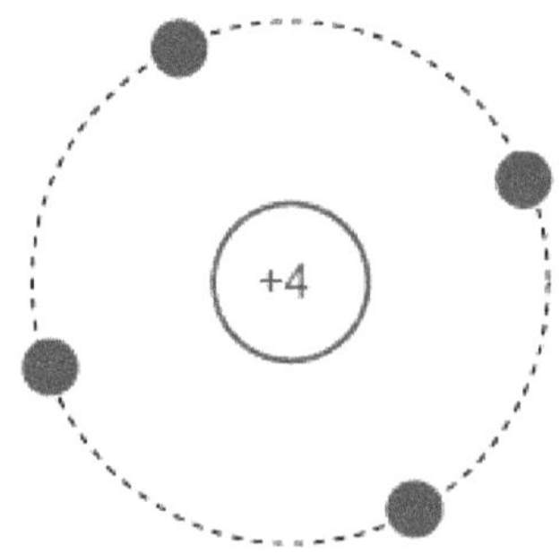

Figure 1.6 Simplified Bohr model of Silicon.

As an alternative, sometimes we will "straighten out" the Bohr model so that it simply shows the energy levels graphically as lines or bands, and without counting specific electrons. This is depicted in Figure 1.7.

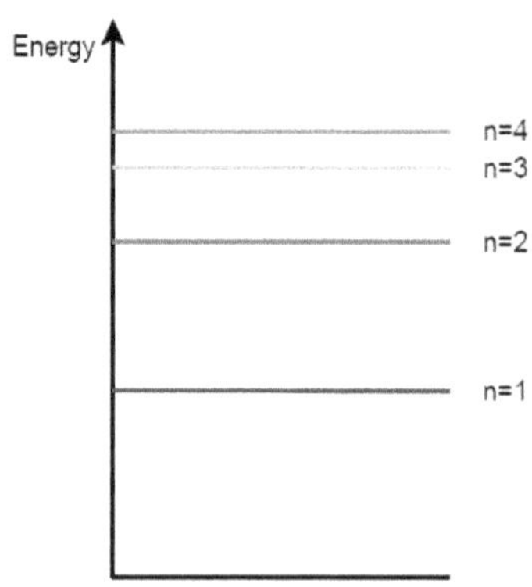

Figure 1.7 Energy level diagram.

1.3 CRYSTALS

We used silicon in the preceding example on purpose. The fact that it has a half- filled valence shell with four electrons puts it in a special place. As is, it's neither a great conductor nor a superior insulator. With some attention to detail, it will become a *semiconductor*. Silicon is not the only material that can be used for semiconductors. In fact, many of the earliest semiconductors were made from germanium and currently we make semiconductors from other materials. Silicon, however, remains the source of most semiconductors today.

It is possible for pure silicon to be arranged in a mono crystalline structure. That is, all of the silicon atoms align in a very specific, well-ordered manner, without any voids or breaks in the pattern. As silicon has only four electrons in its valence shell, four more electrons would be needed to obtain stability (i.e., eight electrons in the outer shell). In the crystal, any given atom of silicon effectively "shares" an electron from its four closest neighbors through a *covalent* bond (meaning "with or among the valence"). Each atom does this, therefore each atom is tightly bound to its neighbors. This is illustrated in Figure 1.8 using simplified Bohr models. Note the color coding that indicates the sharing.

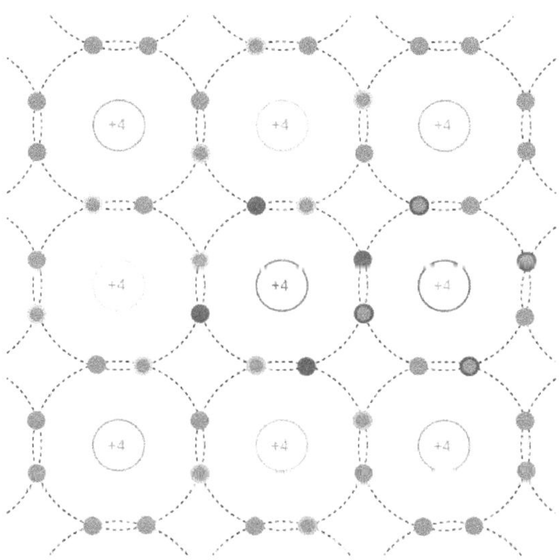

Figure 1.8 Silicon crystal covalent bonding.

Remember, this is an energy diagram. We are not trying to indicate that individual valence electrons are zipping between atoms in figure eight patterns. Indeed, this diagram is drawn flat whereas a real crystal is not a simple sheet, but is three dimensional with varying thickness.

While it would be a practical impossibility to draw a highly realistic representation of atoms in the crystal, given a few liberties we can draw something that at least comes a little closer to reality. We start by representing each silicon atom as a ball and the covalent bond as a connecting tube. Recalling that each atom must be bound to four others in a regular, equal pattern, we come up with the drawing of Figure 1.9. Notice that the overall structure is essentially that of a cube. Further, at the center of each face of the cube there exists an atom of silicon. Therefore, we say that the crystal structure is face centered cubic.

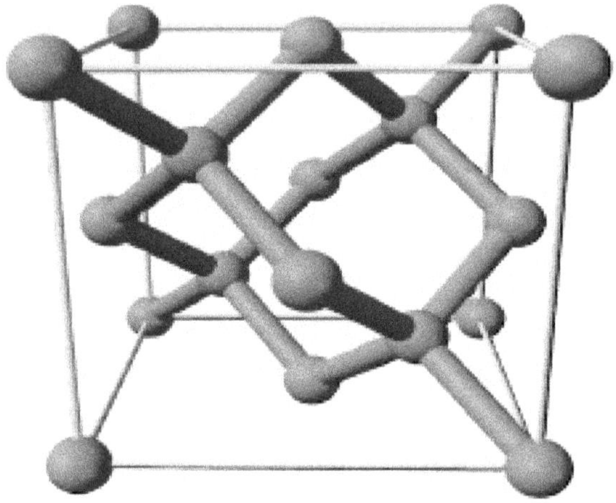

Figure 1.9 Representation of silicon crystal structure.

An interesting thing happens in the crystal when we examine the energy levels. With a single atom we would expect to see an energy diagram like that of Figure 1.7. That is, discrete, permissible steps. Within a crystal, though, each atom is affected by those around it. This causes slight changes in the energy levels. Taken as a whole, all of these individual variations cause the discrete levels to blur into broader bands. If we were to examine the valence and conduction energy levels, instead of discrete, thin lines we'd see the thicker bands as illustrated in Figure 1.10. These bands still represent permissible electron energy levels, it's just that now there is a continuum rather than a discrete level. There will still be non-permissible or forbidden zones between these regions. A forbidden zone is referred to as a band gap.

Associated with this idea is the concept of the *Fermi level*, named after physicist Enrico Fermi. Basically, the Fermi level is the energy level in a given material at which there is a 50% probability that it is filled with electrons. In other words, levels below this value tend to be filled with electrons and levels above tend to be empty. If the Fermi level lies within a band, the material will be good a conductor. On the other hand, if the Fermi level lies between two widely separated bands, the material will be a good insulator. If the Fermi level is between bands that are relatively close, the material is a semiconductor.

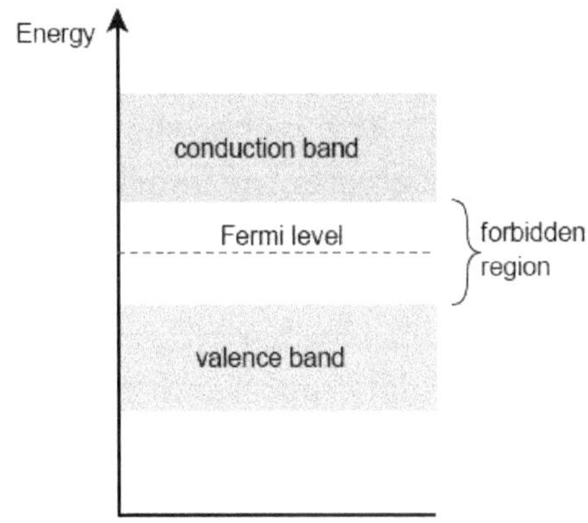

Figure 1.10 Energy diagram for an intrinsic semiconductor.

Figure 1.10 shows the energy bands for an *intrinsic semiconductor*, such as an ideal silicon crystal. The term *intrinsic* simply means that there are no impurities in the crystal. Between the valence band and conduction band is an impermissible or forbidden region. This is a band gap. In practical terms you can think of the band gap as the amount of energy that needs to be applied to an electron in order to move it from the valence band to the conduction band. The value of the band gap will depend on a variety of factors, the precise material being used for the semiconductor is of particular importance.

Without any external energy applied (i.e., isolated and at absolute zero), the crystal lattice is stable and there is no electron movement through the crystal. As we add thermal energy, it is possible for valence electrons to jump up to the *conduction band*. At this point, the electron can "wander" through the crystal in the manner depicted in Figure 1.11.

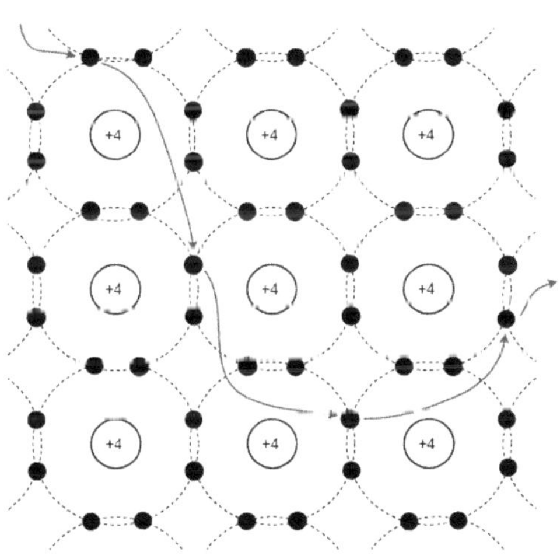

Figure 1.11 Electron movement in a crystal.

Here is how this happens: Because the thermal energy causes the electron to jump to the higher energy level of the conduction band, it leaves behind a "hole", that is, a place devoid of an electron. Now that the hole exists, it provides a place for another electron to "fall into". The higher the temperature, the greater the number of freed electrons and the greater the number of corresponding holes. We now have thermally-induced electron movement. We can also look at this from the opposing perspective, namely that we have an equal magnitude but opposite direction "hole flow". If you find this idea hard to grasp, simply look at Figure 1.12. Each horizontal bar contains four dots representing electrons. In the topmost bar there is an empty space (a hole) to the extreme left. When the leftmost electron moves into this hole it fills it in a process called *electron-hole recombination*, which of course, sounds much more impressive than it really is. The result is the second bar. We repeat this process of moving an electron right to left as we traverse down the diagram.

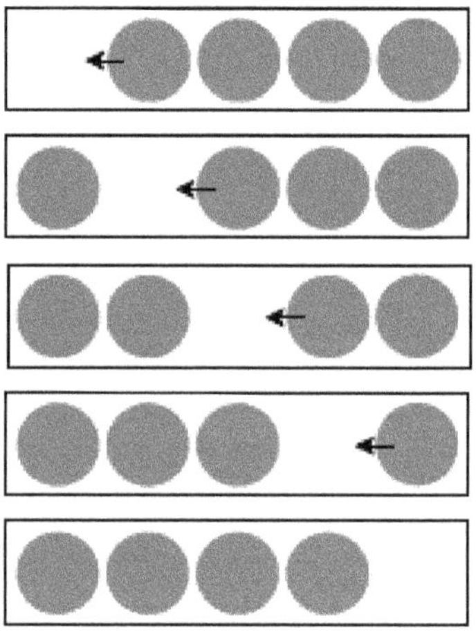

Figure 1.12 Electron versus hole flow.

Eventually we end up with the four electrons packed together toward the left. Finally, instead of focusing on the dots, focus instead on the negative space (the empty white bit). Moving from top to bottom, the hole moves left to right, in the opposing direction.

Just as we think of the movement of electrons as a movement of negative charge, then the movement of holes can be thought of as a movement of positive charge. We can say that the electron is the carrier of negative charge while the hole is the carrier of positive charge.

Before moving on to the next section, it is important to remember that in an intrinsic (pure) semiconductor, the number of thermally produced electrons and holes will be equal. Also, even at room temperature the total number will also be quite small compared to the number of electrons in the crystal.

1.4 DOPED MATERIALS

Doped materials, a crucial aspect of semiconductor technology, involve the intentional introduction of impurities into a semiconductor crystal lattice to modify its electrical and optical properties. This process, known as doping, plays a pivotal role in tailoring semiconductors for specific applications. Two primary types of dopants are utilized: donor dopants, such as phosphorus or arsenic, which introduce extra electrons, and acceptor dopants, like boron or gallium, which create "holes" or electron deficiencies. By strategically incorporating these dopants, semiconductor devices can be engineered with desired conductivity characteristics, allowing for the precise control of electrical currents. This manipulation of charge carriers forms the basis for the creation of transistors, diodes, and other semiconductor components essential to electronic devices and integrated circuits. Doped materials showcase the versatility of semiconductors, enabling the customization of their properties to meet the diverse requirements of modern technology.

By themselves, intrinsic semiconductors are not of particular use. They are neither good conductors nor insulators, and their conduction is largely dependent on temperature. We can alter the properties of the material by introducing foreign substances or impurities into the crystal. These impurities are also known as *dopants*. A crystal with an added dopant is referred to as an *extrinsic* semiconductor or *doped* material. The amount of impurity added is generally small, perhaps in the neighborhood of one part per million. The dopant may be added through a gaseous diffusion process where the crystal is heated in an oven and the dopant added in gaseous form. Over a period of time the impurities will diffuse or "seep into" the target crystal. An alternate approach is ion implantation. In this method the impurities are accelerated and quite literally smash into the target, dislodging and replacing some of the original atoms in the crystal.

There are two different types of semiconductors possible. One is called *N-type material*, and the other, *P-type material*. Unsurprisingly, the N stands for Negative and the P stands for (you guessed it) Positive. N-type material is created by adding pentavalent impurities, that is, a dopant with five electrons in its outer shell.

Examples include phosphorus, arsenic and antimony. In contrast, P-type material is created by adding a trivalent impurity, one with three electrons in its outer shell. Possible trivalent impurities include boron, gallium and indium.

N-Type Material

Figure 1.13 shows a model of a silicon crystal with a pentavalent impurity at its center. Compared to an ordinary silicon atom that would have four electrons in its outer shell, the pentavalent impurity creates an extra, or *donor*, electron. Thus, the crystal has a net negative charge and is referred to as N-type material. The energy level of the donor electrons is just below the bottom of the conduction band. In other words, the difference between the donor level and the bottom of the conduction band is much, much smaller than the band gap itself. Therefore it is relatively easy for these donor electrons to jump into the conduction band, becoming free ionized electrons and leaving behind ionized holes[2].

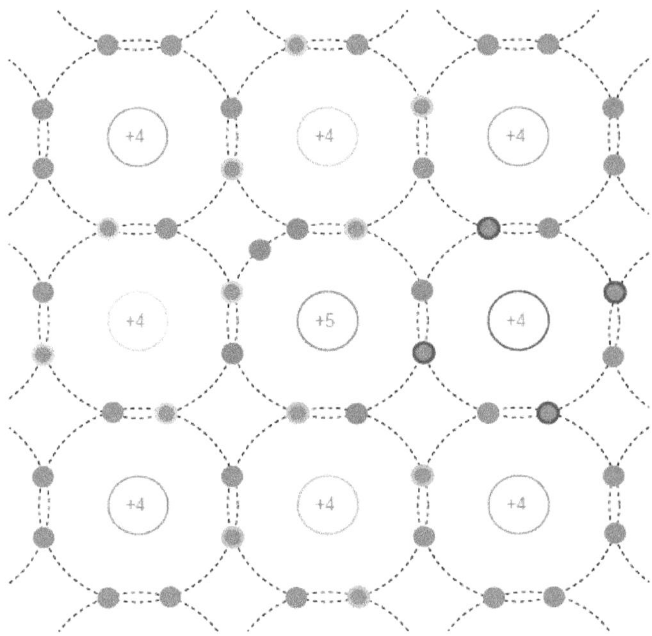

Figure 1.13 Crystal with added pentavalentimpurity (N-type).

Compared to the undoped intrinsic crystal, the doped extrinsic crystal exhibits a relatively high number of free electrons. As you might surmise, this enhances the conductivity of the material, and the greater the doping level, the greater the enhancement. Earlier it was mentioned that both electrons and holes can serve as charge carriers. Because the number of free electrons is significantly larger than the number of holes in N-type material, electrons in N-type material are referred to as the *majority charge carrier* (or more simply, the majority carrier) while holes are referred to as the *minority charge carrier* (or minority carrier).

The extra electrons add to the number of filled energy states and, being of higher energy than the valence electrons, push the Fermi level to a higher value.

Remember, the Fermi level represents the point where 50% of states would be filled, so if we add states above this, then the new 50% point must be higher than the former level. This is illustrated in Figure 1.14. Note how close the donor level is to the conduction band and that the Fermi level has been pushed up, away from the valence band and closer to the conduction band. This will be of great significance in up-coming discussions on semiconductor devices.

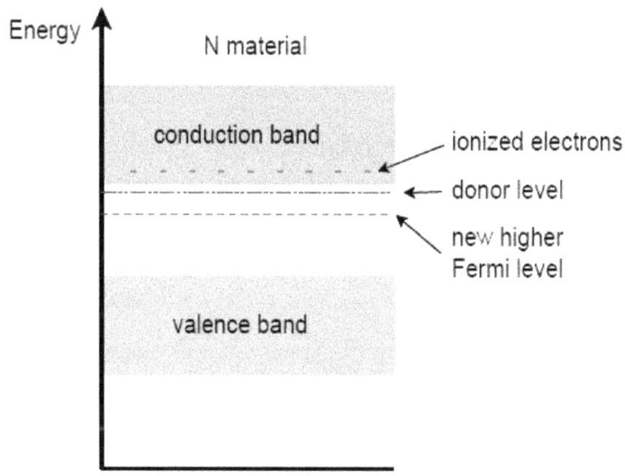

Figure 1.14 Energy band diagram of N-type semiconductor.

P-Type Material

In a similar manner, if we introduce a trivalent impurity, our crystal model now features a hole; a location where an electron is lacking. For this reason, trivalent impurities are sometimes called acceptors. The resulting crystal model is illustrated in Figure 1.15.

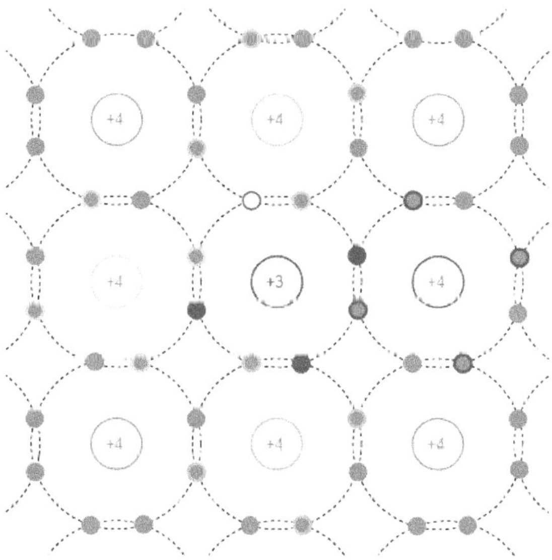

Figure 1.15 Crystal with added trivalent impurity (P-type).

The resulting situation is essentially the reverse of that of the N-type material. Figure shows the energy band diagram for our new P-type material. In this case, the Fermi level has been pushed down, closer to the valence band.

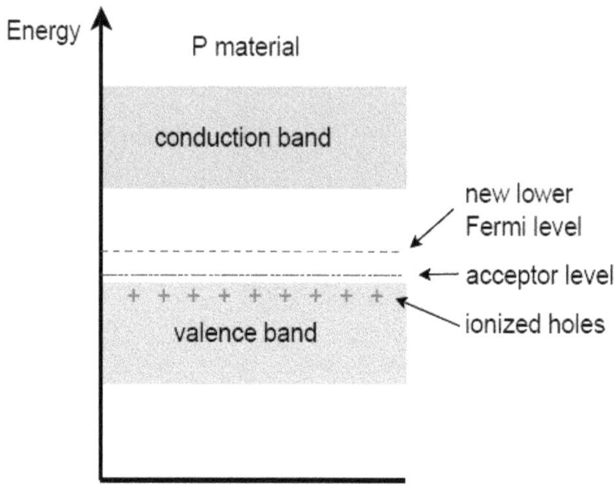

Figure 1.16 Energy band diagram of P-type semiconductor.

In P-type material, holes out number free electrons. Consequently, *holes* are referred to as the *majority carrier in P material* while electrons take on the role of minority charge carrier.

As with N-type material, the greater the amount of trivalent impurity added, the greater the overall effect. By itself, a doped crystal can be used to create a resistor. The resistivity of the material is a function of the doping level. By setting the cross- sectional area, length and doping level, we can create well-defined resistor values. If this was all we could do with semiconductors then we could say two things: first, the solid state semiconductor revolution would not exist; and second, this text would be very short. The interesting bits arrive when we combine both N- and P-type materials into a single device, as we shall begin to see in the next chapter.

CONCISE SUMMARY

- In this chapter we have examined the basic structure of atoms. This includes the concept of electron shells and permissible energy states. We have used both the Bohr model of the atom and the corresponding energy band diagrams.
- Crystals such as silicon show a very ordered three dimensional structure that relies on strong covalent bonds. The crystal tends to "fuzz" or broaden the permissible energy levels into thicker energy bands. Further, the

crystal exhibits a modest energy gap, or band gap, between the valence band and conduction band. This gap is much smaller than the gap seen in insulators, and therefore the material is referred to as a semiconductor, being somewhere between a true conductor and a true insulator.

- The electrical characteristics of a pure, or intrinsic, semiconductor crystal can be altered by adding impurities or dopants. A doped crystal is referred to as an extrinsic crystal. If a pentavalent dopant is added, there will be a surplus of electrons and a raising of the Fermi level. The new crystal is called N-type material. In contrast, if a trivalent dopant is added, there will be a surplus of holes and a lowering of the Fermi level. The new crystal is called P-type material. In N-type material, electrons are the majority charge carrier and holes are the minority charge carrier. In P-type material, holes are the majority carrier while electrons serve as the minority carrier.

CHAPTER-2

PN JUNCTIONS AND DIODES: FUNDAMENTAL COMPONENTS IN SEMICONDUCTOR PHYSICS

2.1 INTRODUCTION: AN OVERVIEW

PN junctions and diodes are fundamental components in semiconductor physics, essential to electronic devices and circuits. A PN junction is formed by joining a region with an excess of positive charge carriers (P-type) and a region with an excess of negative charge carriers (N-type). This junction creates a barrier potential that controls the flow of current. A diode, which incorporates a PN junction, allows current to flow in one direction while blocking it in the opposite direction. When a forward voltage is applied across the diode, the PN junction conducts, enabling the passage of current. In reverse bias, the diode acts as an insulator, preventing the flow of current. Diodes are integral in rectifying alternating current to direct current, serving as electronic switches, and protecting circuits from overvoltage. Understanding the behavior of PN junctions and diodes is foundational to semiconductor electronics and is crucial for designing and optimizing electronic systems.

Having investigated the characteristics of extrinsic N-type and P-type materials in the prior chapter, we shall continue by examining what happens when these two materials are combined into a single device. It is critical to understand that when we combine P- and N-type materials, we do not do so through simple mechanical means. That is, we do not in some way solder, weld, bolt, friction-fit, glue or duct tape3 one type of material to another. Rather, we must maintain a single piece of mono-crystalline silicon, not a poly-crystalline amalgam of individual pieces. This can be achieved via a diffusion or ion implantation technique that is applied repeatedly to a single piece of silicon crystal. This will leave regions or zones in the crystal that are N-type or P-type. In fact, it is quite possible to have a region of one type completely embedded within a region of the opposite type as we shall see in later chapters.

By creating a single zone of N material adjacent to a zone of P material, we wind up with the *PN junction.* The PN junction is arguably the fundamental building block of solid state semiconductor devices. PN junctions can be found in a variety of devices including bipolar junction transistors (BJTs) and junction field effect transistors (JFETs). The most basic device built from the PN junction is the *diode.* Diodes are designed for a wide variety of uses including rectifying, lighting (LEDs) and photodetection (photodiodes). We shall begin by examining the basic structure and operation of the PN junction. This will include a look at the many different kinds of diodes available. To assist with circuit analysis, a series of simplified models will be created and investigated. We shall use these models to solve a number of example circuits that feature the many diode variations available.

2.2 THE PN JUNCTION

If we were to create a region of N material abutting a region of P material in a single crystal, an interesting situation occurs. Assuming the crystal is not at absolute zero, the thermal energy in the system will cause some of the free electrons in the N material to "fall" into the excess holes of the adjoining P material. This will create a region that is devoid of charge carriers (remember, electrons are the majority charge carrier in N material while holes are the majority charge carrier in P material). In other words, the area where the N and P materials abut is depleted of available electrons and holes, and thus we refer to it as a *depletion region.* This is depicted in Figure 2.1. The excess electrons of the N material are denoted by minus signs while the excess holes of the P material are denoted with plus signs. At the interface, the free electrons have recombined with holes. When an electron recombines, it leaves behind a positive ion in the N material (shown here as a circled plus sign) and produces a negative ion in the P material (shown as a circled minus sign).

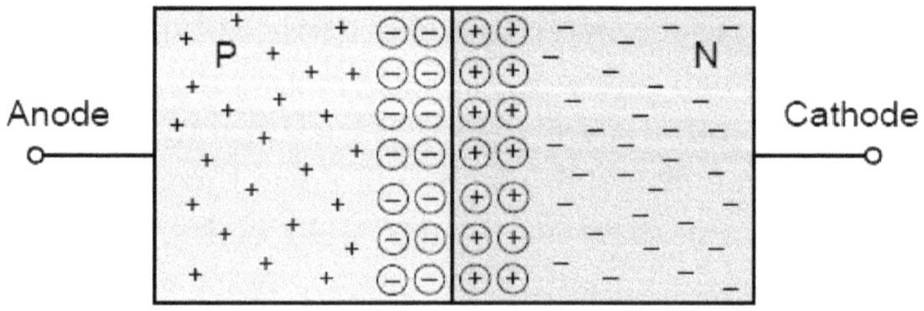

Figure 2.1 PN junction.

We now have a region depleted of charge carriers and this will have an effect on the ability to establish a flow of current through the device. We have, in essence, created an *energy hill* that will need to be overcome.

To understand the concept of the energy hill, recall that in the prior chapter it was discovered that doping an intrinsic crystal would shift the Fermi level. For N material, the Fermi level is shifted up, toward the conduction band. In contrast, for P material the Fermi level is shifted down, nearer to the valence band. When two dissimilar regions adjoin, as in the case here, the energy bands will adjust so that the Fermi levels are consistent. Effectively, this causes the bands of the P material to rise relative to the bands of the N material. The interface between the two appears as a hill, and this is the aforementioned depletion region. This situation is depicted graphically in Figure 2.2. Compare this energy diagram to the energy diagrams for N material and P material presented in the prior chapter. By simply aligning the Fermi levels, it should be clear how we arrive at the new energy diagram.

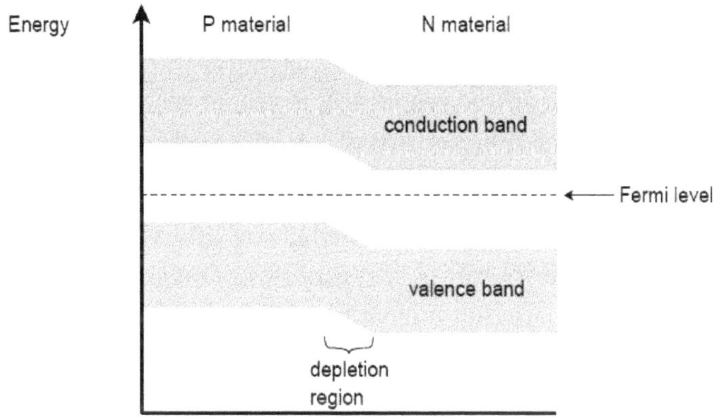

Figure 2.2 Energy bands in PN junction.

Now let's consider what happens if we were to connect this device to an external voltage source as shown in Figure 2.3. Obviously, there are two ways to orient the PN junction with respect to the voltage source. This version is termed forward-bias.

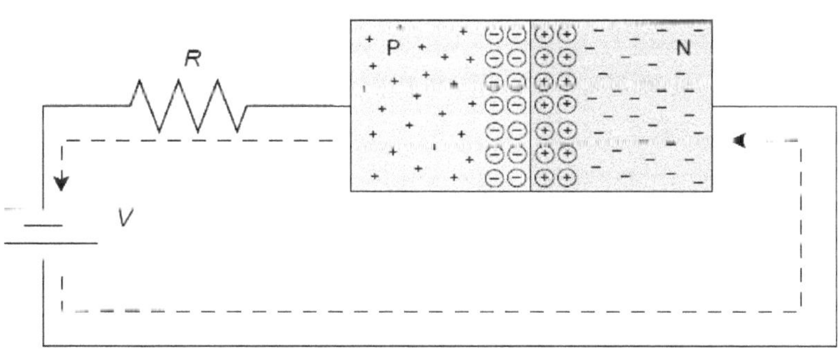

Figure 2.3 PN junction connected to external voltage source.

Forward-Bias

The dotted line of Figure 2.3 shows the direction of electron flow (opposite the direction of conventional flow). First, electrons flow from the negative terminal of the battery toward the N material. In N material, the majority carriers are electrons and it is easy for these electrons to move through the N material. Upon entering the depletion region, if the supplied potential is high enough, the electrons can diffuse into the P material where there are a large number of lower energy holes. From here, the electrons can migrate through to the positive terminal of the source, completing the circuit (the resistor has been added to limit maximum current flow). The "trick" here is to assure that the supplied potential is large enough to overcome the effect of the depletion region. That is, a certain voltage will be dropped across the depletion region in order to achieve current flow. This required potential is called the barrier potential or forward voltage drop. The precise value depends on the material used. For silicon devices the barrier potential is usually estimated at around 0.7 volts. For germanium devices it is closer to 0.3 volts while LEDs may exhibit barrier potentials in the vicinity of 1.5 to 3 volts, partly depending on the color.

Another way of thinking about this is that the addition of the voltage source "flattens" the inherent energy hill of the junction. Once the applied forward-bias voltage is at least as big as the hill, current can flow easily.

Reverse—Bias

If the voltage source polarity is reversed in Figure 2.3, the behavior of the PN junction is altered radically. In this case, the electrons in the N material will be drawn toward the positive terminal of the source while the P material holes will be drawn toward the negative terminal, creating a small, short-lived current. This has the effect of widening the depletion region and once it reaches the supplied potential, the flow of current ceases. In essence, we have increased the size of the energy hill. Further increases in the source voltage only serve to make the situation worse. The depletion region simply expands to fill the void, so to speak. Ideally, the PN junction acts like an open circuit with an applied reverse-bias voltage.

This asymmetry in response to a supplied potential turns out to be extraordinarily useful. Perhaps the simplest of all semiconductor devices is the diode. In its basic form a diode is just a PN junction. It is a device that will allow current to pass easily in one direction but prevent current flow in the opposite direction.

Shockley Equation

We can quantify the behavior of the PN junction through the use of an equation derived by William Shockley.

$$I = I_S \left(e^{\frac{V_D q}{n k T}} - 1 \right) \quad (2.1)$$

Where

I is the diode current,

IS is the reverse saturation current,

VD is the voltage across the diode,

q is the charge on an electron, 1.6E-19 coulombs,

n is the quality factor (typically between 1 and 2),

k is the Boltzmann constant, 1.38E-23 joules/kelvin,

T is the temperature in kelvin.

At 300 kelvin, q/kT is approximately 38.6. Consequently, for even very small forward (positive) voltages, the "-1" term can be ignored. Also, IS is not a constant. It increases with temperature, approximately doubling for each 10 C° rise (more on this in a moment).

If we plot the Shockley equation using typical values for a silicon device, we arrive at the curve shown in Figure 2.4. This plots the junction current as a function of the forward (positive) device voltage. It is a representative curve only. While all silicon diodes will exhibit this same general shape, the precise value of current for a specific voltage will vary depending on the device design.

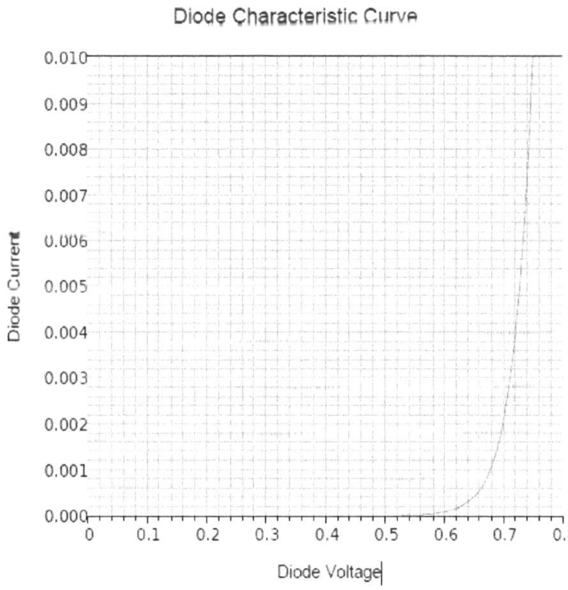

Figure 2.4 Characteristic curve of forward-biased silicon PN junction.

For potentials below about 0.5 volts, the current is virtually non-existent. Above this value, the current rises rapidly, becoming nearly vertical after approximately 0.7 volts. If the plot was recreated using a higher temperature, the effect would be to shift the curve to the left (i.e., a higher current for a given voltage).

If we were to alter the graph to use a logarithmic current scale rather than a linear scale, the graph of Figure 2.5 results. The resulting straight line plot shows clearly the logarithmic relationship between the diode's voltage and current.

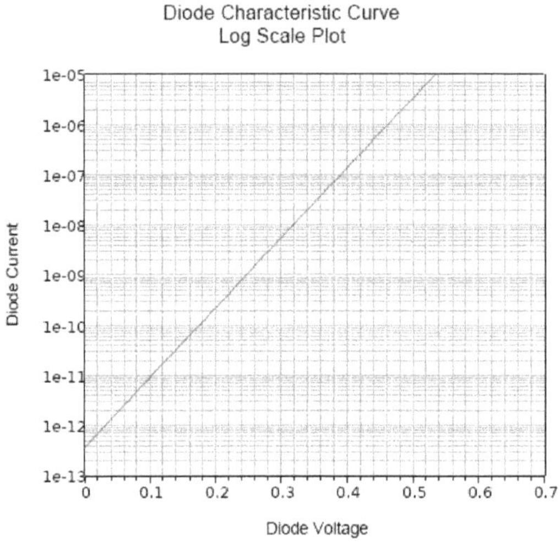

Figure 2.5 Characteristic curve of forward-biased silicon PN junction using log scale.

For negative voltages (reverse-bias) the Shockley equation predicts negligible diode current. This is true up to a point. The equation does not model the effects of *breakdown*. When the reverse voltage is large enough, the diode will start to conduct. This is shown in Figure 2.6. In the first quadrant we see the same general shape we found in Figure 2.4. *VF* is the forward "knee" voltage (roughly 0.7 volts for silicon). *IR* is the reverse saturation current (ideally zero but in reality a very small amount of current will flow). *VR* is the reverse breakdown voltage. Note that the current increases rapidly once this reverse voltage is reached.

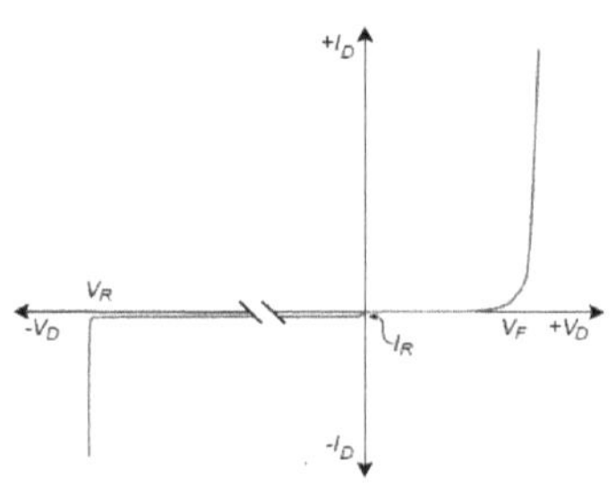

Figure 2.6 Simplified forward and reverseI-V curve for diode.

In general, diodes should not be operated in the breakdown region (the exception being Zener diodes). There are two mechanisms behind this phenomenon. The Zener effect, named after Clarence Zener, predominates when the doping levels are high and produces breakdown voltages below roughly five or six volts. It is due to the production of a very high electric field across the depletion region which then results in the production of a high current through electron tunneling. In devices using lower levels of doping, avalanche dominates. In this instance, a high electric field accelerates the free electrons to the point where they can impact surrounding atoms and create new electron-hole pairs, thus creating new free electrons that can repeat the process, resulting in a rapid increase of current.

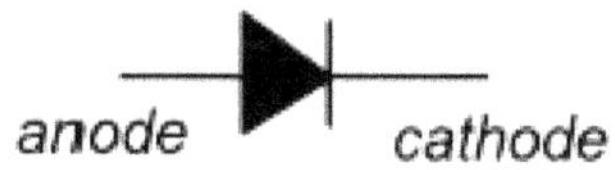

Figure 2.7 Diode schematic symbol (ANSI).

The schematic symbol for a basic switching or rectifying diode is shown in Figure2.7. This is the ANSI standard which predominates in North America. The P material is the anode while the N material is the cathode4. As a general rule for semiconductor schematic symbols, arrows point toward N material. In this case, the arrow also points in the direction of easy conventional current flow. Figure 2.8 shows an alternate schematic symbol, the IEC international standard, the difference being that it is in outline form without the body being filled in.

Figure 2.8 Alternate diode schematic symbol (IEC).

When it comes to physical device packaging, small and medium current and power devices for through-hole mounting include the DO-35 and DO-204, with the model number stamped on the body. The typical size is comparable to a 1/4 to 1/8 watt resistor. As seen in Figure 2.9 the cathode end is denoted by a band, reminiscent of the bar on the schematic symbol. Surface mount packages are also available.

Devices handling higher currents and powers often come in stud or bolt styles such as the DO-4 shown in Figure 2.10. These packages facilitate mounting to a metal plate or heat sink to help dissipate the excess heat.

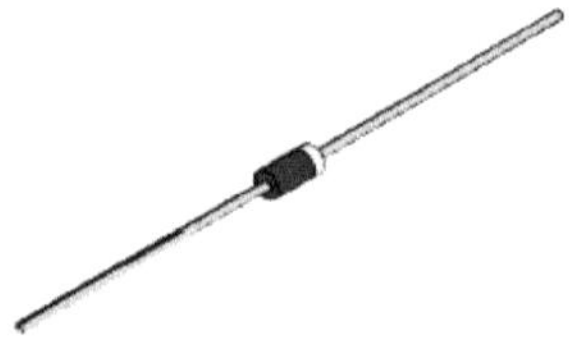

Figure 2.9 DO-204 case.

2.3 DIODE DATA SHEET INTERPRETATION

A data sheet for the popular 1N4148 switching diode is shown in Figure 2.11. The 1N4148 is designed for high speed operation required in high frequency signal applications but also finds use in a variety of general purpose applications that do not require very high current or power handling.

Some of the key features include a four nanosecond switching speed, a maximum reverse voltage of 100 volts and a 450 milliamp maximum forward current (with short single pulses as high as four amps being possible). Power dissipation is 500 milliwatts.

Referring to Figure 2.11b, the variation in reverse current with regard to temperature is obvious. This also verifies the "doubles every 10 C°" rule-of-thumb.

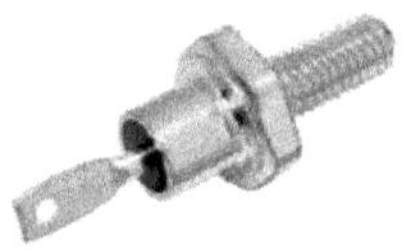

Figure 2.10 DO-4 case. Courtesy of Vishay Intertechnology, Inc.

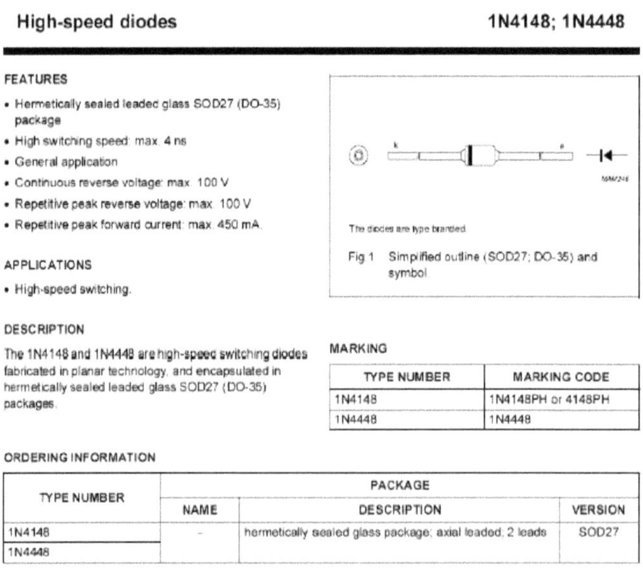

High-speed diodes 1N4148; 1N4448

FEATURES

- Hermetically sealed leaded glass SOD27 (DO-35) package
- High switching speed: max. 4 ns
- General application
- Continuous reverse voltage: max. 100 V
- Repetitive peak reverse voltage: max. 100 V
- Repetitive peak forward current: max. 450 mA.

APPLICATIONS

- High-speed switching.

The diodes are type branded.

Fig.1 Simplified outline (SOD27; DO-35) and symbol.

DESCRIPTION

The 1N4148 and 1N4448 are high-speed switching diodes fabricated in planar technology, and encapsulated in hermetically sealed leaded glass SOD27 (DO-35) packages.

MARKING

TYPE NUMBER	MARKING CODE
1N4148	1N4148PH or 4148PH
1N4448	1N4448

ORDERING INFORMATION

TYPE NUMBER	PACKAGE		
	NAME	DESCRIPTION	VERSION
1N4148	–	hermetically sealed glass package; axial leaded; 2 leads	SOD27
1N4448			

Figure 2.11a 1N4148 data sheet. Courtesy of NXP Semiconductors.

NXP Semiconductors — Product data sheet

High-speed diodes — 1N4148; 1N4448

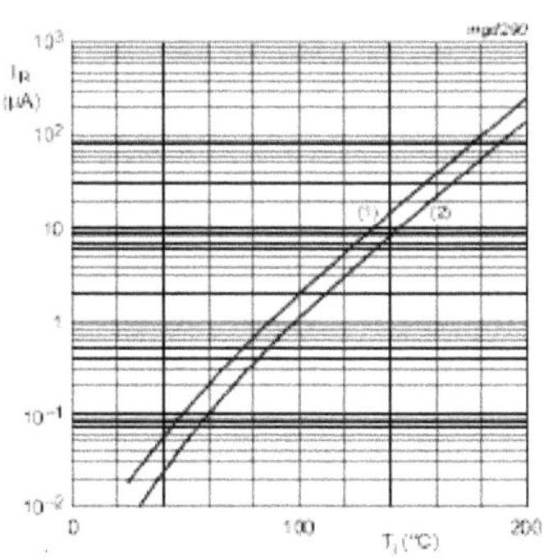

(1) V_R = 75 V; typical values.
(2) V_R = 20 V; typical values.

Fig.5 Reverse current as a function of junction temperature.

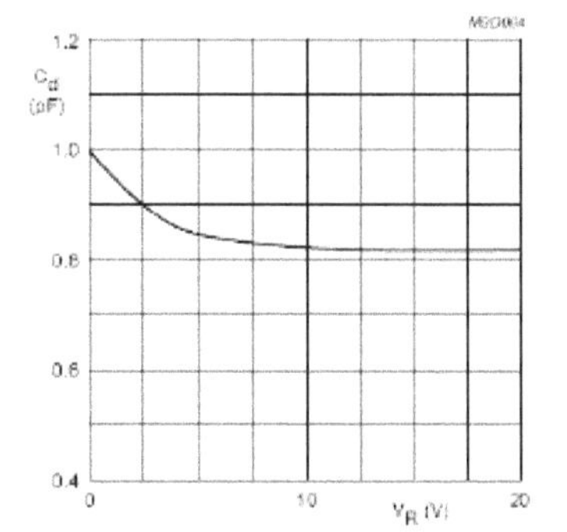

f = 1 MHz; T_j = 25 °C.

Fig.6 Diode capacitance as a function of reverse voltage; typical values.

Figure 2.11b 1N4148 data sheet (continued).

NXP Semiconductors — Product data sheet

High-speed diodes — 1N4148; 1N4448

LIMITING VALUES

In accordance with the Absolute Maximum Rating System (IEC 60134).

SYMBOL	PARAMETER	CONDITIONS	MIN.	MAX.	UNIT
V_{RRM}	repetitive peak reverse voltage		–	100	V
V_R	continuous reverse voltage		–	100	V
I_F	continuous forward current	see Fig.2; note 1	–	200	mA
I_{FRM}	repetitive peak forward current		–	450	mA
I_{FSM}	non-repetitive peak forward current	square wave; T_j = 25 °C prior to surge; see Fig.4			
		t = 1 μs	–	4	A
		t = 1 ms	–	1	A
		t = 1 s	–	0.5	A
P_{tot}	total power dissipation	T_{amb} = 25 °C; note 1	–	500	mW
T_{stg}	storage temperature		−65	+200	°C
T_j	junction temperature		–	200	°C

Note

1. Device mounted on an FR4 printed-circuit board; lead length 10 mm.

ELECTRICAL CHARACTERISTICS

T_j = 25 °C unless otherwise specified.

SYMBOL	PARAMETER	CONDITIONS	MIN.	MAX.	UNIT
V_F	forward voltage	see Fig.3			
	1N4148	I_F = 10 mA	–	1	V
	1N4448	I_F = 5 mA	0.62	0.72	V
		I_F = 100 mA	–	1	V
I_R	reverse current	V_R = 20 V; see Fig.5		25	nA
		V_R = 20 V; T_j = 150 °C; see Fig.5	–	50	μA
I_R	reverse current, 1N4448	V_R = 20 V; T_j = 100 °C; see Fig.5	–	3	μA
C_d	diode capacitance	f = 1 MHz; V_R = 0 V; see Fig.6	–	4	pF
t_{rr}	reverse recovery time	when switched from I_F = 10 mA to I_R = 60 mA; R_L = 100 Ω; measured at I_R = 1 mA; see Fig.7	–	4	ns
V_{fr}	forward recovery voltage	when switched from I_F = 50 mA; t_r = 20 ns; see Fig.8	–	2.5	V

THERMAL CHARACTERISTICS

SYMBOL	PARAMETER	CONDITIONS	VALUE	UNIT
$R_{th(j-tp)}$	thermal resistance from junction to tie-point	lead length 10 mm	240	K/W
$R_{th(j-a)}$	thermal resistance from junction to ambient	lead length 10 mm; note 1	350	K/W

Note

1 Device mounted on a printed-circuit board without metallization pad.

Figure 2.11c 1N4148 data sheet (continued).

NXP Semiconductors Product data sheet

High-speed diodes 1N4148; 1N4448

GRAPHICAL DATA

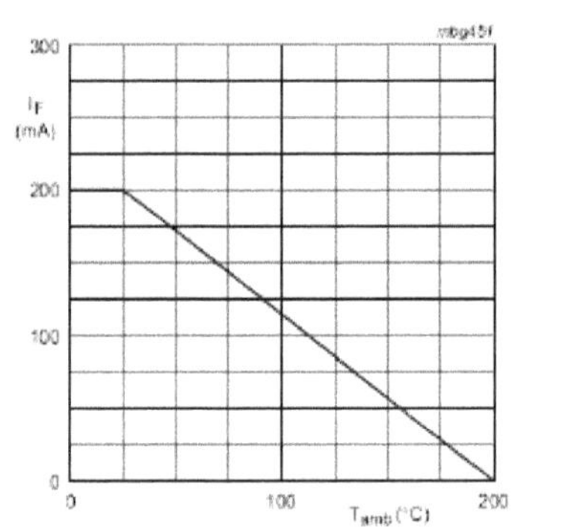

Device mounted on an FR4 printed-circuit board; lead length 10 mm.

Fig.2 Maximum permissible continuous forward current as a function of ambient temperature.

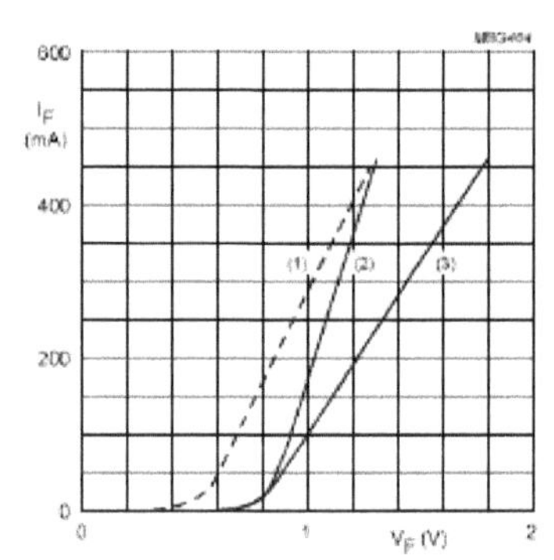

(1) T_j = 175 °C; typical values.
(2) T_j = 25 °C; typical values.
(3) T_j = 25 °C; maximum values.

Fig.3 Forward current as a function of forward voltage.

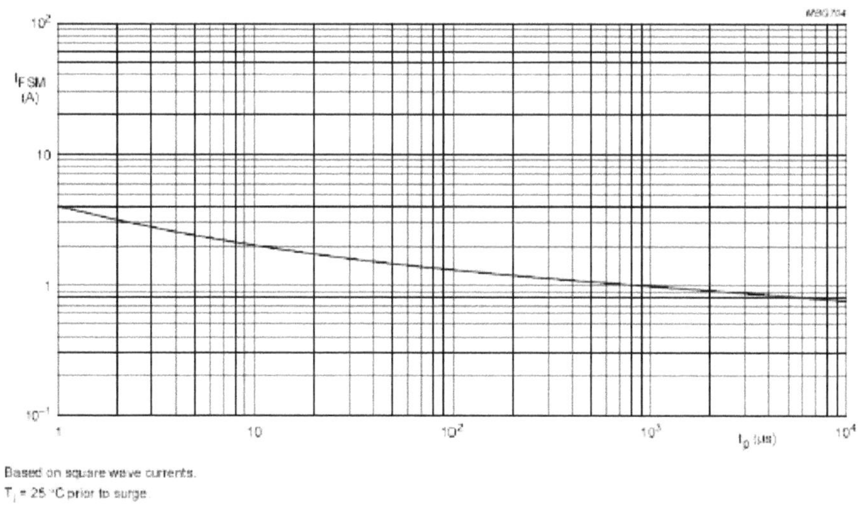

Fig.4 Maximum permissible non-repetitive peak forward current as a function of pulse duration.

Figure 2.11d 1N4148 data sheet (continued).

Power derating and permissible pulse amplitudes can be seen in Figure 2.11d. Finally, note the variation in the forward voltage curves due to temperature. As stated previously, for a given current, an increase in temperature results in a lower forward voltage. Also, at room temperature, we see a knee voltage of approximately 0.7 volts.

2.4 DIODE CIRCUIT MODELS

One thing is very clear from the characteristic curve of the diode: It is not a linear bilateral[5] device, quite unlike a resistor. Consequently, we cannot use the superposition technique to solve diode circuits unless we have *a priori* knowledge about it, that is, whether or not it is forward- or reverse-biased. For example, we can imagine a circuit comprised of two voltage sources, resistors and a diode. By itself, one of the voltage sources might forward-bias the diode while the other would reverse-bias it. Obviously, a diode cannot be both forward and reverse-biased at the same time.

A second problem we face with circuit analysis is the added complexity of the Shockley equation. For speed and ease of computation we find it useful to model the diode with simpler circuit elements. Three diode models are shown in Figure 2.12.

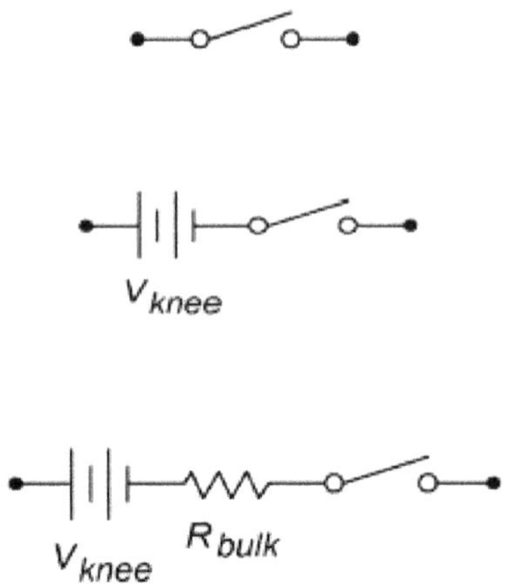

Figure 2.12 Simplified diode models. Top to bottom: first, second and third approximations, increasing in accuracy.

The first approximation is the simplest of the three. It treats the diode as a simple dependent switch: the switch is closed if the diode is forward-biased and open if it is reverse-biased. The second approximation adds the effect of the forward voltage.

V_{knee} is the "turn-on" potential required to overcome the energy hill. It would be 0.7 volts for a silicon device. The third approximation is the most accurate of the three. A close look at the characteristic curve of Figure 2.4 shows that once the knee voltage is reached, the curve does not transition to a perfect vertical line. Instead, there remains some positive, non-infinite slope. That is, the voltage continues to increase, although modestly, with further increases in current. We can approximate this effect as a small resistive value, *Rbulk*. The three corresponding I-V plots are shown in Figure 2.13. Compare these to Figure 2.6 and note the increasing accuracy.

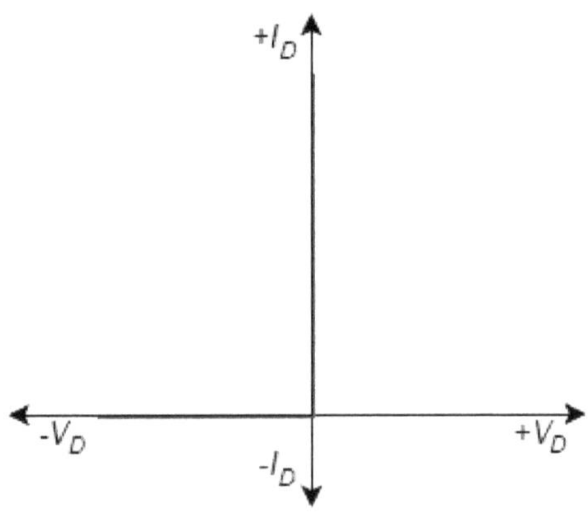

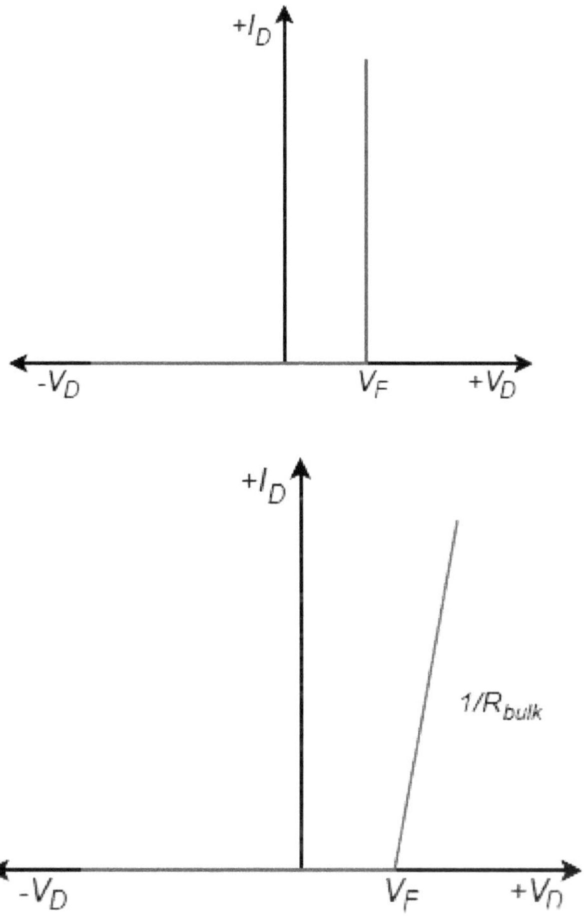

Figure 2.13 I-V curves for simplified diode models. Top to bottom: first, second and third approximations.

In many applications the second approximation will yield sufficiently accurate results and we will tend to make greatest use of it. Just remember that these are behavioral models; don't think that there are literally 0.7 volt sources or little resistors in the diodes.

It should be noted that *Rbulk* does not represent the "diode resistance" per se, rather, it models a minimum value. There really is no such thing as a singular diode resistance. We can, however, talk about the *effective resistance* of a diode in a particular circuit in both DC and AC terms.

The key to understanding this concept is to remember that resistance is a linear function, a straight line on a I-V graph. Therefore, we need to find a straight line "fit" for the diode curve. Two possibilities are shown in Figure 2.14.

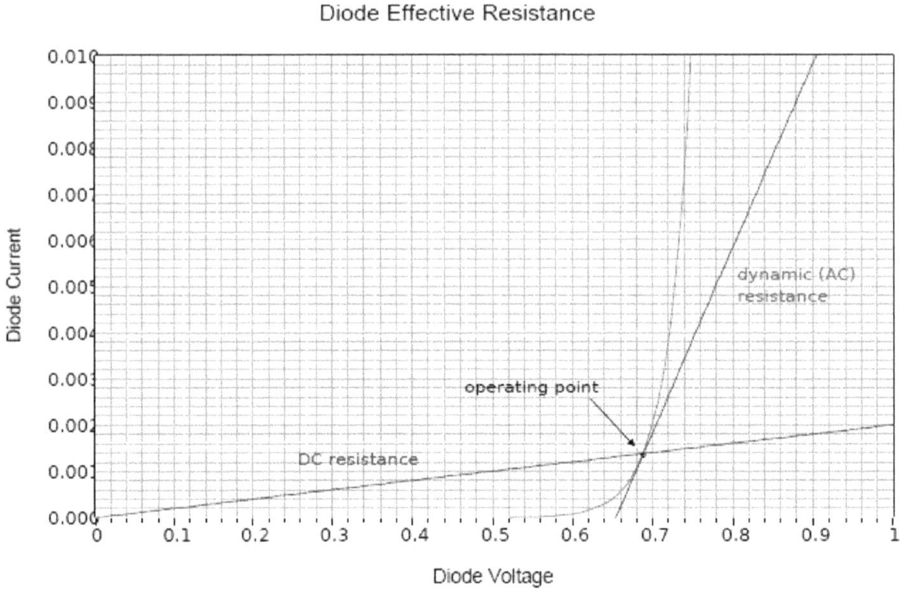

Figure 2.14 Diode effective resistance for DC and AC.

The red curve is the diode's characteristic curve (arbitrary current values are shown). For some particular DC circuit, a specific current will flow through the diode which will produce a particular voltage, denoted on the graph as the *operating point*. If we simply compute the ratio of that voltage to the driving current, we wind up with a resistance. This is the effective DC resistance of the diode under these circuit conditions and is represented by the blue line. That is, the reciprocal of the slope of the blue line is the effective DC resistance. Obviously, if we shift the operating point along the red diode curve, the slope of the intersecting blue line changes and therefore we arrive at a new DC resistance. The higher the current, the lower the effective DC resistance.

Instead of just DC, the diode might see a combination of DC and AC signals. Visualize this as adding a small AC variation on top of the DC. We can imagine the operating point moving along the red diode curve, back and forth about the operating point. If we divide the small AC voltage variation by its associated AC current variation, we wind up with the AC equivalent resistance, also known as the dynamic resistance. Graphically, we can think of this as finding the slope of a line that is tangent to the operating point (the purple line). This will in fact, be an average value across the AC variation. It should also be apparent that the effective AC resistance must be smaller than its DC counterpart because the AC approximation (purple line) must be steeper than the DC approximation (blue line)6.

It is time for a few illustrative examples.

Example 2.1

Consider the resistor-diode circuit of Figure 2.15. Assume the voltage source is 12 volts and the resistor is 2 kΩ. Further, assume the diode is silicon and its bulk resistance is 10 Ω. Using the three diode approximations, compute the circulating current.

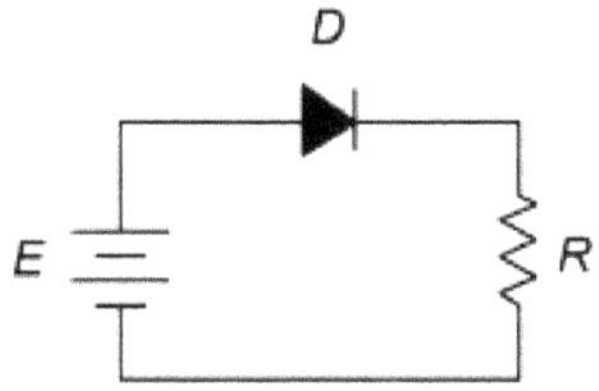

Figure 2.15 Schematic for Example 2.1.

First, note that the diode is forward-biased. This must be the case because there is a single voltage that is larger than the knee voltage and its positive terminal is attached to the diode's anode. No matter which approximation we use, Kirchhoff's voltage law (KVL) must be true so it will be a matter of summing the available voltage drops versus resistance(s).

Using the first approximation:

Here we assume the diode is a closed switch. Consequently all of the source voltage must drop across the single resistor.

$$I = \frac{E}{R}$$
$$I = \frac{12\,\text{V}}{2\,\text{k}\Omega}$$
$$I = 6\,\text{mA}$$

Using the second approximation:

In this instance we include the knee voltage.

$$I = \frac{E - V_{knee}}{R}$$
$$I = \frac{12\,\text{V} - 0.7\,\text{V}}{2\,\text{k}\Omega}$$
$$I = 5.65\,\text{mA}$$

Using the third approximation:

The most accurate of the three, we include both the knee voltage and bulk resistance.

$$I = \frac{E - V_{knee}}{R + R_{bulk}}$$
$$I = \frac{12\,\text{V} - 0.7\,\text{V}}{2\,\text{k}\Omega + 10\,\Omega}$$
$$I = 5.622\,\text{mA}$$

In this particular case the difference between the second and third approximations is less than 1%. It is also worth noting that the third approximation predicts a diode voltage of slightly more than 0.7 volts (approximately 0.756 volts) due to the additional potential across the bulk resistance.

Example 2.2

Determine the circulating current for the circuit in Figure 2.16. Also find the diode and resistor voltages. Assume the power supply is 20 volts, the diode is silicon and the resistor is 2 kΩ.

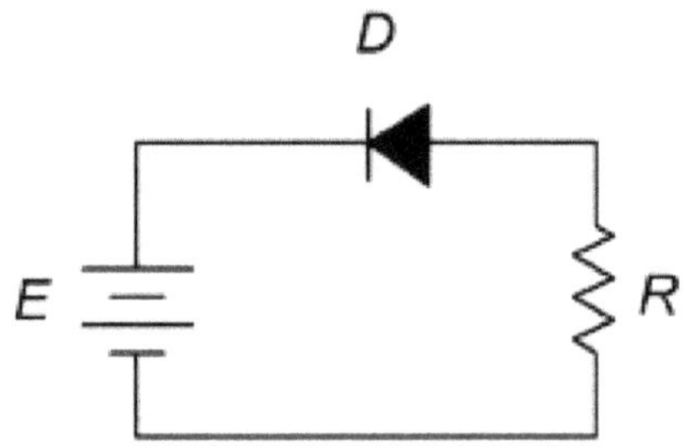

Figure 2.16 Schematic for Example 2.2.

This problem is deceptively easy. Note that the positive terminal of the source is connected to the cathode. As there are no other sources in the circuit, the diode must be reverse-biased. The model for a reverse-biased diode is an open switch and the circulating current in an open circuit is zero. Therefore, the resistor voltage must also be zero and values for the knee voltage and bulk resistance are not needed. In order to satisfy KVL, the diode voltage will equal the source of 20 volts (+ to - from cathode to anode).

The only time this would not be the case is if the reverse breakdown voltage of the diode is less than the 20 volt source. In that case the diode voltage would equal the breakdown voltage with the remainder of the source voltage dropping across the resistor.

Example 2.3

Determine the circulating current for the circuit in Figure 2.17. Also find the diode and resistor voltages. Assume the power supply is 9 volts, the diodes are silicon and $R1$ = 1 kΩ, $R2$ = 2 kΩ.

$$I = \frac{E - V_{knee1} - V_{knee2}}{R_1 + R_2}$$

$$I = \frac{9\text{V} - 0.7\text{V} - 0.7\text{V}}{1\,\text{k}\Omega + 2\,\text{k}\Omega}$$

$$I = 2.533\,\text{mA}$$

According to KVL, the applied source must equal the sum of the voltage drops across the resistors and diodes as this is a single loop. Both diodes are forward-biased (conventional current entering the anodes).

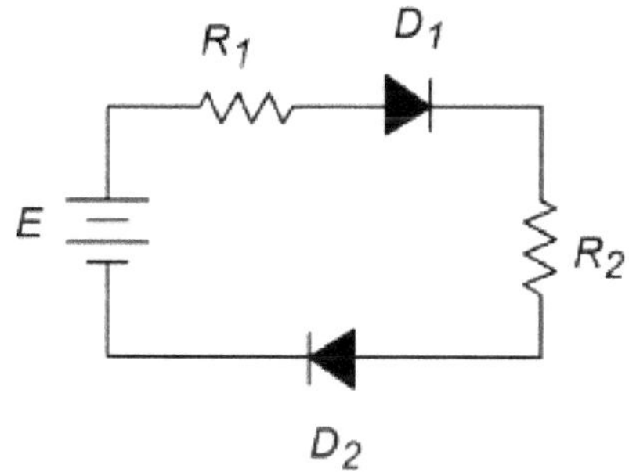

Figure 2.17 Schematic for Example 2.3.

Note that if either diode was reversed, there would be no current flow and all of the source potential would drop across the reversed diode.

Example 2.4

Determine the source current and resistor voltages for the circuit in Figure2.18. Also find the resistor voltages if the diode polarity is reversed. Assume the power supply is 10 volts, the diode is silicon and the resistors are 1 kΩ each.

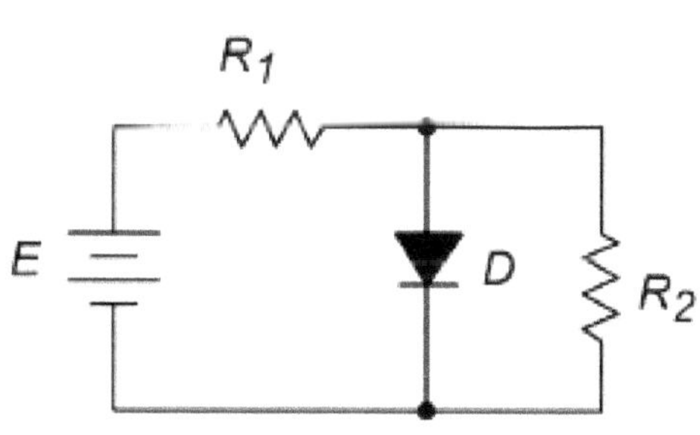

Figure 2.18 Schematic for Example 2.4.

As *D* and *R2* are in parallel they must have the same voltage drop. Also, the diode is forward-biased. Therefore, the voltage across *R2* must be approximately 0.7 volts, leaving 9.3 volts to drop across *R1*. The current through *R1* is the source current.

$$I = \frac{E - V_D}{R_1}$$

$$I = \frac{10\,\text{V} - 0.7\,\text{V}}{1\,\text{k}\Omega}$$

$$I = 9.3\,\text{mA}$$

If the diode is reversed it behaves as an open switch. The circuit reduces to a simple 1:1 voltage divider, each resistor dropping half of the supply, or 5 volts each.

Computer Simulation

To verify our results, Example 2.4 is simulated. The circuit is captured as shown in Figure 2.19a. This particular example is shown in Multisim although any decent quality simulator will do. The very common 1N4148 switching diode is used here. Another popular choice would be the 1N914 switching diode or a 1N400X series rectifier.

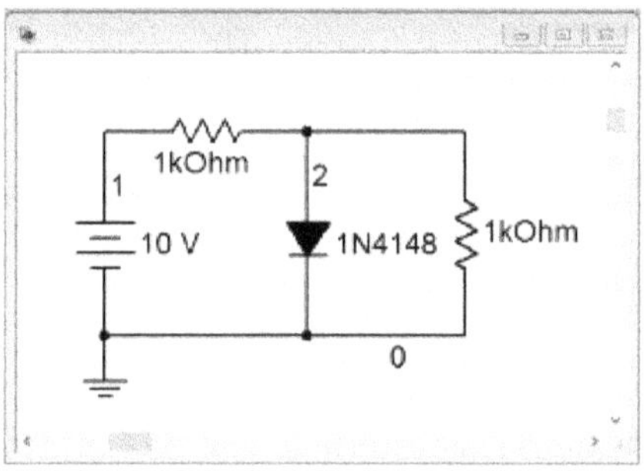

Figure 2.19a The circuit of Example 2.4 in Multisim.

Next, a DC Operating Point analysis is performed. The results are shown in Figure 2.19b. Note that the diode potential is just under the 0.7 volt approximation. From this we can deduce that the voltage drop across the first resistor must be slightly more than 9.3 volts, producing a current slightly more than 9.3 mA.

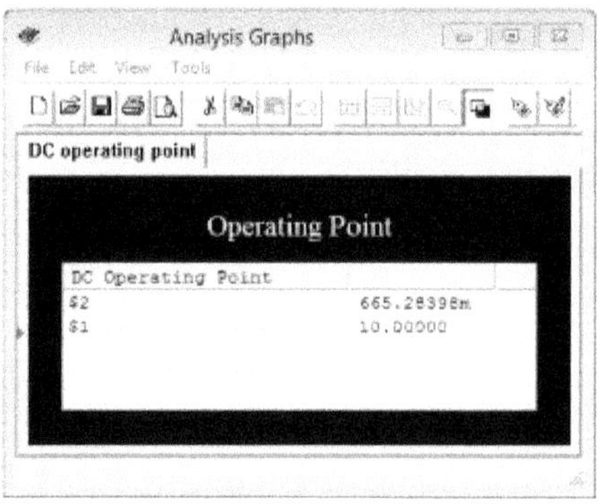

Figure 2.19b DC Operating Point simulation results for the circuit of Example 2.4.

Finally, Figure 2.19c shows the results when the diode is reversed in the circuit. The second resistor (node 3 to ground) shows 5 volts as expected. Therefore, the first resistor must also be dropping 5 volts.

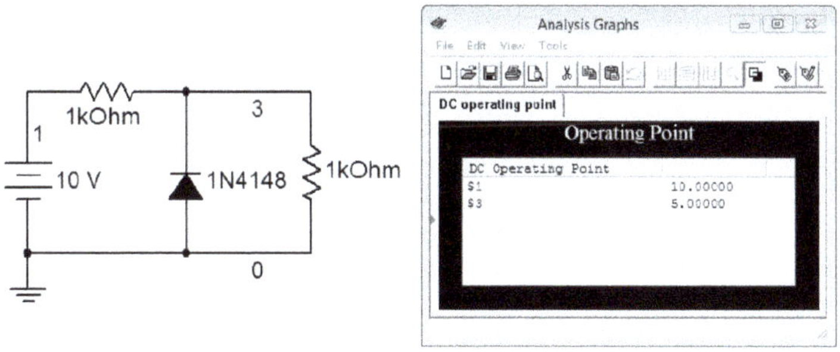

Figure 2.19c: Simulation of Example 2.4 using reversed diode orientation.

Before moving on to another topic, let's take a look at a somewhat more involved example using multiple diodes.

Example 2.5

Determine the diode and resistor voltages for the circuit in Figure 2.20. Assume the diodes are silicon.

$$I = \frac{E - V_{D1}}{R_1 + R_2}$$

$$I = \frac{20\,\text{V} - 0.7\,\text{V}}{1\,\text{k}\Omega + 2\,\text{k}\Omega}$$

$$I = 6.433\,\text{mA}$$

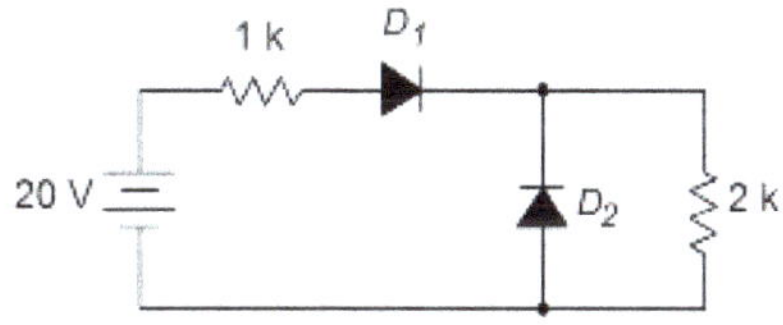

Figure 2.20 Schematic for Example 2.5.

The first thing to notice is that *D1* is forward-biased while *D2* is reverse-biased. Therefore, the 20 volt source must equal the drop across *D1* and the two resistors. *D2* will take on whatever the drop across the 2 kΩ works out to as they are in parallel.

Note that virtually no current flows down through D2 as it is reverse-biased. Using Ohm's law, the drop across the first resistor is 6.433 volts and for the second resistor, 12.867 volts.

2.5 OTHER TYPES OF DIODES

Diodes have been designed to exploit different aspects of PN junctions. Besides the basic use as a switching or rectifying device, diodes are available for voltage regulation, variable capacitance, illumination and light sensing. The schematic symbols for a number of popular diode types are shown in Figure 2.21. Note the similarities of the symbols. The "bar" portion represents the cathode for all of them.

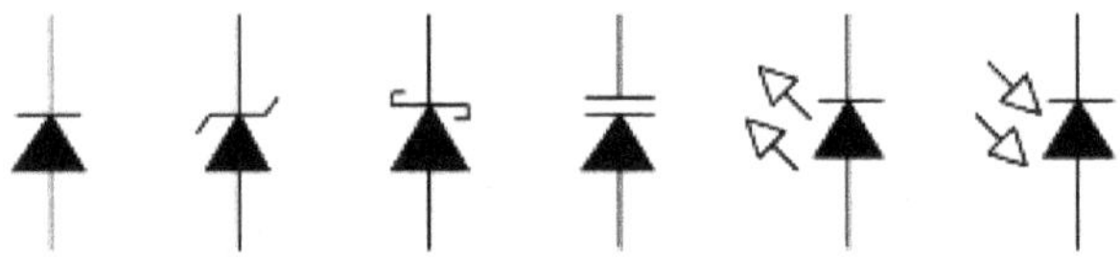

Figure 2.21 Diode schematic symbols: a) switching or rectifying b) Zener c) Schottky d) varactor e) LED f) photodiode

Zener Diode

The Zener diode behaves like an ordinary signal diode when forward-biased. Normally, though, Zeners are used in a reverse-bias condition. From our previous discussion, recall that if the reverse potential is high enough, a diode can go into breakdown, causing a rapid increase current. This was caused by either of two effects, Zener conduction or avalanche. The Zener diode takes advantage of this in order to produce a stable voltage7. Zeners are specified by their reverse potential (generically referred to as the "Zener voltage") and are designed to handle larger currents and powers than the average signal diode. Zener voltages are standardized in much the same manner as resistors so values such as 3.9 volts, 5.1 volts and 6.8 volts are to be expected. The Zener voltage is measured at IZT, the Zener test current. A lower current may not fully push the diode into conduction resulting in a lower than expected diode potential.

Figure 2.22 Zener diode schematic symbol.

Instead of modeling the Zener as an open switch when reverse-biased, instead we model it as an open when its voltage is less than the rated voltage, and as a voltage source equal to the rated value if its voltage tries to exceed that value. When analyzing Zener-based circuits, the first thing to do is determine if the diode is forward-biased. If it is then treat it like an ordinary switching diode. If, on the other hand, it is reverse-biased then treat it like an open switch. If the resulting diode voltage is greater than the Zener voltage then recompute the circuit but this time mentally replace the Zener with a voltage source equal to the Zener voltage. Our next example will illustrate this method.

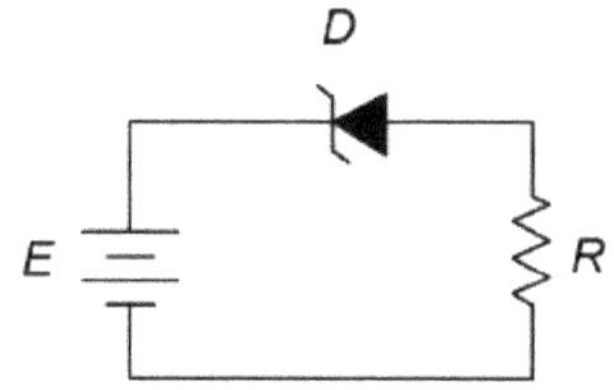

Figure 2.23 Schematic for Example 2.6.

Determine the circulating current for the circuit in Figure 2.23. Also find the diode and resistor voltages. Assume the power supply is 9 volts, the Zener voltage is 5.1 volts and the resistor is 3.3 kΩ.

The diode is reverse-biased. If we treat it as an open then it would drop the entire source voltage, or 9 volts. This is greater than the Zener potential so the device must be in Zener conduction. This means that conventional current will flow relatively easily in a clockwise direction. The voltage across the diode will equal the rated value of 5.1 volts, + to - from cathode to anode. By KVL the resistor drop must be 9 V - 5.1 V, or 3.9 volts.

$$I = \frac{E - V_{Zener}}{R}$$

$$I = \frac{9\text{V} - 5.1\text{V}}{3.3\text{k}\Omega}$$

$$I = 1.182\text{ mA}$$

If the diode was flipped in orientation then it would be forward-biased and show the expected 0.7 volts with 8.3 volts across the resistor.

Computer Simulation

A Zener diode circuit is simulated as shown in Figure 2.24.

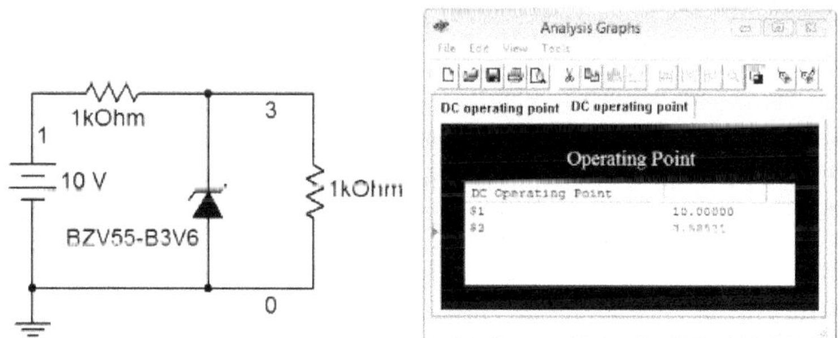

Figure 2.24 Zener diode circuit simulation.

Without the Zener, the two resistors would simply split the supply voltage equally, each receiving 5 volts. If an ordinary diode was used it would be reverse-biased and act as an open. The resulting voltages would be the same. In this case, however, the Zener is activated at 3.6 volts (it is typical to include the Zener voltage as part of the model number with the letter "V" replacing the decimal point when needed).

Therefore we see roughly 3.6 volts across the Zener and the parallel second resistor (node 3 to ground). The precise value of voltage will depend on the magnitude of the diode current. If the simulation is rerun with a higher voltage source, the increased current will produce a slightly higher voltage at node 3. This is because the breakdown curve is not infinitely steep once it goes past the rated Zener voltage. The effect is similar to that of Rbulk in a forward-biased diode. On a data sheet this value is referred to as the differential resistance, or Rdif.

An excellent use of the Zener is to limit or regulate a voltage. When a Zener is placed in parallel with other components we can ensure that those components will not see a potential higher than the rated Zener voltage. We will take a much closer look at this in the next chapter.

Light Emitting Diode ((LED)) and Photodiode

The light emitting diode (LED) and photodiode are complements. While the LED produces light with an electrical input, the photodiode produces a current when exposed to light. Both devices can operate within the human visible spectrum and can also be designed to operate at wavelengths outside this range, in the infrared (IR) and ultraviolet (UV). In fact, most TV remote controls rely on IR emitter/detector pairs for communication[8].

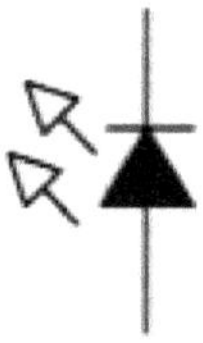

Figure 2.25 Light emitting diode (LED)schematic symbol.

The LED has displaced traditional incandescent (filament-based) light sources in many applications due to its high efficiency in turning an electrical energy input into a light output. They are small, physically robust, operate relatively cool and are available in a number of different colors. The schematic symbol is shown in Figure2.25. The basic idea behind its operation is fairly simple. In a forward-biased PN junction, when free electrons recombine and "fall" into lower energy valence holes they must give up this energy differential in some manner. In most diodes, this energy is emitted as heat. In LEDs, the energy transition is designed such that it is radiated at shorter wavelengths (i.e., visible light). In order to achieve this, LEDs are not formed just using silicon as in a typical switching diode. Instead, somewhat more exotic materials are used. From an analysis or design standpoint, the important thing to remember is that the forward voltage tends to be noticeably higher than silicon's 0.7 volt drop. The precise value will depend on the material, which in turn effects the color. A generic red LED will likely exhibit a forward drop of around 1.8 volts or so. Other colors tend to be somewhat higher as we move through the rainbow, ending with blue and UV LEDs (and also high brightness versions) up around 3 to 4 volts. In a lab it is easy to determine the approximate forward drop of a given diode by connecting it in series with a voltage source and current limiting resistor. The supply is increased until the desired brightness is achieved and then the diode drop can be measured with a DMM. When reverse-biased the LED behaves like a switching diode, that is, it looks like an open switch. Unlike switching and rectifying diodes, LED maximum reverse potentials tend to be relatively low, perhaps just a few volts.

A datasheet for the Cree C566D series LED is presented in Figure 2.26. Notice that the colors are specified in terms of wavelength (in nanometers) and luminous intensity (brightness) is given in millicandella (mcd).

CREE

PRODUCT FAMILY DATA SHEET

Cree® Screen Master® 5-mm Oval LED
C566D-RFF/GFF/BFF/AFF
C566D-RFE/GFE/BFE/AFE

PRODUCT DESCRIPTION

The oval LED is specifically designed for variable-message signs and passenger information signs. The oval-shaped radiation pattern and high luminous intensity ensure that these devices are excellent for wide-field-of-view outdoor applications where a wide viewing angle and readability in sunlight are essential.

These lamps are tinted and diffused. The encapsulation resin contains anti-UV material in order to reduce the effects of long-term exposure to direct sunlight.

FEATURES

- Size (mm): 5
- Color and Typical Dominant Wavelength:
  Red (621nm)
  Green(527nm)
  Blue(470nm)
  Amber(591nm)
- Luminous Intensity (mcd)
  C566D-RFF/RFE:(2130-5860)
  C566D-GFF/GFE:(4180-12000)
  C566D-BFF/BFE:(1100-3000)
  C566D-AFF/AFE:(2130-5860)
- Lead - Free
- RoHS Compliant

APPLICATIONS

- Electronic Signs & Signals (ESS)
- Full Color video screen
- Motorway Signs
- Variable Message Sign (VMS)
- Advertising signs
- Petrol Signs

Cree, Inc.
4600 Silicon Drive
Durham, NC 27703
USA Tel: +1.919.313.5300

Figure 2.26a LED datasheet. Courtesy of Cree, Inc.

These devices are RoHS compliant which stands for Restriction of Hazardous Substances. It is an EU directive that limits usage of toxic materials such as lead, cadmium and mercury. Continuing, we find the device maximum ratings:

CREE

Cree 5-mm Oval LED
C566D-RFF/GFF/BFF/AFF
C566D-RFE/GFE/BFE/AFE

ABSOLUTE MAXIMUM RATINGS (T_A = 25°C)

Items	Symbol	Absolute Maximum Rating		Unit
		Red and Amber	Blue and Green	
Forward Current	I_F	50 Note1	35	mA
Peak Forward Current Note2	I_{FP}	200	100	mA
Reverse Voltage	V_R	5	5	V
Power Dissipation	P_D	130	140	mW
Operation Temperature	T_{opr}	-40 ~ +95		°C
Storage Temperature	T_{stg}	-40 ~ +100		°C
Lead Soldering Temperature	T_{sol}	Max. 260°C for 3 sec. max. (3 mm from the base of the epoxy bulb)		
Electrostatic Discharge Classification (MIL-STD-883E)	ESD	Class 2		

Note:

1. For long term performance the drive currents between 10mA and 30mA are recommended. Please contact CREE sales representative for more information on recommended drive conditions.
2. Pulse width ≤0.1 msec, duty ≤1/10.

TYPICAL ELECTRICAL & OPTICAL CHARACTERISTICS (T_A = 25°C)

Characteristics	Color	Symbol	Condition	Unit	Minimum	Typical	Maximum
Forward Voltage	Red/Amber	V_F	I_F = 20 mA	V		2.1	2.6
	Blue/Green	V_F	I_F = 20 mA	V		3.4	4.0
Reverse Current	Red/Amber	I_R	V_R = 5 V	µA			100
	Blue/Green	I_R	V_R = 5 V	µA			100
Dominant Wavelength	Red	λ_D	I_F = 20 mA	nm	619	621	624
	Green	λ_D	I_F = 20 mA	nm	520	527	535
	Blue	λ_D	I_F = 20 mA	nm	460	470	475
	Amber	λ_D	I_F = 20 mA	nm	584	591	596
Luminous Intensity	Red	I_V	I_F = 20 mA	mcd	2130	3000	
	Green	I_V	I_F = 20 mA	mcd	4180	7000	
	Blue	I_V	I_F = 20 mA	mcd	1100	2000	
	Amber	I_V	I_F = 20 mA	mcd	2130	3000	

Figure 2.26b LED datasheet (continued). Figure 2.26c

Note the different values between the various colors. The forward current is specified as 50 mA for red/amber with 35 mA for green/blue. Nominal operating currents are between 10 and 30 mA. Reverse voltage is 5 volts, typical for many LEDs although much lower than the average switching diode. Forward voltage is typically 2.1 volts for the red end of the spectrum and, as expected, 3.4 volts for the green/blue end. The expected luminous intensities also vary with color. Further, it Figure 2.26b LED datasheet (continued)should be noted that LEDs do not produce "pure color" light in the manner of a laser. Rather, they produce a range of wavelengths clustered in a specific area. The wavelength that produces the highest output in this area is referred to as the peak or dominant wavelength. Human vision covers the range of roughly 400 nanometers (violet) to 700 nanometers (red)9.

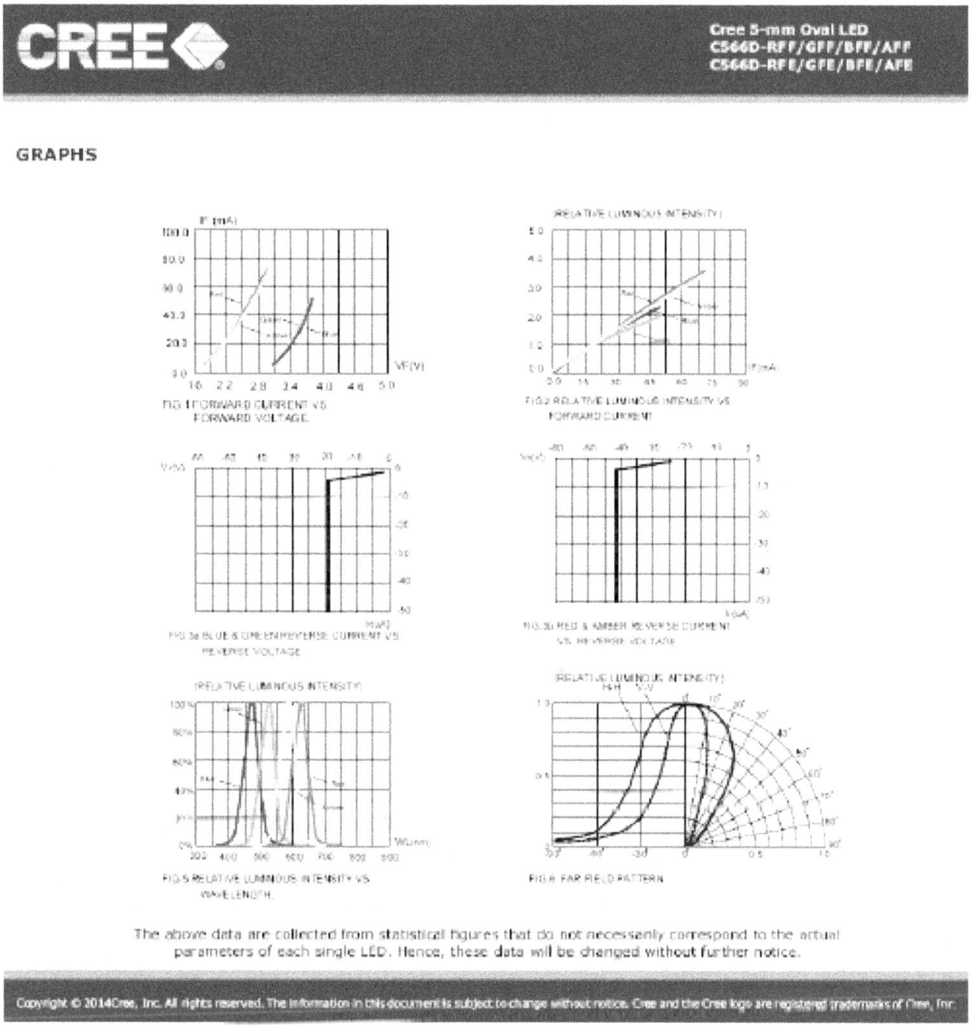

Figure 2.26c LED datasheet (continued).

Figure 2.26c presents pertinent graphical data. We observe a roughly linear increase in luminous intensity with increasing current. Also, note the difference in the reverse voltage/current plots between blue/green and red/amber. Of particular interest is the final graph which shows the beam pattern or beam angle. You can think of this in terms of how narrow or broad the illumination pattern is. When comparing different model LEDs it is useful to remember that on-axis brightness can be increased by narrowing the angle. This graph is split in half using two different ways of showing the data. On the left side we have a linear graph depicting the relative brightness as we move off of the center axis (zero degrees). On the right side we see a polar plot version of the same data.

Example 2.7

Determine the circulating current for the circuit in Figure 2.27. Assume the power supply is 5 volts, the LED forward voltage is 2.1 volts and the resistor is 330 Ω.

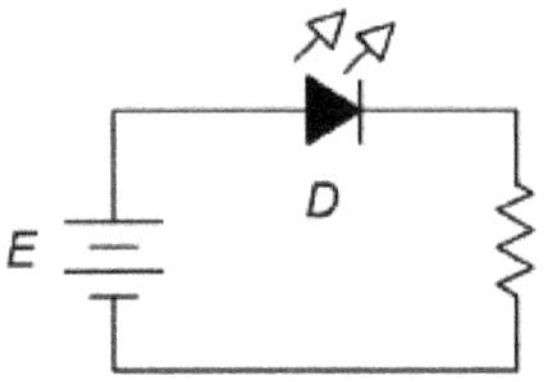

Figure 2.27 Schematic for Example 2.7.

The LED is forward-biased and as the source is greater than the LED potential, it should light. Using KVL,

$$I = \frac{E - V_{LED}}{R}$$

$$I = \frac{5\,\text{V} - 2.1\,\text{V}}{330\,\Omega}$$

$$I = 8.788\,\text{mA}$$

This should result in a relatively bright LED. The resistor can be used to effectively program the brightness by changing the current level (a smaller resistance yields a higher current and therefore a brighter LED). Given the 2.1 volt forward potential, it is likely that this is an amber or yellow LED. If a different color had been used, say a 1.6 volt red or 3.2 volt blue, there would be a change in current and most likely a change in brightness. The change in brightness might not perfectly echo the change in current because the conversion efficiency for the two diodes may not be the same.

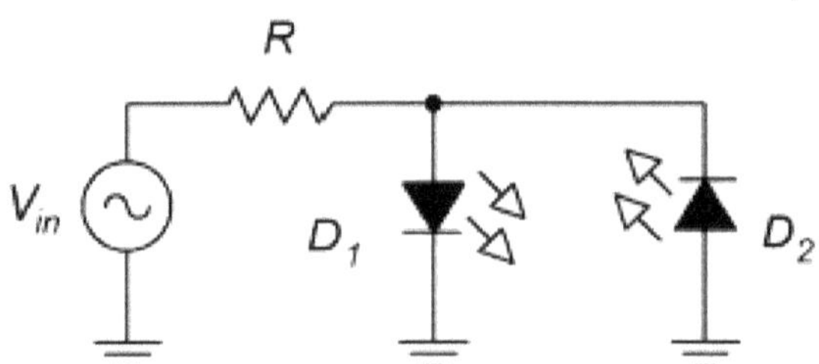

Figure 2.28 Dual LED application.

An interesting circuit using two differently colored LEDs is shown in Figure 2.28. An AC source is used to drive the LEDs. Only one of the two will be forward-biased at any given time. For positive source voltages D_1 will be on while D_2 will be off.

For negative source voltages the opposite will be true. Resistor *R* serves to limit current for both of them. Assume D_1 is red and D_2 is blue. Further, suppose the source frequency is relative low, say 1 hertz. For the positive half cycle (.5 seconds)the red LED will light and for the negative half cycle the blue LED will light. This alternating pattern continues for as long as the source is applied but a curious thing happens as we increase the frequency. At first, the blink rate will increase with the red and blue flickering back and forth faster and faster. At some point, perhaps around 30 hertz or so, it will appear as though both LEDs are lit continuously. This is because the human visual sense will tend to integrate the rapid motion and we effectively see the "average" intensity. In fact, this "on-off" trick is often used in digital circuits to control the brightness of LEDs or the speed of motors. Bi-color LEDs are available in a single package. Using a common lead and two control leads (one for each color), it is possible to achieve color mixing.

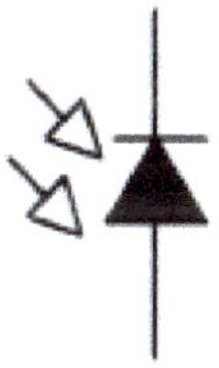

Figure 2.29 Photodiode schematic symbol.

The logical inverse of the LED is the photodiode, the schematic symbol of which is illustrated in Figure 2.29. The photodiode includes some manner of port that allows light to hit the junction. A sufficiently energetic photon of light can knock lose an electron. This creates an electron-hole pair which results in current flow. As more light energy is added to the system, an increasing current or voltage will result[10].

Photodiodes can operate in one of two modes. The first mode is *photovoltaic* mode. It uses zero-bias (that is, no external bias potential). In this mode, the photodiode operates as a voltage source. This is the mode used by photovoltaic solar cells. They can be thought of as very large photodiodes. The second mode of operation is *photoconductive* mode. This mode requires reverse-biasing the diode with an external potential. In this mode the diode acts more like a current source. The advantage is that the response is faster than photovoltaic mode. The downside is that noise and *dark current* are worse. Dark current is the current produced even when no light is shining on the photodiode. Ideally this would be zero. A large dark current reduces the effective dynamic range of the device.

Schottky and Varactor Diode

The Schottky diode is a special purpose device. It is named after Walter Schottky, a German physicist. Unlike other diodes that rely on a semiconductor-to-semiconductor junction, the schottky diode is comprised of a semiconductor-to-metal contact. The Schottky diode exhibits two major advantages over traditional diodes. First, they have very fast switching times, perhaps orders of magnitude of improvement. Second, they exhibit relatively low turn on voltages. Instead of the 0.6 to 0.7 volts seen with a silicon junction diode, a Schottky diode may turn on with as little as 0.2 or 0.3 volts. Consequently, Schottky diodes are used when very fast switching speed and/or minimizing forward voltage drops are important. Examples include shunting diodes in switch mode power supplies and RF detector circuits. Its schematic symbol is shown in Figure 2.30.

Figure 2.30 Schottky diode schematic symbol.

The varactor diode is another special purpose device. Its schematic symbol is illustrated in Figure 2.31. It is used as an electrically controlled capacitance (note that the schematic symbol appears as a hybrid of normal diode and capacitor symbols).

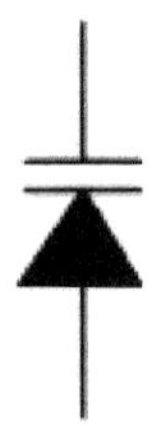

Figure 2.31 Varactor diode schematic symbol.

Varactors are used in reverse-bias mode. The key to understanding their operation is to consider the structure of a diode, comparing it to the construction of a capacitor. Consider the depletion region to be the dielectric of a capacitor with the anode and cathode being the capacitor plates. Consequently, all junction diodes exhibit some capacitance. Normally, designers try to minimize this effect but it is exploited with varactors. As noted in our earlier discussion, increasing the reverse-bias potential on a diode causes its depletion region to widen. All else being equal, increasing the plate separation of a capacitor decreases its capacitance.

Thus, by increasing the reverse-bias potential, we increase the effective plate spacing and decrease the diode junction capacitance. We now have a capacitance the value of which is determined by a DC bias voltage. This capacitance can be used as part of electronic tuning circuits for applications such as oscillators and filters. Compared to fixed capacitors the values tend to be small, in the tens to hundreds of picofarads, but it is sufficient for much radio frequency work. The advantages over mechanically adjustable capacitors are manifold, including small size, high reliability, low cost and the ability to rapidly change the capacitance[11].

CONCISE SUMMARY

- In this chapter we have examined the structure and functioning of the PN junction. APN junction produces a depletion region which is an area devoid of free charges. This leads to an energy hill or barrier voltage, the precise value of which depends on the material used as well as other factors such as temperature. The PN junction is the basis for most diodes. Its current-voltage characteristic is described by the Shockley equation and shows a logarithmic characteristic (i.e., the voltage is proportional to the log of the current).
- The terminals of a diode are identified as the anode (P material) and the cathode (N material). If a positive potential which is greater than the barrier voltage is applied from anode to cathode, the diode will conduct current. If the polarity is reversed, the diode will not conduct. Therefore a simple model of the diode is a polarity sensitive switch. Improved models include the forward barrier voltage and the bulk resistance of the diode. Another refinement includes the effect of reverse breakdown, that is, the tendency of a diode to suddenly begin conducting if the reverse-bias potential is large enough. For ordinary diodes, the reverse potential should not be allowed to reach breakdown.
- Besides the common switching and rectifying diodes, other types are also available. These include the Zener which is normally used in reverse-bias mode. It is commonly used to set or limit a specific voltage. In forward-bias, a Zener behaves like an ordinary diode. LEDs produce light from an electrical input. Their forward potentials tend to be in the neighborhood of a few volts. The photodiode is the complement of the LED and produces a current or voltage that scales with incident light. The Schottky diode is notable for its fast switching speeds and low barrier potential. Finally, the varactor is used as an electrically controlled capacitance. It is used in reverse-bias mode.

CHAPTER-3

APPLICATIONS OF DIODE

3.1 INTRODUCTION: AN OVERVIEW

Diodes, fundamental semiconductor devices with a PN junction, find widespread applications across the realm of electronics, serving as crucial building blocks in various circuits and systems. These semiconductor components, which allow current to flow in one direction while restricting it in the opposite direction, are indispensable in shaping the behavior of electrical signals. From rectifying alternating current to direct current in power supplies to acting as electronic switches in digital circuits, diodes play pivotal roles in ensuring the proper functioning and efficiency of electronic devices. Their applications extend to voltage regulation, signal demodulation in communication systems, and protection against voltage spikes. Understanding the diverse applications of diodes is essential for engineers and designers working in fields ranging from telecommunications and power electronics to consumer electronics and beyond. This exploration of diode applications underscores their versatility and significance in shaping the landscape of modern electronic technology.

The preceding chapter was concerned primarily with introducing the practical considerations of diodes while presenting them in DC circuits. This chapter will extend the discussion by focusing on AC circuit applications. A prime example is AC to DC conversion, the concept behind most electronic power supplies. It also includes the basics behind regulation and limiting/level shifting circuits such as clippers and clampers. The inherent asymmetry in the conductance of diodes, that is, their sensitivity to the direction of current flow, is what makes these circuits possible. Non-ideal effects such as a diode's forward voltage drop might be ignored in some instances but may be quite important in others.

3.2 RECTIFICATION PROCESS

Rectification is the process of turning an alternating current waveform into a direct current waveform, i.e., creating a new signal that has only a single polarity. In this respect it's reminiscent of the common definition of the word, for example where "to rectify the situation" means "to set something straight". Before continuing, remember that a DC voltage or current does not have to exhibit a constant value (like a battery). All it means is that the polarity of the signal never changes. To distinguish between a fixed DC value and one that varies in amplitude in a regular fashion, the latter is sometimes referred to as pulsating DC.

The concept of rectification is crucial to the operation of modern electronic circuits. Most electronic devices such as a TV or computer require a fixed, unchanging DC voltage to power their internal circuitry. In contrast, residential and commercial power distribution is normally AC. Consequently, some form of AC to DC conversion is required[12]. This is where the asymmetry of the diode comes in.

Half-wave Rectification

To understand the operation of a single diode in an AC circuit, consider the diagram of Figure 3.1. This is a simple series loop consisting of a sine wave source, a diode and a resistor that serves as the load. That is, primarily we will be interested in the voltage developed across the resistor.

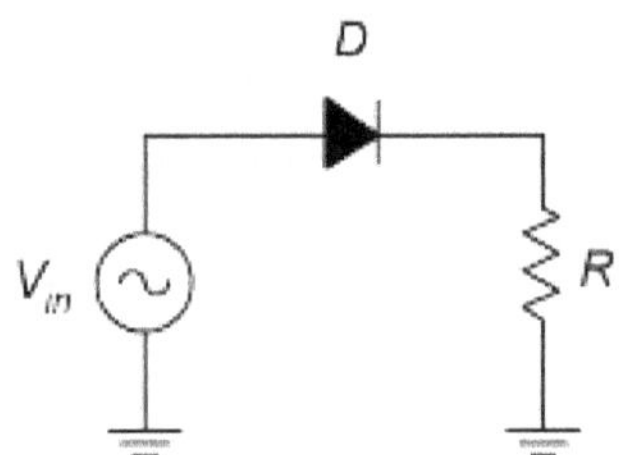

Figure 3.1 Basic AC diode-resistor circuit.

For positive portions of the input wave, the diode will be forward-biased. To a first approximation it will appear as a closed switch. Consequently, all of the input signal will drop across the resistor. In contrast, when the input signal switches to a negative polarity on the other half of the waveform, the diode will be reverse-biased.

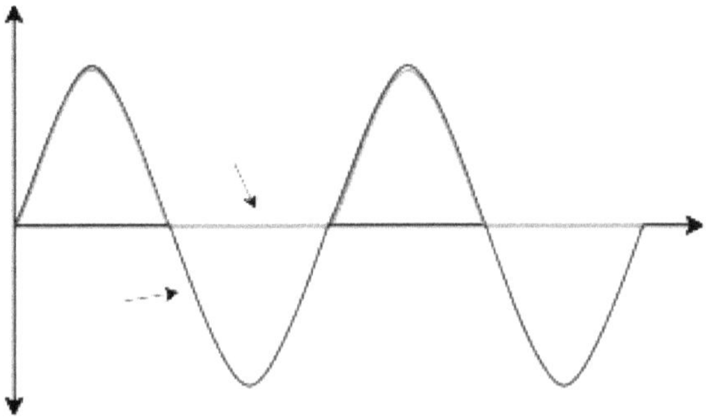

Figure 3.2 Half-wave rectification waveforms.

Therefore, the diode acts as an open switch. The circulating current drops to zero thereby producing no voltage across the resistor. All of the applied potential drops across the diode, as indicated by Kirchhoff's voltage law (KVL). The input and load resistor's voltage waveforms can be seen in Figure 3.2.

The resulting signal seen across the load resistor is a pulsating DC waveform. We have effectively removed the negative half of the waveform leaving just the positive portion. Because only half of the input waveform makes it to the load, this is referred to as half-wave rectification.

It is worth noting that if the AC peak input voltage is not particularly large, there can be an obvious discrepancy between the peak levels of the input and load signals. For example, if the peak input voltage is in the range of three or four volts and a silicon diode is used, the resulting waveforms would look more like Figure 3.3.

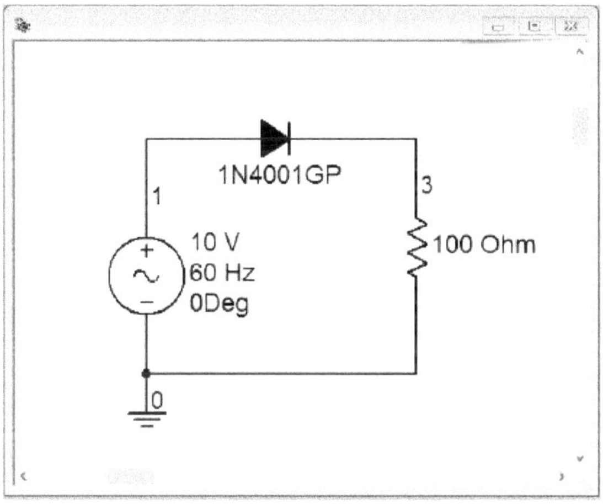

Figure 3.4 Simulation schematic for half-wave rectifier.

In this case the 0.7 volt forward drop cannot be ignored as it represents a sizable percentage of the input peak. The positive pulses are also slightly narrowed as current will not begin to flow at reasonable levels until the input voltage reaches 0.6 to 0.7 volts.

If the diode was oriented in reverse, it would block the positive portion of the input and allow only the negative portion through. In this instance the load waveform would appear flipped top to bottom compared to Figures 3.2 and 3.3.

Computer Simulation

A simulation schematic for a simple half-wave rectifier is shown in Figure 3.4. A sine wave source of 10 volts peak is used to feed a popular 1N4000 series rectifier diode connected to a 100 Ω load. The source frequency is 60 hertz, the North American standard for power distribution.

A transient analysis is run resulting in the waveforms shown in Figure 3.5. The source voltage waveform is shown in red while the load voltage waveform is depicted in blue. While the half-wave rectification is obvious, the loss due to the forward voltage drop of the diode is clearly evident. Based on the vertical scale, a value just under one volt would be a reasonable estimate. The simulation agrees nicely with the expected result as drawn in Figure 3.3, although not as extreme due to the increased source voltage.

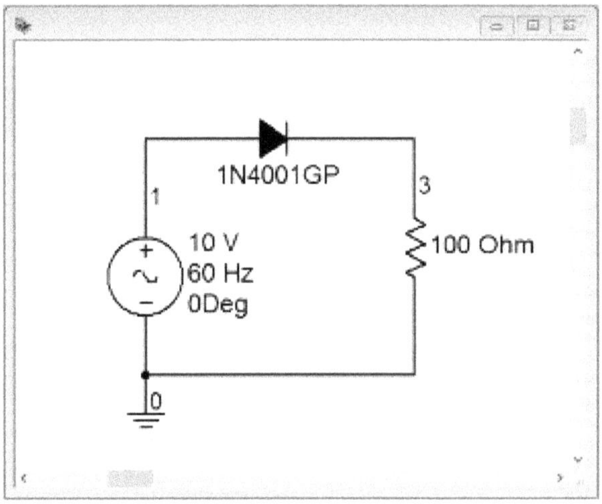

Figure 3.4 Simulation schematic for half-wave rectifier.

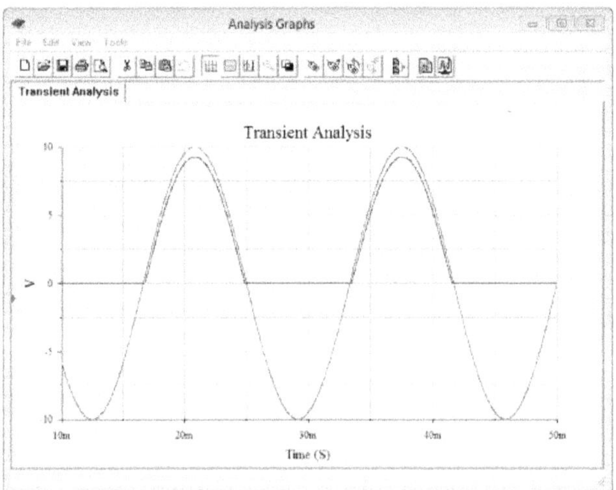

Figure 3.5 Transient analysis for half-wave rectifier.

On a practical note, there are still two items to consider when it comes to converting AC to DC. The first item is the issue of scaling the 120 VAC RMS outlet voltage to a more useful level. In many cases this means lowering the voltage although there are some applications such as high power amplifiers where the voltage will need to be increased. The second item involves smoothing the pulsating DC to produce a constant value, much like a battery.

A Note Regarding Transformers

The aforementioned voltage scaling issue can be addressed through the use of a transformer. While a complete exploration of transformers is beyond the scope of this chapter, we can present the basics. In simple terms, a transformer has an input side, or *primary*, and an output side, or *secondary*. Each side is made up of a coil of wire and these coils are wound around a common magnetic core. The current in the primary-side coil creates a magnetic flux in the core. This flux induces a current in the secondary coil. Ideally, the voltage is decreased and the current is increased by the ratio of the number of loops between these coils. For example, if the secondary-side coil has half as many turns as the primary-side coil then the secondary voltage will be half of the primary voltage and its current will be twice as large as the primary current. This implies that in the ideal case there is no power lost within the transformer. It simply transforms the power from high-voltage/low-current to low- voltage/high-current (or vice versa), hence the name. In reality, transformers do have voltage and current limits, and they are specified in terms of a volt-amp or VA rating which is simply the product of the nominal secondary voltage and maximum allowed secondary current. Transformers that decrease the voltage are referred to as *step-down* while those that increase the voltage are referred to as *step-up*. Finally, it is possible to create transformers with multiple primaries and secondaries (via either separate coils or multi-tapped coils). The resulting series and parallel coil configurations make them much more flexible.

The second issue we have is smoothing and leveling the pulsating DC. The most straightforward method to achieve this is to add a capacitor in parallel with the load. The capacitor will charge up during the conduction phase, thus storing energy. When the diode turns off, the capacitor will begin to discharge, thus transferring its stored energy into the load. The larger the capacitor, the greater its storage capacity and the smoother the load voltage will be. It turns out that there is a down side to large capacitors, as we shall see. Consequently, the goal will not be to use as large of a capacitor as possible but rather to use an optimal size for a given application. A half- wave rectifier with transformer and capacitor is shown in Figure 3.6.

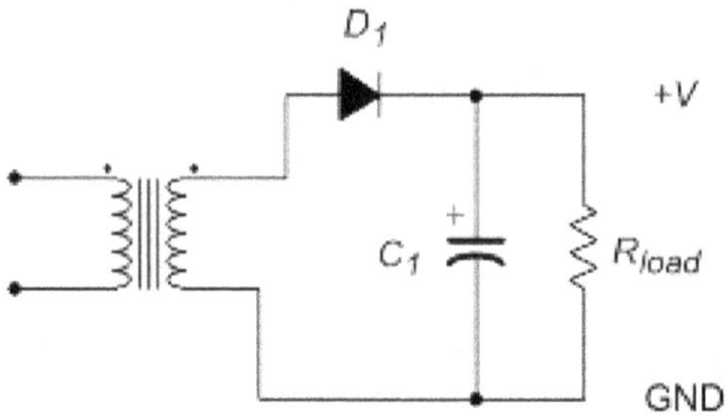

Figure3.6 Half-wave rectifier with transformer and filter capacitor.

One way of looking at the inclusion of the smoothing capacitor is to consider that it, along with the load resistance, make up an *RC* discharge network. To achieve a smooth load voltage the discharge time constant should be much longer than the gap produced when the diode turns off. For 60 hertz operation, this gap is half of the period, or roughly 8.3 milliseconds. The time constant equation is

$$\tau = RC$$

Recalling that in one time constant the capacitor voltage will fall to well below half of the starting value (roughly 37%), we will need a time constant several times larger than 8.3 milliseconds. For example, suppose our effective load resistance is 100 Ω. If we use a 1000 μF capacitor, the resulting time constant would be 100 milliseconds, or over ten times the gap duration. A much smaller capacitor, say around 50 μF, would not be nearly so effective at keeping the voltage constant.

The variation in output voltage due to capacitor discharge is referred to as *ripple*. It can be modeled as an AC voltage riding on a larger DC output. The magnitude of the ripple worsens as the load current increases. Under light load conditions, the output will tend to float to the peak voltage of the secondary with very little ripple. As load current demand goes up, the ripple magnitude increases and the nominal output voltage begins to drop.

Computer Simulation

Two variations on a filtered half-wave rectifier are simulated below. Both versions use a 100 Ω load with a 10 volt source, similar to the prior simulation. The first version uses a 50 μF filter capacitor while the second ups this to 1000 μF. In both cases a 1 Ω resistor is added in series with the capacitor to serve as a current sensor. The first version is shown in Figure 3.7.

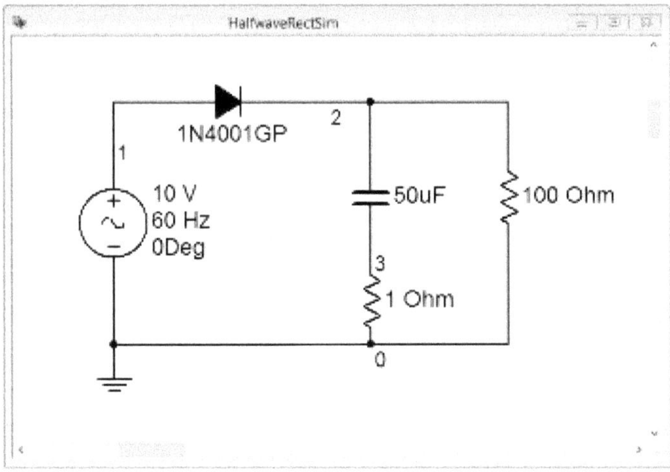

Figure 3.7 Simulation schematic for half- wave rectifier with 50 μF filter capacitor.

A transient analysis simulation graph is shown in Figure 3.8. The input waveform is colored blue while the load voltage is red. Comparing this waveform to that depicted in Figure 3.5 shows the effect of the capacitor stretching out the pulse and partially filling in the gap. It is obvious that this capacitor is too small given the load resistance and the resulting current demand. Indeed, by the time the next pulse arrives the capacitor is nearly depleted and the output voltage has dropped to around one volt.

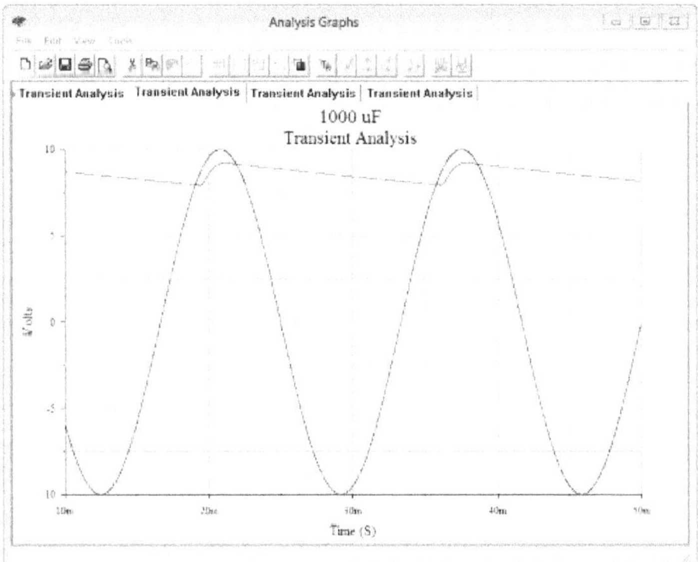

Figure 3.8 Transient analysis simulation for half-wave rectifier using a 50 μF filter capacitor.

In Figure 3.9 the simulation is rerun, but this time using a 1000 μF capacitor in place of the 50 μF. As expected, the increased RC time constant results in a much more stable load voltage. In this version the output has dropped from a little over nine volts to about eight volts yielding a peak-to-peak ripple of a volt and a half or so. The peak voltage of just over nine volts versus the applied ten volts is largely due to the voltage drop across the rectifying diode.

One thing that may not be apparent immediately is that the charge time for the larger capacitor is much shorter than for the smaller unit. This is perhaps counterintuitive. With a larger capacitor, the diode turns on for a shorter time because its cathode is held at a high voltage due to the capacitor. That is, it will only turn on when the input voltage exceeds the capacitor voltage by roughly 0.7 volts. It is only during this time that the capacitor will be replenished, and this can lead to very large current spikes.

To investigate this effect, the simulations are rerun, but this time adding the voltage across the 1 Ω sensing resistor. This relatively small value will have only a modest effect on the charging and discharging, and conveniently scales to the current value (i.e., 100 millivolts signifies 100 milliamps). First, examine the transient simulation of Figure 3.10 using the 50 μF capacitor.

The red sweep is the output voltage while the blue sweep represents the capacitor current. The output voltage plot uses the left vertical axis while the current plot uses the right vertical axis. As the load voltage begins to rise, we see an abrupt spike in the capacitor current. This is current charging the capacitor and it peaks at about 180 milliamps. The total time for the charge phase is around 4 milliseconds. Once the output voltage peaks, the capacitor starts to discharge into the load. During the discharge phase note that the capacitor current's polarity has reversed. It is negative, peaking at roughly -80 milliamps, and delivering current to the load.

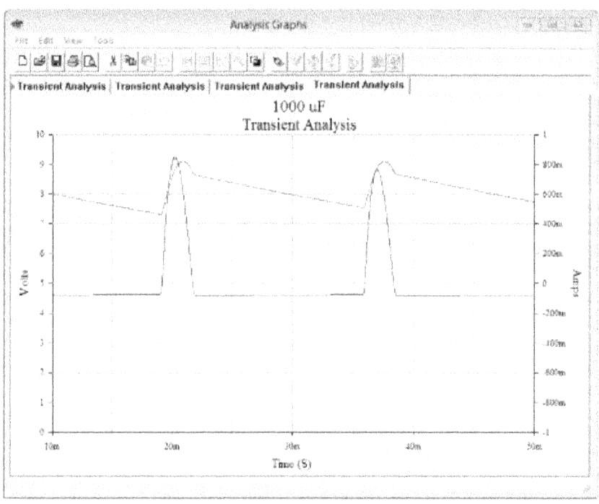

Figure 3.9 Transient analysis simulation for half-wave rectifier using a 1000 μF filter capacitor.

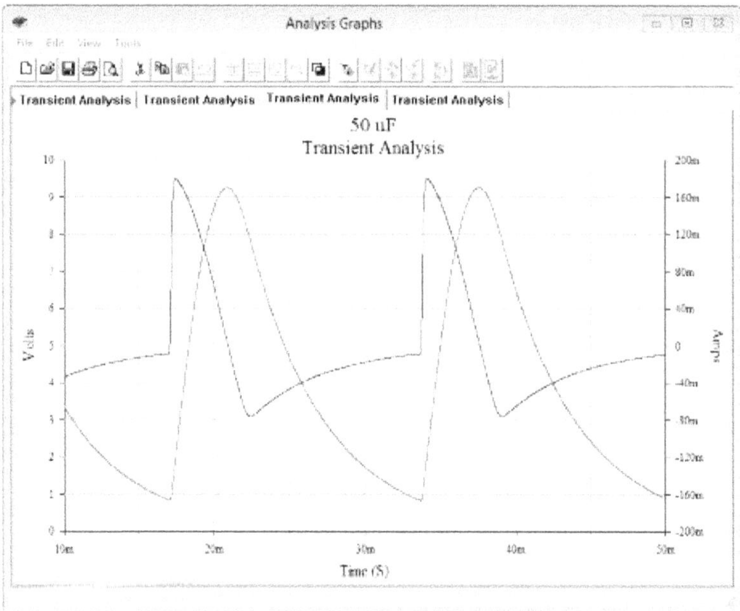

Figure 3.10 Transient analysis current waveform using a 1000 μF filter capacitor.

The blue current waveform peaks at approximately 800 milliamps, or over four times the value compared to using the smaller capacitor. Also, the width of the positive pulse has decreased to about 2.5 milliseconds. The discharge phase is nearly flat, implying that the output voltage must be more stable as this capacitor is the only source for load current during this phase.

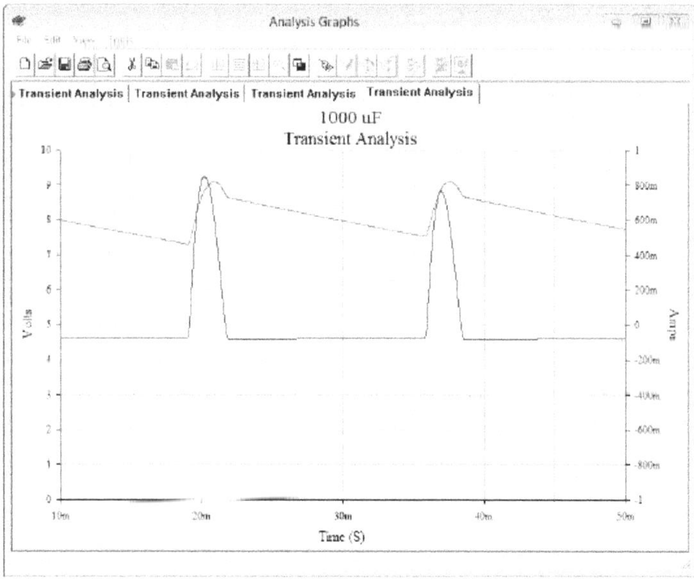

Figure 3.11 Transient analysis current waveform using a 1000 μF filter capacitor.

Full-wave Rectification

An improvement on half-wave rectification is *full-wave rectification.* Half-wave rectification is inefficient because it essentially throws away the negative portion of the input. In contrast, full-wave rectification makes use of the negative portion by inverting or flipping its polarity. The resulting circuit is modestly larger and more complicated but results in large performance improvements. For example, filter capacitor size is greatly reduced.

There are two popular methods to achieve full-wave rectification. The first method uses a pair of diodes with a center-tapped (i.e., split) secondary. The second method uses a four diode bridge network. The diode bridge form is also capable of producing a bipolar output (i.e., a positive output along with a negative output, typically of the same magnitude).

The two diode center-tapped secondary circuit is shown in Figure 3.12. This schematic also includes the filter capacitor.

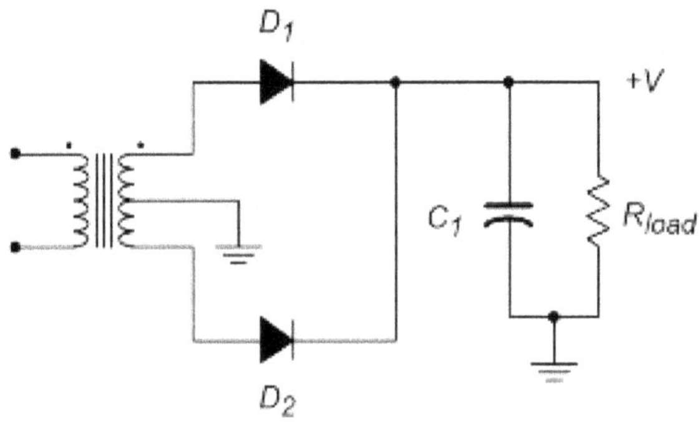

Figure 3.12 Full-wave center-tapped rectifier with capacitor.

The operation is as follows. During the positive half of the source voltage diode D1 is forward-biased while D2 is reverse-biased. Therefore the upper half of the secondary behaves like a simple half-wave rectifier allowing current to flow through D1 and into the load. Due to the reverse-bias on D2, the lower half presents an open circuit and is effectively removed. In mirror fashion, when the applied potential switches polarity D1 will be reverse-biased while D2 becomes forward-biased. Current is now free to flow through D2 into the load. Thus, both halves of the input waveform are used. The resulting waveforms are illustrated in Figure 3.13. For clarity, the filtering effect of the capacitor is not shown and Vin represents one half of the total secondary voltage.

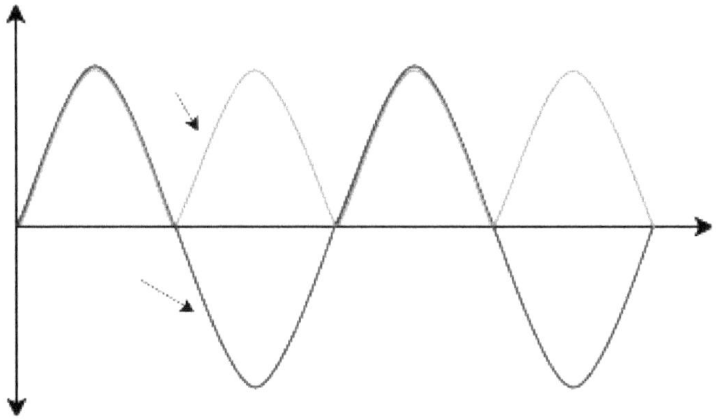

Figure 3.13 Full-wave rectifier waveforms

An important point to remember about this configuration is that the load only "sees" half of the secondary at any given time. Therefore, the load voltage will only be half of the total secondary voltage (minus one forward diode drop). For example, if the transformer has a 10:1 turns ratio and is being fed from a standard 120 volt source, the secondary will produce 12 volts RMS. Ignoring the diode drop, the load would see half of this, or 6 volts RMS (about 8.5 volts peak). Typically, transformers are rated by their total secondary voltage so this transformer would be referred to as having a "12 volt center-tapped secondary".

A four diode bridge rectifier is shown in Figure 3.14. A filter capacitor is included. Also, note the usage of a standard, non center-tapped secondary. As this is a very common configuration, the four diode bridge is available as a single four-lead part in a variety of sizes and current capacities.

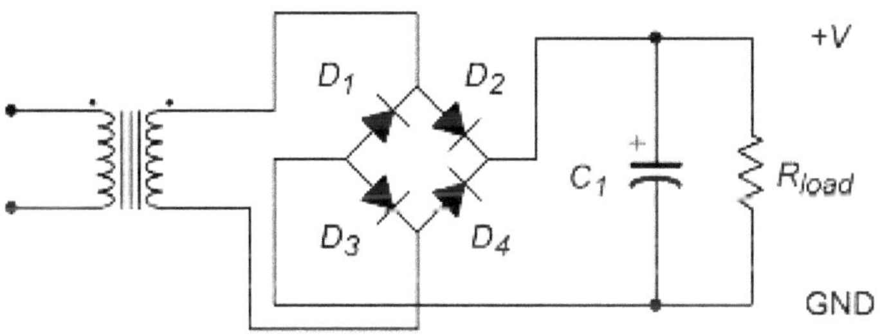

Figure 3.14 Full-wave bridge rectifier with capacitor.

The operation of this circuit is illustrated in Figure 3.15 for the positive portion of the input. First, current flows from the top of the secondary to the D1/D2 junction. Only D2 offers a forward-bias path so current flows through D2 to the junction with D4 and the load. As D4 presents a reverse-bias path, current must flow down through the load. From ground, current continues to the D1/D3 junction. Although at first glance it appears that current could flow through either diode, remember that the cathode of D1 is tied to the high side of the secondary. Therefore, its potential must be higher than the anode side, making it reverse-biased. Consequently, the current flows down through D3. A similar situation occurs at D4 and current is directed back to the low side of the secondary. In short, D2 and D3 are forward-biased while D1 and D4 are reverse-biased. The load sees the entire secondary voltage minus two forward diode drops.

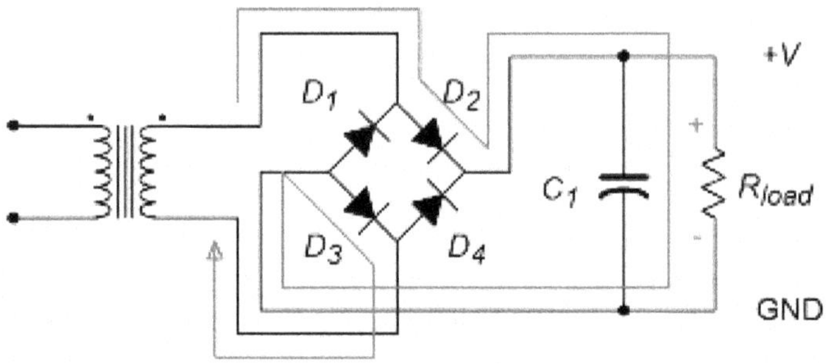

Figure 3.15 Full-wave bridge rectifier analysis, positive input.

During the negative polarity portion of the input the situation is reversed as illustrated in Figure 3.16. Current will flow from the bottom of the secondary through *D4*, down through the load, and finally back to the top of the secondary via *D1*. Thus, *D1* and *D4* are forward-biased while *D2* and *D3* are reverse-biased. The important thing is that in both cases, the current flows down through the load, top to bottom, resulting in a positive output voltage.

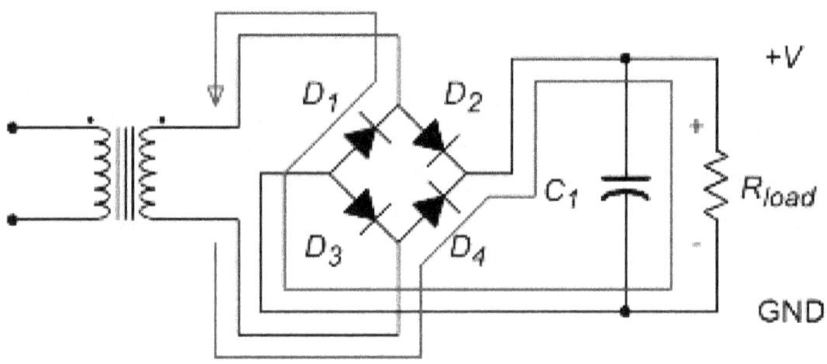

Figure 3.16 Full-wave bridge rectifier analysis, negative input.

Example 3.1

Design a rectifier/filter that will produce an output voltage of approximately 30 volts with a maximum current draw of 300 milliamps. It is to be fed from a 120 VAC RMS source. The ripple voltage should be less than 10% of the nominal output voltage at full load.

For this design we shall focus on using common off-the-shelf parts. As we have seen, the full-wave rectifiers are more efficient at converting AC to DC so we shall go that route, specifically, a four diode bridge arrangement. We will use the circuit of Figure 3.14 as a guide.

The first item to consider is the size of the transformer. A 30 volt output would require a peak secondary voltage of at least 32 volts as we must add in two forward diode drops. The equivalent RMS value is 32/√2 or 22volts. At full load the filtered output voltage will droop somewhat so a somewhat larger value is called for. A standard 24 volt secondary should suffice. Given the 300 milliamp load current rating, the transformer must be at least 0.3 amps · 24 volts or 7.2 VA.

As far as the capacitor is concerned, it must be rated for the peak voltage. The peak equivalent is 24 VAC RMS · √2 or 34 volts. Although a 35 volt rated capacitor might be tried, a standard 50 volt rating would leave a generous safety margin and increase reliability. To find the capacitance value we must first find the effective worst case load impedance.

$$R = \frac{V_{out}}{I_{max}}$$

$$R = \frac{30\text{ V}}{0.3\text{ A}}$$

$$R = 100\ \Omega$$

It will be useful to compare this back to the simulation depicted in Figure 3.9. Our ripple specification is somewhat tighter than that achieved in the prior simulation. This is apparent by noting how far the output voltage has dropped by midway through the off portion of the cycle. Consequently, we will need a larger time constant, perhaps by a factor of two. That puts us at 200 milliseconds.

$$\tau = RC$$

$$C = \frac{\tau}{R}$$

$$C = \frac{0.2\text{ s}}{100\ \Omega}$$

$$C = 2000\ \mu\text{F}$$

A 2200 μF standard value should be sufficient.

Computer Simulation

To verify our results, the design from Example 3.1 is simulated. The schematic is shown in Figure 3.17. To simplify the simulation, a 24 volt RMS source is used in place of the transformer. The worst case load is simulated via a 100 Ω resistor. For the initial test the filter capacitor is omitted so that we can ensure the proper peak voltage and waveforms are created. The results of a transient analysis are shown in

Figure 3.18. The secondary voltage is shown in red while the load voltage is shown in blue. The full-wave waveform is exactly as expected, including a slight reduction in the peak voltage value due to two forward diode drops. The output peak is just above 30 volts, as desired.

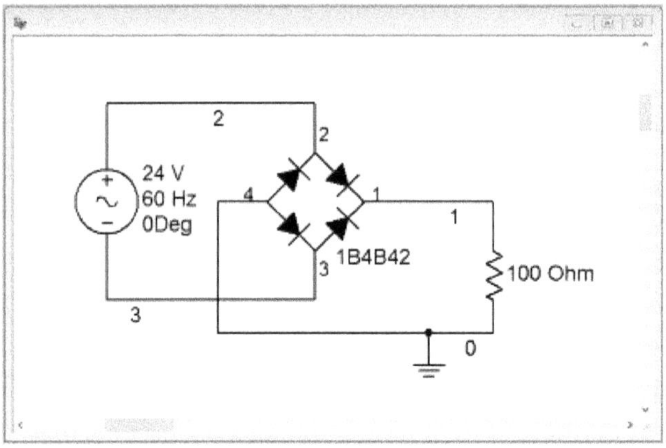

Figure 3.17 Simulation schematic for the design of Example 3.1 without capacitor.

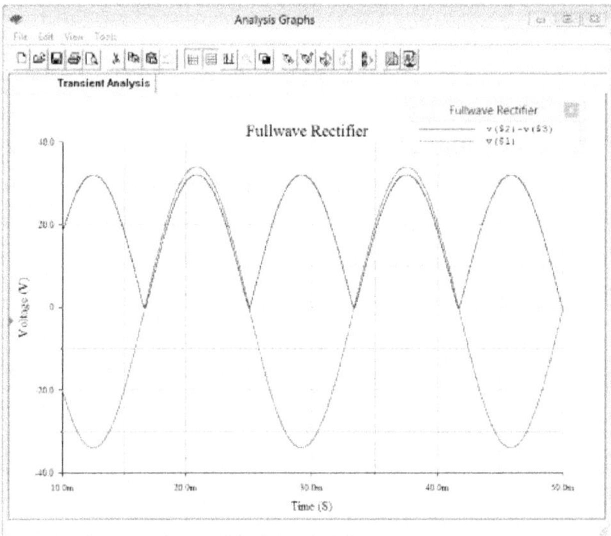

Figure 3.18 Transient analysis of the design of Example 3.1 without capacitor.

Now that we have confidence in the voltage level and waveform, the output filter capacitor is added as shown in Figure 3.19. A transient analysis is run again with the resulting input and load voltage waveforms depicted in Figure 3.20. The load voltage is shown in red. The average value is just over 30 volts and the peak-to-peak ripple is less than two volts, as desired. Note that the full-load peak voltage with the capacitor is slightly less than what was seen in the capacitor-less version. If the load current demand were to increase, both droop and ripple would get worse.

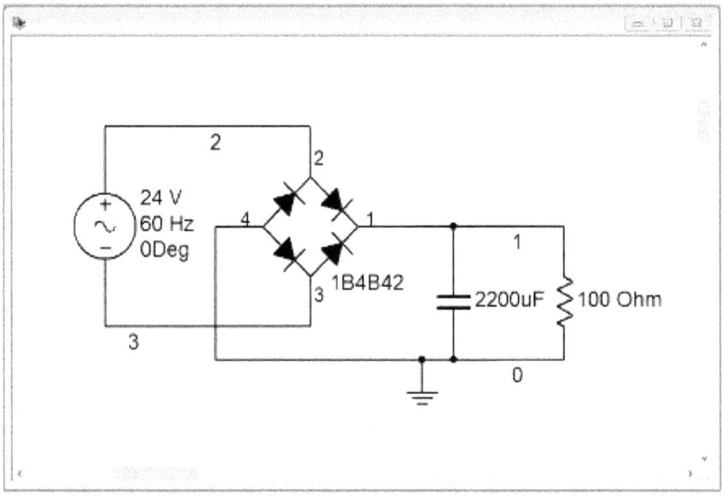

Figure 3.19 Simulation schematic for the design of Example 3.1 with capacitor.

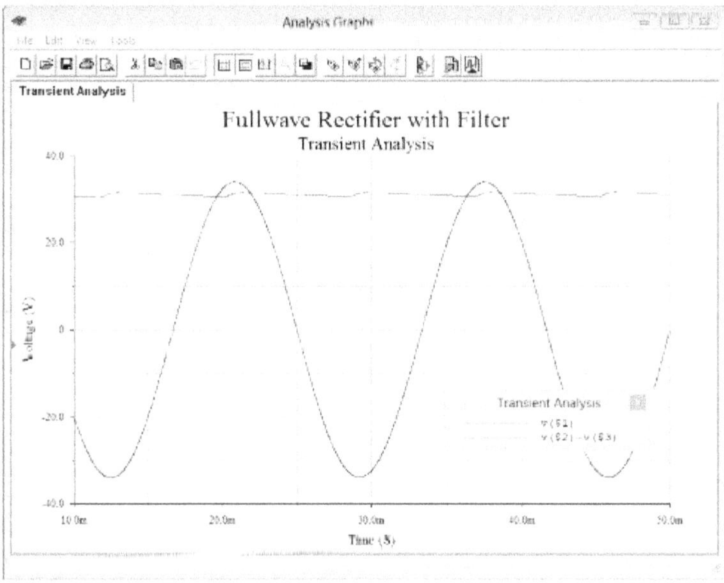

Figure 3.20 Transient analysis of the design of Example 3.1 with capacitor.

Full-wave Bridge with Dual Outputs

As mentioned, the full-wave bridge can be configured to create a dual output bipolar supply. This is shown in Figure 3.21. Note the inclusion of the center tap on the secondary of the transformer and the location of the ground connection between the two loads and their associated capacitors.

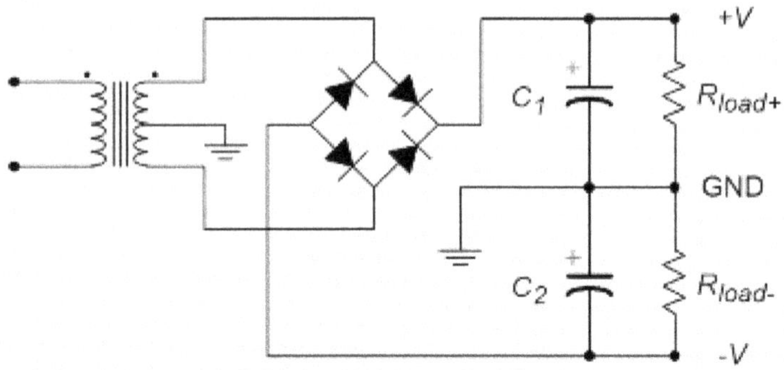

Figure 3.21 Dual output full-wave bridge rectifier.

One way of thinking of this is that we have simply created a new reference point, splitting in half the total output potential of the circuit presented in Figure 3.14. Alternately, it can be thought of as the upper half of the secondary driving Rload+ while the bottom half drives Rload-, as if the bridge and two-diode versions were somehow combined in a transporter accident, as in the 1958 movie The Fly, although it doesn't scream "Help me! Help me!" in a tiny little voice at the end.

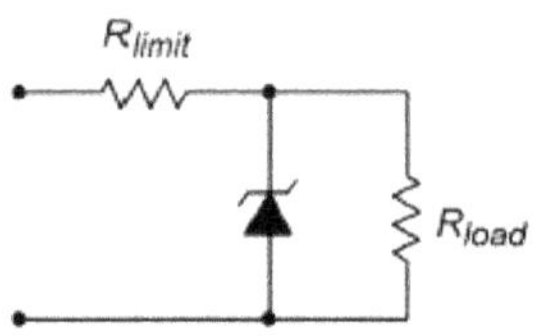

Figure 3.22 Simple Zener regulator

Zener Regulation

Adding a large capacitor to a rectifier is necessary to store and transfer energy so that a smooth, ideally non-varying voltage results. As noted previously, under heavy load the ripple would increase in amplitude and the average voltage would drop.

This issue can be greatly reduced by adding a Zener diode and current limiting resistor to the output, following the capacitor. This is called a Zener regulator and is shown in Figure 3.22.

The operation of the Zener regulator is fairly straightforward. Recall that when reverse-biased with a sufficiently large potential, the normal reverse diode behavior of an open switch abruptly changes to maintain a fixed voltage; the Zener potential. The current through the diode begins to increase dramatically once this potential is reached. If we place a Zener diode across the output of our filtered rectifier, the Zener will attempt to limit the output voltage to the Zener potential. To prevent excessive and possibly destructive current draw by the Zener diode, the voltage difference between the capacitor voltage and the Zener potential is dropped across a series current limiting resistor. This limiting resistor will set the maximum amount of output current. This current is then split between the Zener diode and the load.

Under light load conditions, most of this current will flow through the Zener diode. Under heavy load conditions, most of the current will be drawn by the load with little flowing through the Zener diode. If the load current demand is too heavy, no current is available for the Zener diode and it stops conducting. Regulation is lost and the limiting resistor forms a voltage divider with the load.

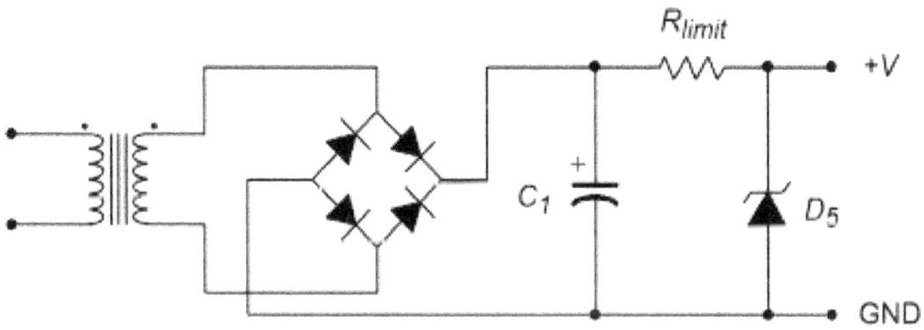

Figure 3.22 Simple Zener regulator.

A complete rectifier/filter/Zener regulator circuit is show in Figure 3.22. Let's examine how R limit interacts with the load.

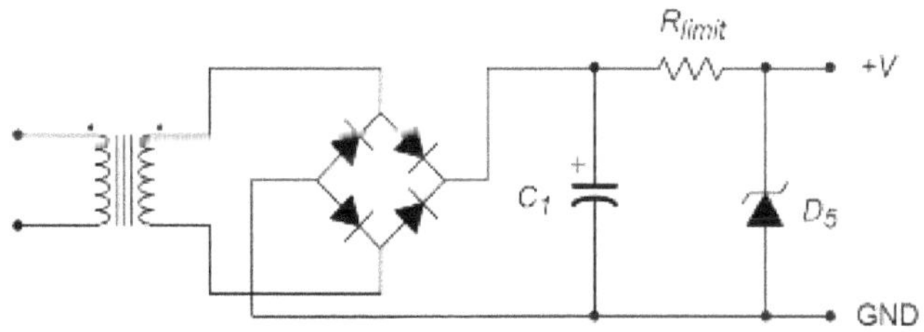

For proper operation, the Zener potential (D5) is the desired DC output voltage and the peak secondary voltage is set somewhat higher. We wish to guarantee that under full load conditions the lowest capacitor voltage due to ripple is still greater than the desired DC output voltage. The difference between the capacitor voltage and the Zener potential drops across Rlimit. Therefore

$$I = \frac{V_{cap} - V_Z}{R_{limit}}$$

Under no-load conditions all of this current flows down through the Zener diode. The maximum load current is equal to this value (at which point no current flows through the Zener diode).

Example 3.2

Determine the maximum load current for a DC supply such as that found in Figure 3.22. The capacitor voltage is 15 volts average with ±1 volt of ripple (i.e., 16 volts dropping to 14 volts). The Zener potential is 12 volts and Rlimit is 4.7 Ω.

The highest possible continuous load current is the current through Rlimit (ignoring IZT). The limiting case for continuous draw will occur when the capacitor voltage is at its lowest value, or 14 volts.

$$I = \frac{V_{cap} - V_Z}{R_{limit}}$$

$$I = \frac{14\,\text{V} - 12\,\text{V}}{4.7\,\Omega}$$

$$I = 426\,\text{mA} \;\; \text{(actually a few mA less due to } I_{ZT}\text{)}$$

The highest peak current through the Zener diode is found at the maximum capacitor voltage and assumes no current is drawn by the load.

Note that this worst case current times the Zener potential results in a power dissipation of about 10 watts. Of course, during normal operation with a load drawing current, the diode dissipation is much reduced. It is interesting to note that the Zener dissipates maximal power when the load current is zero. Consequently, we can think of this circuit as shifting current from the Zener diode to the load as the load demands more current13.

3.3 CLIPPERS

Sometimes it is useful to limit the maximum amplitude of a signal. This might be done for protection, for example when too large of an input signal might damage the following circuit. It might also be employed as a means of wave shaping, that is, morphing a signal into another shape. A good example is the purely aesthetic desire to emulate the sound of "fuzz" guitar. In the early days of rock music it was discovered that over-driving a guitar amplifier in an attempt to make it louder created considerable distortion and this produced a new and interesting sound quality. Technically, this is largely caused by the power stage of the amplifier reaching its maximum output level. Any portion of the waveform above this level is simply cut off or *clipped*[14]. The practical problem here is that the only way to achieve this sound is to crank up the guitar amplifier's volume to ten[15] and live with the attendant high loudness level. Not too popular with the neighbors, that's for sure. In contrast, if the signal could be limited before the power amplifier stage in an attempt to mimic the clipping, the effect could be achieved without the resulting loudness. This proved to be so popular among guitarists that by the 1970s numerous companies were making "fuzz boxes" and "distortion pedals", each with their own twist on the concept.

The simplest form of clipper places a diode (or two parallel diodes of opposing polarity) in parallel with the load. The diode will limit the output voltage swing to its forward turn-on potential; 0.7 volts for a silicon device. This circuit is somewhat limiting (pun intended) as you are stuck with a 0.7 volt limit value. What if we need to limit at some other potential, say 12 volts? While it is possible to stack a bunch of diodes in series to increase the limit point, a more flexible and practical approach involves biasing the diode with a DC source. This is called a biased diode clipper.

A biased diode clipper used to limit positive signals is shown in Figure 3.23. The operation is as follows. For any input signal that is less positive than the bias potential Vclip, the diode will be in reverse-bias. This means that the diode branch presents a high resistance and is effectively removed from the circuit. Therefore, Vin flows through R to the output unimpeded. If the input signal exceeds by Vclip by approximately 0.7 volts, the diode turns on resulting in a very low internal resistance. As the internal resistance of the DC source is also very low, this creates a low impedance path to ground and results in a voltage divider with R. As R is a much greater resistance value, virtually all of the input signal above the turn-on voltage will be dropped across R, never reaching the output. Therefore, we can control or program the clip point by adjusting the bias voltage. Clipping will occur at approximately Vclip plus 0.7 volts, assuming a silicon diode is used.

A negative biased clipper is shown in Figure 3.24. Both the diode and DC bias voltages have been flipped to the opposite polarity. The operation of this circuit is similar to the positive clipper. In this variant, the negative bias potential ensures that the diode is reverse-biased as long as the input level is more positive than Vclip minus0.7 volts. Once the input signal goes below this voltage, the diode turns on creating the shorting path and limiting the output voltage.

A bipolar or dual-polarity clipper can be created by combining the positive and negative clippers. It is possible to limit the positive and negative swings independently, as illustrated in the following Example.

Example 3.3

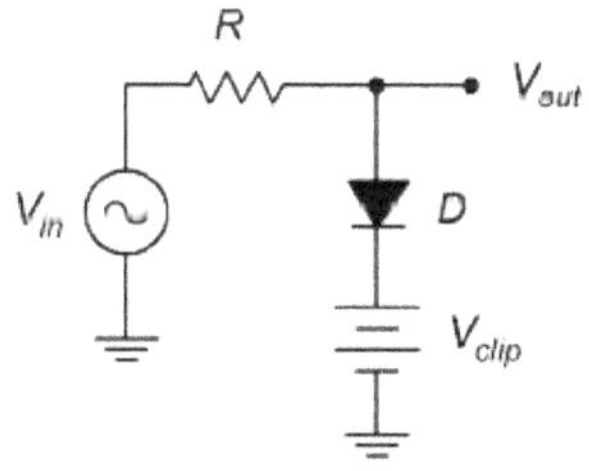

Figure 3.23 Biased positive clipper.

Determine the output signal for the circuit of Figure 3.25. The input voltage is a 12 volt peak sine wave at 100 Hz. D1 and D2 are silicon switching diodes. R = 10 kΩ, V1 = 4 volts and V2 = 8 volts.

First off, note that the precise value of *R* is unimportant here. It simply needs to be significantly larger than the on-resistance of the diodes. Based on our prior study of diode resistance in Chapter 2 (see the discussion around Figure 2.14) it is likely that the dynamic diode resistance will be under 100 Ω once full turn- on is reached. This resistance ratio is more than sufficient to create an effective voltage divider.

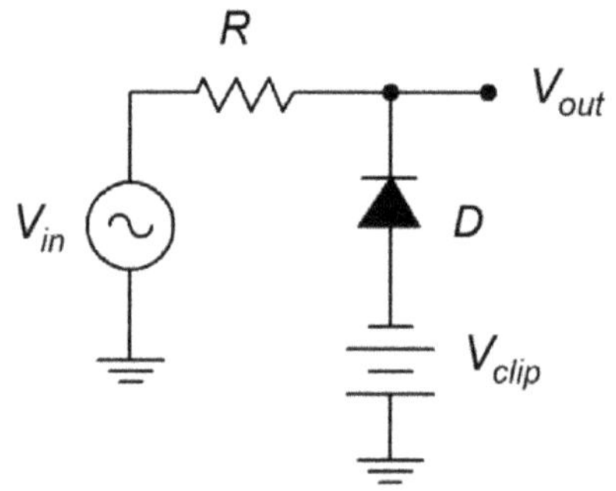

Figure 3.24 Biased negative clipper.

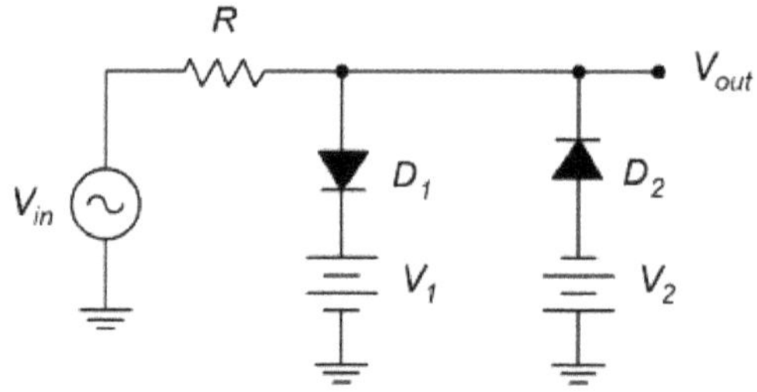

Figure 3.25 Dual clipper for Example 3.3.

The positive clip level will be set by *V1*. Adding in the forward potential of *D1* we arrive at 4.7 volts. The negative clip level will be set by *V2*. Including in the forward potential of *D2* we arrive at -8.7 volts.

Thus, we expect to see a sine wave that is clipped at +4.7 volts and -8.7 volts. It should appear as a sort of lopsided cross between a sine wave and a square wave.

Computer Simulation

To verify and visualize our computations, the circuit of Example 3.3 is simulated with a transient analysis. The circuit schematic is shown in Figure 3.26 (a). For the diodes, common 1N914 switching diodes are used.

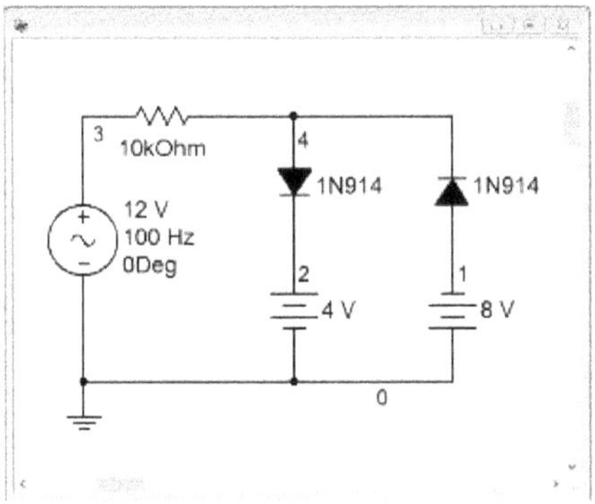

Figure 3.26 (a) Simulation schematic for dual clipper of Example 3.3.

The results of the transient analysis are shown in Figure 3.26 (b). The input waveform is shown in red while the output waveform is shown in blue. The input sine wave appears as expected with the specified 12 volt peak. The output follows the input perfectly for all values that are within the clip points. Beyond the clip points, the output voltage flattens. That is, it is limited to just below +5 volts (the programmed

+4.7) and to just above -9 volts (the programmed -8.7 volts).

Careful inspection of the output waveform plot reveals that it is not perfectly flat at the voltage limits. In fact, there is a slight rounding that is most noticeable toward the transitions. This is due to the fact that the diodes do not turn on immediately. The dynamic resistance of the diodes change with the size of the signal. That is, the greater the input signal is above the clip point, the more current that will flow, and thus the dynamic resistance decreases, strengthening the effect of the voltage divider.

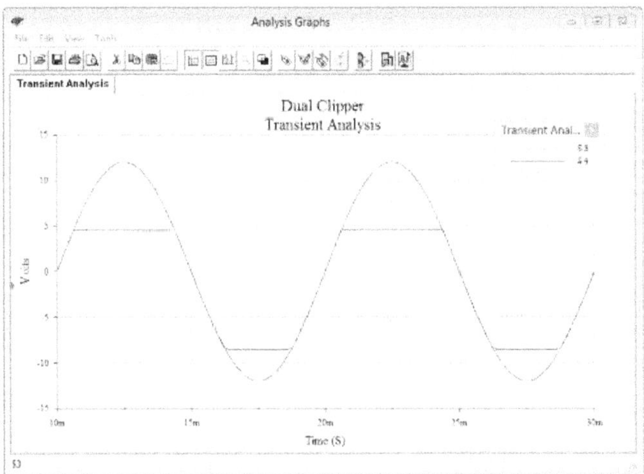

Figure 3.26 (b) Transient analysis for dual clipper of Example 3.3.

3.4 CLAMPERS

A *clamper* is circuit that adds a DC offset to an AC signal in such a way that the resulting voltage is uni-polar. A positive clamper adds a positive offset such that the former negative peak now sits at zero volts. In like fashion, a negative clamper adds in a negative offset such that the former positive peak now sits at zero volts.

Clampers are also referred to as *DC restorers*. Clampers can also be biased so that the new peak point is something other than zero volts.

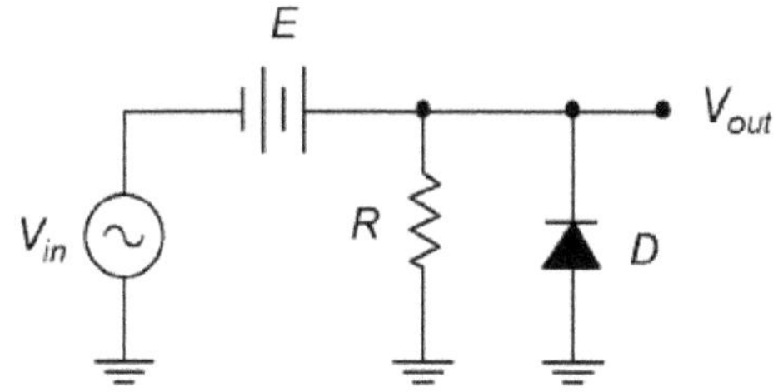

Figure 3.27 Prototype DC offset circuit.

The concept of a clamper is fairly simple; we just add a DC voltage to the existing AC signal. The trick is in getting the circuit to automatically determine what the DC shift needs to be. This way, if the amplitude of the input signal changes, the offset can track with it.

First, let's consider the prototype circuit in Figure 3.27. This is a fixed DC offset circuit. The DC source E adds a positive offset to the input signal. If the offset is equal to the peak value of the input, the negative peak will rise up to zero volts and the diode will never turn on (meaning that it will not load the input and change the wave shape).

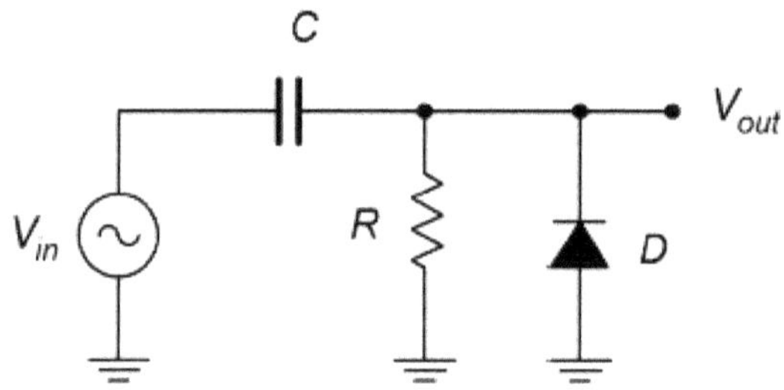

Figure 3.27 Prototype DC offset circuit.

In Figure 3.27 the fixed DC source has been replaced with a capacitor. This capacitor is used to create the DC offset. Unlike the fixed source, the capacitor voltage will vary with the peak value of the input and therefore precisely compensate to produce an ideally clamped output signal. As long as the time constant for this capacitor and the surrounding resistance is much longer than the period of the input waveform, it will achieve proper clamping action.

Here is how the circuit operates. For the initial positive cycle, the capacitor is uncharged and the diode is reverse-biased. As the RC time constant is much longer than the input period, the output voltage merely follows the input voltage. Once the input signal swings negative, the diode turns on. This bypasses the parallel resistor and drastically reduces the charge time constant. This means that the capacitor voltage will begin to track the negative portion of the input signal while the output stays near zero volts.

Note that the capacitor voltage will have a polarity of minus-to-plus from left to right, in keeping with Kirchhoff's voltage law. The capacitor voltage will track the negative input voltage all the way down to the negative peak. Once the input begins to reverse slope and rise toward zero, the diode will be turned off due to the potential now held on the capacitor. At this point, the capacitor has a voltage across it that is equivalent to the negative peak value of the input signal and it will behave just like the fixed DC voltage source in the prototype. The input is just now starting to track in the positive direction from its negative peak while the capacitor holds this same magnitude of voltage. The result is that the output is at zero volts and as the input continues to swing positive, the output will track it, thus producing the desired level shift.

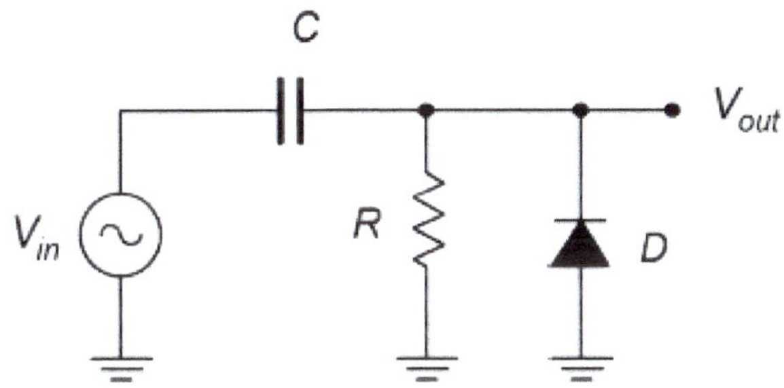

Figure 3.28 Biased positive clamper.

Of course, circuits are never perfect. First, the forward voltage drop of the diode will result in a negative peak that's not precisely at zero volts but is instead about -0.7 volts. Second, it may take more than one cycle of the input to "grab" the peak value, all depending on the period and the precise charge and discharge time constants. As you might guess, flipping the polarity of the diode will result in a negative clamper instead of a positive clamper. Also, if we add a DC source in series with the diode, like we did with the biased clipper, we can create a biased clamper. This is shown in Figure 3.28.

Example 3.4

Determine the output signal for the circuit of Figure 3.29. The input voltage is a 10 volt peak sine wave at 1 kHz. C = 10 μF, R = 10 kΩ, Vclamp = 5.7 volts and D is a silicon switching diode.

The configuration is a positive biased clamper. First, we need to ensure that the discharge time constant is much longer than the period. The period is 1/f, or 1 millisecond. The discharge time constant is

$$\tau = RC$$
$$\tau = 10\,k\Omega \times 10\,\mu F$$
$$\tau = 100\ \text{milliseconds}$$

The DC clamping source will produce a positive offset of 5 volts (5.7 volts minus the 0.7 volt forward diode drop). This means that we should see a 20volt peak-to-peak sine wave that swings between +5 volts and +25 volts.

Computer Simulation

To verify the analysis of Example 3.4, the circuit is captured as shown in Figure3.30. A common 1N914 switching diode is used

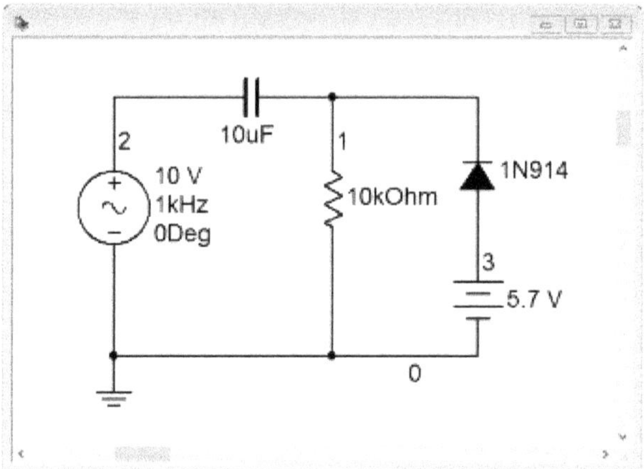

Figure 3.29 Simulation schematic for biased clamper of Example 3.4.

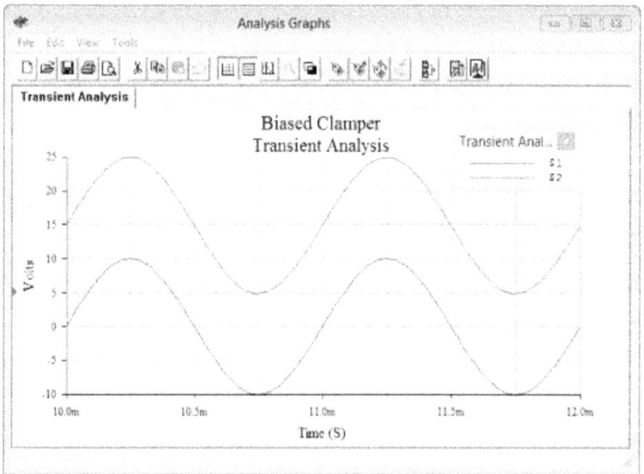

Figure 3.30 Transient analysis for biased clamper of Example 3.4.

The transient analysis is shown in Figure 3.31 and is precisely as predicted. The input waveform is blue and the clamped output is shown in red. The analysis was delayed ten milliseconds in order to get past the initial charge phase and observe the steady state operation.

CONCISE SUMMARY

- Numerous AC diode applications have been examined in this chapter. A single diode may be used to create half-wave rectification, producing pulsating DC from AC. This is achieved by simply blocking one of the two polarities with a diode. A more efficient form of rectification is full-wave rectification. In this scheme, one of the two polarities is effectively flipped. This may be achieved via a two diode circuit that employs a split secondary transformer or via a four diode bridge circuit using a non-tapped secondary. The addition of a split secondary to the bridge circuit enables a dual polarity output.
- In order to smooth the pulsating DC into a relatively constant level, a filter capacitor is added in parallel with the load. The larger the capacitor, the greater the filtering and smoothing action, however, this will also increase peak charging current. A Zener diode may be employed to further stabilize the output voltage.
- Clippers are used to limit the range of an input signal. They may be designed to clip the positive portion, the negative portion or both polarities of the input waveform. The positive and negative clip levels may be adjusted independently.
- Clampers are used to create a DC level shift that is dependent on the peak level of the input waveform. The shift may be positive or negative, and may also include an optional bias. The operation of the clamper hinges upon the charge versus discharge time constants for a series capacitor and associated diode.

CHAPTER-4

BASICS OF BIPOLAR JUNCTION TRANSISTORS (BJTS)

4.1 INTRODUCTION: AN OVERVIEW

Bipolar Junction Transistors (BJTs) are fundamental semiconductor devices with crucial roles in electronic circuits and amplification. Comprising three layers—P-type and N-type semiconductor materials—the BJT operates based on the movement of charge carriers across these layers. There are two main types: NPN (negative-positive-negative) and PNP (positive-negative-positive). In operation, a small current at the transistor's input terminal controls a larger current flow between the other two terminals. This controlled amplification mechanism forms the basis for BJT applications in signal amplification, switching, and voltage regulation. BJTs are integral to analog electronics, serving in audio amplifiers, radio frequency circuits, and various signal processing applications. A thorough understanding of the basics of BJTs is essential for electronics engineers and designers, forming the cornerstone for the development of diverse electronic systems and technologies.

The bipolar junction transistor, or BJT, is a foundational electronic component. It serves as the basis for a variety of applications ranging from simple amplifiers to device control to complex digital computing circuitry.

Variations exist for applications spanning very low to very high frequency work; low, medium and high power; inexpensive general purpose through highly specialized niche items; and so forth. No matter what a BJT has been optimized for, all BJTs can be considered to be current boosting devices. Of course, if you can boost current, then you can also boost voltage and power, depending on the associated impedances. Further, all BJTs share the same basic structure: three alternating layers of N-type and P-type material with one external lead attached to each layer. In this manner, the BJT can be thought of as an extension of the basic diode: just add another segment of oppositely doped material to one end of the diode creating a second PN junction. The configuration could be either PNP or NPN. There are uses for both types and circuits often work best when the two types are used together.

4.2 THE BIPOLAR JUNCTION TRANSISTOR

In prior work we discovered that the PN junction is the foundation of the basic diode. Under normal operating conditions the interface between the N-type and P-type materials is devoid of free charges and is referred to as a depletion region. The dissimilar Fermi levels of N-type and P-type materials lead to an "energy hill" between them, and without an external potential of the proper polarity, the junction will not allow current to flow. The required magnitude is a function of the material used but it is always the case that the P material (anode) must be positive with respect to the N material (anode). We extend this idea by adding a second portion of N material to the other side of the P material, creating an N-P-N "sandwich" of sorts. This is shown in Figure 4.1.

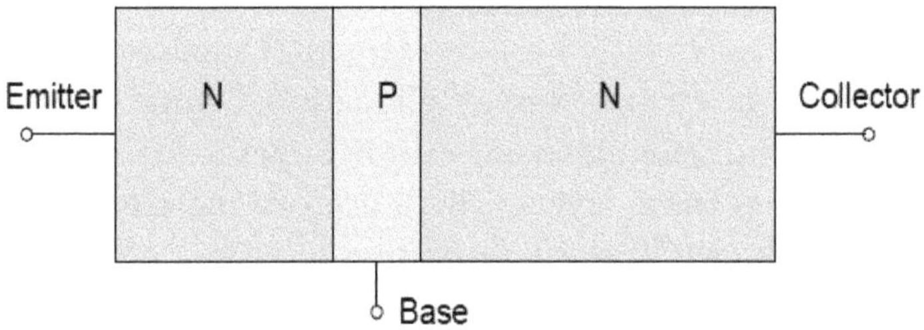

Figure 4.1 Basic configuration of NPN bipolar junction transistor.

This diagram is drawn to ease the understanding of the operation of the device, extending our earlier diode work. In contrast, real BJTs are built in more of a "layer cake" fashion, N-P-N bottom to top16. Of course, the spatial orientation of the device has no bearing on its operation so this is not a major issue for our purposes. The three terminals are named the emitter, base and collector. The collector is the largest of the three regions while the base is relatively thin and lightly doped.

Above absolute zero there will be recombination and two depletion regions will form as shown in Figure 4.2. Compare this figure to the basic PN junction drawing found at the beginning of Chapter 2, Figure 2.1.

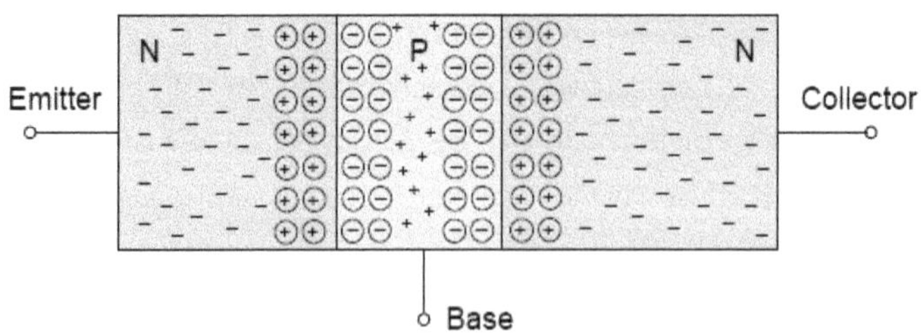

Figure 4.2 Charges in NPN BJT (base region widened to show detail).

A Simple Two-Diode Model

Because this device contains two depletion regions, a much simplified model can be created using two diodes as shown in Figure 4.3. Please keep in mind that this is a very limited model (as we shall soon see).

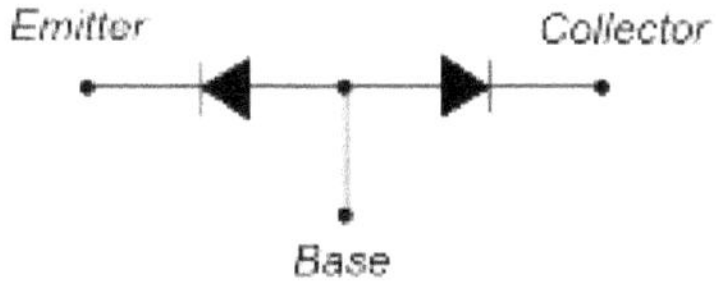

Figure 4.3 Diode model of NPN BJT.

If you were to test an NPN BJT with an ohmmeter, two leads at time, this model would successfully predict the results. If the red (positive) lead of the ohmmeter is connected to the base and the black (negative) lead is connected to either the emitter or collector, a low resistance will be indicated. This is because the ohmmeter will modestly forward-bias the base-emitter or base-collector junction. Similarly, if the leads are reversed, the meter will indicate high resistance because the junction under consideration will be reverse-biased. If the two leads are connected to the emitter and collector, a high reading will result regardless of the polarity. This is because one of the two junctions will be reverse-biased which results in no current flow through either of them due to the series connection.

Biasing the BJJT

Now let's consider adding external sources to bias the transistor. We begin by adding two DC sources with associated current limiting resistors as shown in Figure 4.4.

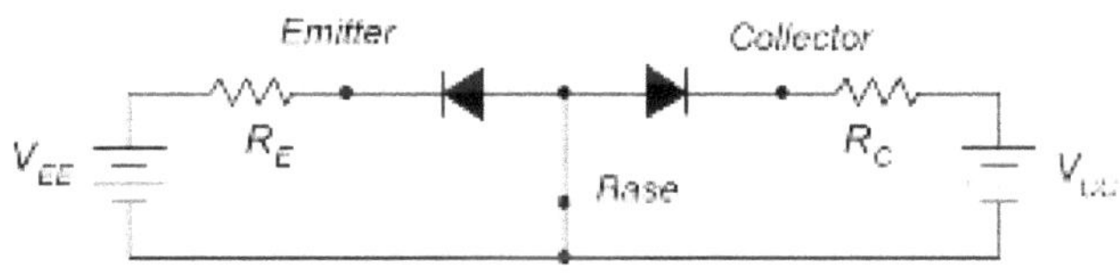

Figure 4.4 Double reverse-bias.

This circuit is comprised of two loops, one between the base emitter and the second between the base-collector. In the B-E loop, the emitter supply *VEE* reverse biases the base-emitter diode. A similar situation occurs in B-C loop where the collector supply reverse-biases the base-collector diode. The result is that virtually no current flows anywhere in the circuit. If the two supplies are reversed in polarity then both diodes become forward-biased and we see currents flowing in both loops dependent on the precise values of the supplies and associated resistors.

No surprises so far. Now consider if we forward-bias the base-emitter diode while simultaneously reverse- biasing the base-collector diode, as shown in Figure 4.5.

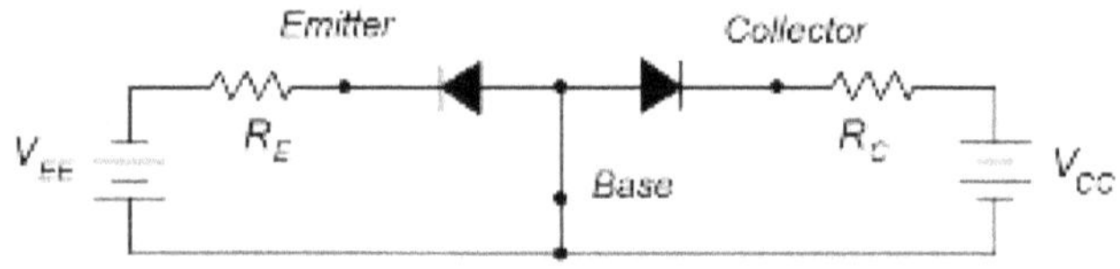

Figure 4.5 Forward-reverse bias.

With a simple pair of diodes we'd expect the B-E loop to show a high current and the B-C loop show negligible current. With a BJT, this is not what happens. Instead, what we see is a high current in both loops, and those currents are very nearly equal in magnitude. How does this come about?

The key to understanding this situation is that the base of the BJT is thin and lightly doped. In contrast, the dual diode model splits the base into two separate pieces of material and that makes all the difference. To get a better handle on what's happening here, let's take a closer look at this forward-reverse bias circuit but this time substituting the transistor diagram of Figure 4.2. Refer to Figure 4.6.

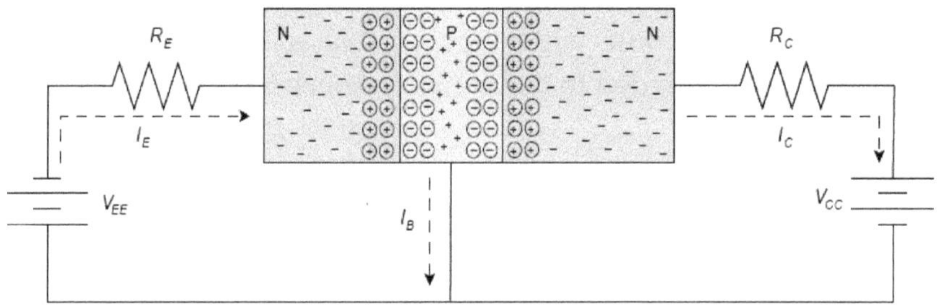

Figure 4.6 Forward-reverse bias, electron flow.

Electron flow will facilitate this explanation so we'll draw the current directions using dashed lines. From the left side of the diagram, electrons exit the emitter supply and enter the N emitter. Here they are the majority carrier. The base-emitter depletion creates an energy hill just as it did with a single PN junction. As long as there is sufficient potential from the emitter supply, the electrons will be pushed into the base. These electrons will attempt to recombine with the majority base holes, however, because the base is physically thin and lightly doped, only a small percentage of the injected electrons will recombine with base holes and exit the base terminal back to ground. This current is called the base current or the recombination current. Meanwhile, the vast majority of the remaining electrons (95% to over 99%) will find their way to the base-collector depletion region and then to the collector. Once in the collector, the electrons are again the majority carrier and flow back to the positive terminal of the collector power supply. The energy diagram of the transistor is depicted in Figure 4.7. Compare this to the single PN junction energy diagram found at the beginning of Chapter 2, Figure 2.2.

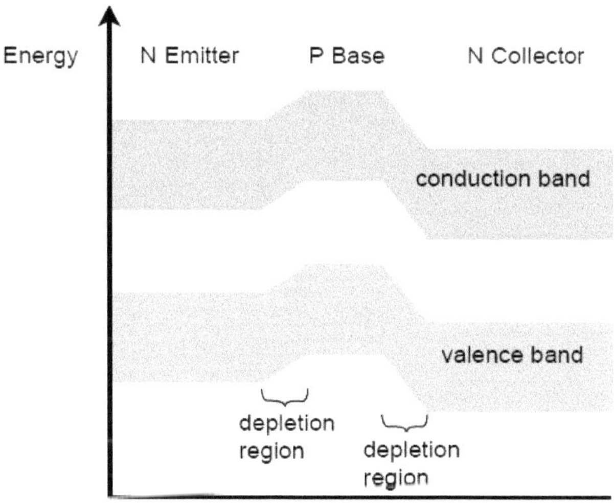

Figure 4.7 Energy diagram of BJT.

At first glance, it might appear as though the emitter and collector leads can be swapped with no change in operation. With real-world devices this is not possible generally because the emitter and collector regions are optimized and not physically identical. Thus, placing transistors into a circuit backwards, with emitter and collector leads swapped, will usually result in unpredictable behavior.

Based on the foregoing discussion and what we already know about PN junctions, we can summarize transistor performance as follows:

- From KCL, $IE = IC + IB$.
- $IC >> IB$, therefore $IE \approx IC$.
- The base-emitter junction is forward-biased, therefore $VBE \approx 0.7$ V (silicon).
- The base-collector junction is reverse-biased, therefore VCB is large.
- Conventional current flows into the collector and base, and out of the emitter.

We can also define a couple of transistor performance parameters. The ratio of collector current to emitter current is called α (alpha). α typically is greater than 0.95. A somewhat more useful parameter is the ratio of collector current to base current. This is called β (beta) and can also be found on transistor spec sheets as hFE (hFE is one of four hybrid parameters). It is also referred to generically as current gain (if IB is in the input signal and IC is the output signal then β represents the amount of signal boost or gain). For small signal transistors β typically is in the range of 100 to 200, although it can be larger. For power transistors, β tends to be smaller, more like 25 to 50. Presented as formulas we have:

$\alpha = IC / IE$ (4.1)

$\beta = IC / IB$ (4.2)

And with a little math,

$\alpha = \beta / (\beta+1)$

$\beta = \alpha / (1-\alpha)$

$IC = \beta IB$

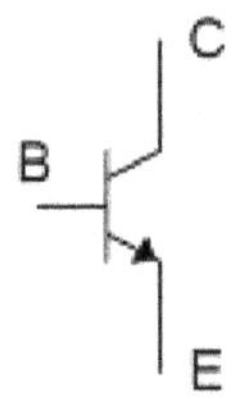

Figure 4.8 NPN Schematic symbol

Finally, we come to the schematic symbol of the NPN BJT, as shown in Figure 4.8. A common variation places the body of the device within a circle. Following the standard, the arrow points to N material and in the direction of easy conventional current flow.

The PNP Bipolar Junction Transistor

The PNP version of the BJT is created by swapping the material for each layer. The outcome is the logical inverse of the NPN regarding current directions and voltage polarities. That is, conventional current flows into the emitter, and out of the collector and base (echoing the electron flow of the NPN). Further, voltages across the device have reversed polarity, for example, $VBE \approx -0.7$ V. All of the other characteristics remain unchanged so equations such 4.1 and 4.2 are still applicable. Just about any NPN-based circuit has its PNP counterpart. The schematic symbol of the PNP reverses the emitter arrow. As the base is now the N material, the arrow points toward the base. This is illustrated in Figure 4.9.

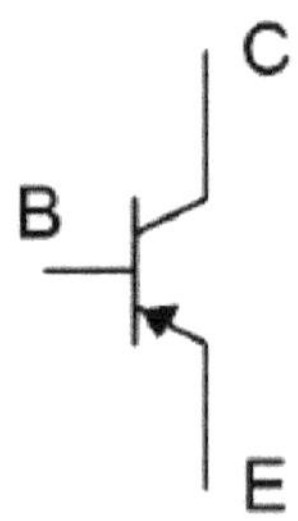

Figure 4.9 PNP Schematic symbol

4.3 BJT COLLECTOR CURVES

One of the more useful BJT device plots is the family of collector curves. This is a series of plots of collector current, IC, versus collector-emitter voltage, VCE, at varying levels of base current, IB. To generate these curves we drive the base terminal with a fixed current source establishing IB. A DC power supply is attached from the collector to emitter and then swept from zero volts to some upper value. This establishes VCE. Simultaneously, we track the resulting collector current and plot the result. This results in one trace. The base current is then increased and the DC supply swept again for a second trace. This process is repeated to create a family of curves. An example is shown in Figure 4.10.

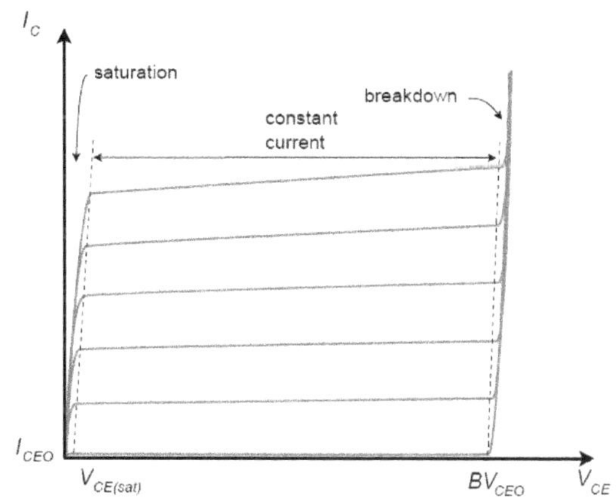

Figure 4.10 A family of collector curves.

The bottom curve results when *IB* = 0. Ideally, the corresponding collector current would be 0 but a small leakage current occurs. This is usually referred to as *ICEO*, meaning the Collector-Emitter current with the base terminal Open (i.e., no base current). The curves above this correspond to increasing levels of base current; each new base current stepping up a fixed amount for each subsequent trace (e.g., 0 μA, 10 μA, 20 μA, 30 μA, etc.).

The most striking thing about this set of curves is that there are three distinct regions or zones of operation. To the extreme left of the curve the current rises rapidly. This is known as the *saturation region*. The break-over point is fairly small at just a few tenths of a volt. This can be found on a data sheet as *VCE(sat)*. The saturation region is used in transistor switching applications.

At the extreme right is another region where the collector current rises rapidly. This is called the *breakdown region*. This is the same effect we saw with individual diodes. We do not wish to operate devices in this region as damage may result. The breakdown voltage is denoted on most data sheets as *BVCEO* (Collector to Emitter voltage with an Open base). For general purpose devices this will be in the range of 30 to 60 volts or so, but it can be much higher.

In between these two extremes is a region where the collector current is relatively constant, showing only a modest positive slope. This is the *constant current region*. This is where we want the transistor to operate for applications such as linear amplifiers.

A device called a *curve tracer* can be used to generate this family of curves in the lab. A very good approximation for β can be determined using these curves. First, we determine the approximate circuit values for *IC* and *VCE*, and locate this point on the graph. We then find the nearest plot line to that point. From the intersection of *VCE* and and this plot line we track back to the vertical axis to find the precise value of *IC* for that trace. We count the number of traces and multiply by the base current step size to determine the corresponding base current. We then divide the two values and arrive at β.

Example 4.1

Assume we have a BJT operating at *VCE* = 30 V and *IC* = 4 mA. If the device is placed in a curve tracer and the resulting family of curves appears as in Figure 4.11, determine the value of β. Assume the base current is increased 10 µA per trace.

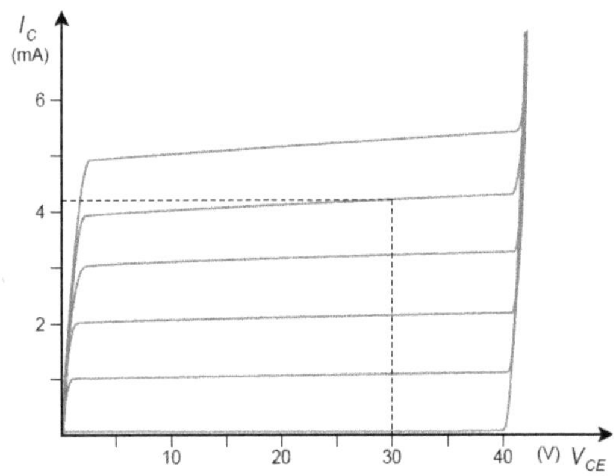

Figure 4.11 Curve tracer display for Example 4.1.

First, find a trace close to the operating point of 30 volts and 4 mA. Draw a vertical line at 30 volts and stop when that line intersects the trace nearest to4 mA. In this example that's the second trace from the top. From that intersection point, track back to the vertical axis to determine the precise collector current. That's roughly 4.2 mA here. To determine the base current count up the number of traces to the selected trace. The selected trace is the fourth one up (do not include the bottom trace where IB is 0). The base current is increased by 10 µA per trace so that leaves us with IB = 40 µA.

$$\beta = \frac{I_C}{I_B}$$

$$\beta = \frac{4.2\ \text{mA}}{40\ \mu\text{A}}$$

$$\beta = 105$$

When looking at the collector curves, a good question we might ask is why the collector current rises as *VCE* increases. This is due to the fact that the increased collector-emitter voltage is responsible for an increase in collector-base voltage (by definition, *VCE* = *VCB* + *VBE*). *VCB* is the reverse-bias potential on the collector-base PN junction. As this reverse potential increases, the collector-base depletion region widens. As it widens, it penetrates further into the base layer. Because the base is effectively narrowed, the chances for recombination are reduced, thus reducing base current and effectively increasing β.

If we extend the constant current region traces back into the second quadrant they intersect at a point called the Early Voltage, named after James Early, and denoted as *VA*. This is illustrated in Figure 4.12.

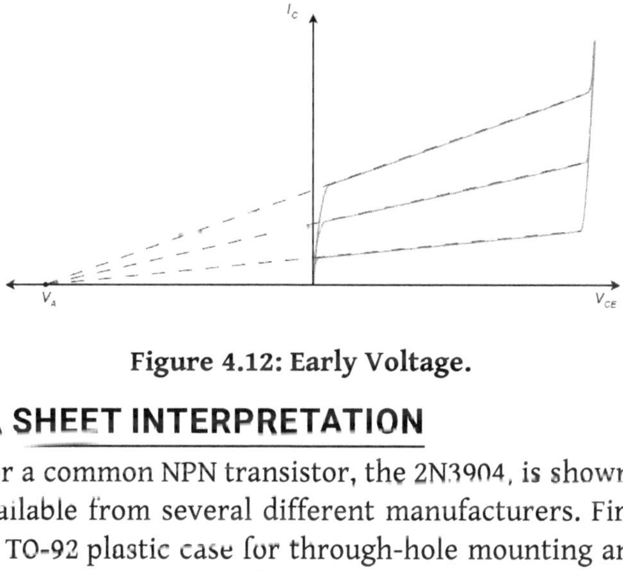

Figure 4.12: Early Voltage.

4.4 BJT DATA SHEET INTERPRETATION

The data sheet for a common NPN transistor, the 2N3904, is shown in Figure 4.13. This model is available from several different manufacturers. First off, note the case style. This a TO-92 plastic case for through-hole mounting and is commonly used for small signal transistors. Under the maximums we find the device has a maximum power dissipation of 625 mW in free air (ambient temperature of 25°C), a maximum collector current of 200 mA and a maximum collector-emitter voltage of 40 V. Obviously, the device cannot withstand maximum current and voltage simultaneously.

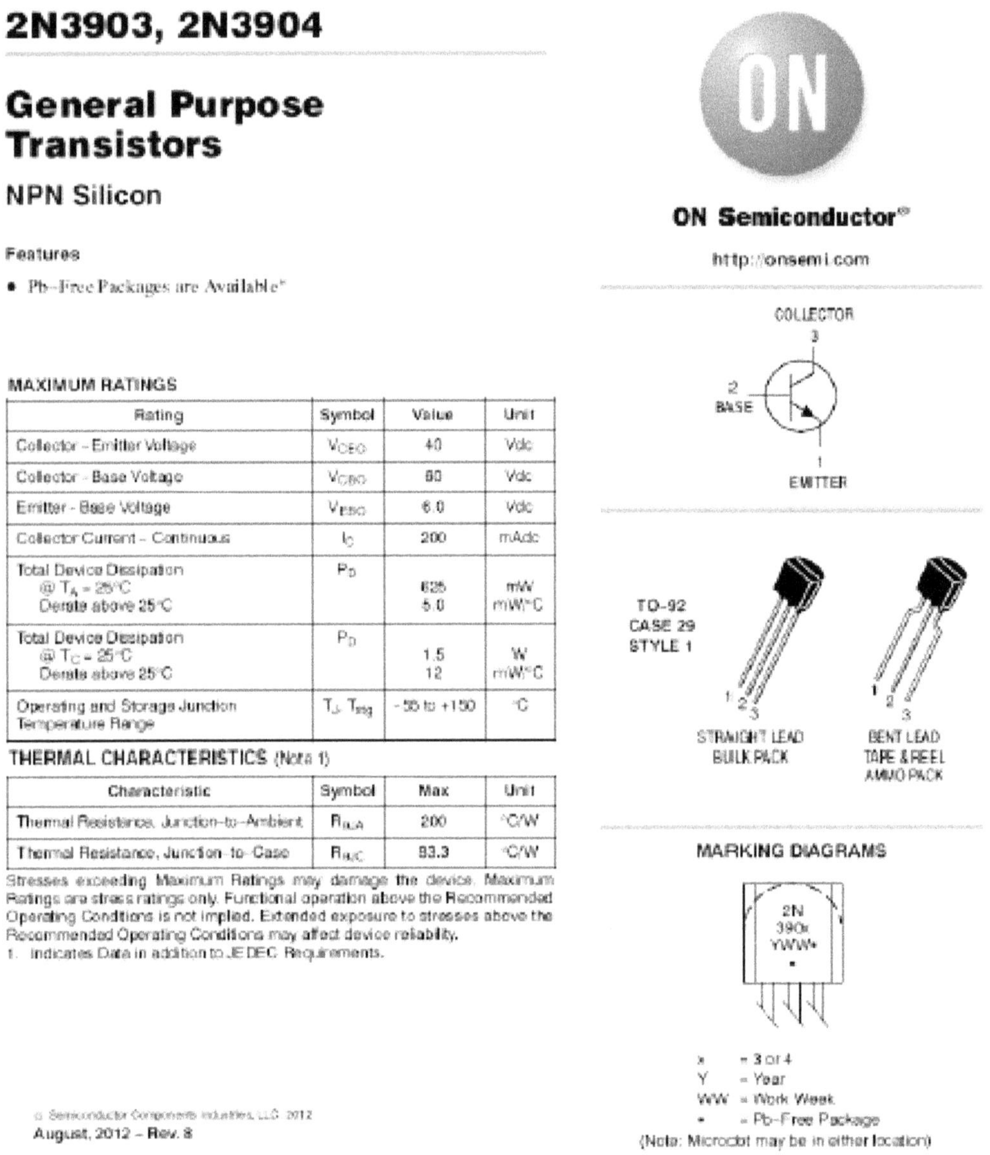

2N3903, 2N3904

General Purpose Transistors

NPN Silicon

Features

- Pb−Free Packages are Available*

MAXIMUM RATINGS

Rating	Symbol	Value	Unit
Collector − Emitter Voltage	V_{CEO}	40	Vdc
Collector − Base Voltage	V_{CBO}	60	Vdc
Emitter − Base Voltage	V_{EBO}	6.0	Vdc
Collector Current − Continuous	I_C	200	mAdc
Total Device Dissipation @ $T_A = 25°C$ Derate above 25°C	P_D	625 5.0	mW mW/°C
Total Device Dissipation @ $T_C = 25°C$ Derate above 25°C	P_D	1.5 12	W mW/°C
Operating and Storage Junction Temperature Range	T_J, T_{stg}	−55 to +150	°C

THERMAL CHARACTERISTICS (Note 1)

Characteristic	Symbol	Max	Unit
Thermal Resistance, Junction−to−Ambient	$R_{\theta JA}$	200	°C/W
Thermal Resistance, Junction−to−Case	$R_{\theta JC}$	83.3	°C/W

Stresses exceeding Maximum Ratings may damage the device. Maximum Ratings are stress ratings only. Functional operation above the Recommended Operating Conditions is not implied. Extended exposure to stresses above the Recommended Operating Conditions may affect device reliability.
1. Indicates Data in addition to JEDEC Requirements.

August, 2012 − Rev. 8

Figure 4.13a 2N3904 data sheet. Used with permission from SCILLCdba ON Semiconductor.

In Figure 4.13b we find a variety of characteristics including nominal values for β (listed here as *hFE*) under various conditions. At particularly small or large collector currents β tends to drop off. Also, note the wide 3:1 variance at 10 mA. Perhaps more illustrative are the graphs from the third page, Figure 4.13c.

2N3903, 2N3904

ELECTRICAL CHARACTERISTICS (T_A = 25°C unless otherwise noted)

Characteristic		Symbol	Min	Max	Unit
OFF CHARACTERISTICS					
Collector-Emitter Breakdown Voltage (Note 2) (I_C = 1.0 mAdc, I_B = 0)		$V_{(BR)CEO}$	40	–	Vdc
Collector-Base Breakdown Voltage (I_C = 10 μAdc, I_E = 0)		$V_{(BR)CBO}$	60	–	Vdc
Emitter-Base Breakdown Voltage (I_E = 10 μAdc, I_C = 0)		$V_{(BR)EBO}$	6.0	–	Vdc
Base Cutoff Current (V_{CE} = 30 Vdc, V_{EB} = 3.0 Vdc)		I_{BL}	–	50	nAdc
Collector Cutoff Current (V_{CE} = 30 Vdc, V_{EB} = 3.0 Vdc)		I_{CEX}	–	50	nAdc
ON CHARACTERISTICS					
DC Current Gain (Note 2)		h_{FE}			–
(I_C = 0.1 mAdc, V_{CE} = 1.0 Vdc)	2N3903		20	–	
	2N3904		40	–	
(I_C = 1.0 mAdc, V_{CE} = 1.0 Vdc)	2N3903		35	–	
	2N3904		70	–	
(I_C = 10 mAdc, V_{CE} = 1.0 Vdc)	2N3903		50	150	
	2N3904		100	300	
(I_C = 50 mAdc, V_{CE} = 1.0 Vdc)	2N3903		30	–	
	2N3904		60	–	
(I_C = 100 mAdc, V_{CE} = 1.0 Vdc)	2N3903		15	–	
	2N3904		30	–	
Collector-Emitter Saturation Voltage (Note 2)		$V_{CE(sat)}$			Vdc
(I_C = 10 mAdc, I_B = 1.0 mAdc)			–	0.2	
(I_C = 50 mAdc, I_B = 5.0 mAdc)			–	0.3	
Base-Emitter Saturation Voltage (Note 2)		$V_{BE(sat)}$			Vdc
(I_C = 10 mAdc, I_B = 1.0 mAdc)			0.65	0.85	
(I_C = 50 mAdc, I_B = 5.0 mAdc)			–	0.95	

Figure 4.13b 2N3904 data sheet (cont).

The upper-most graph depicts the variation of β with both collector current and temperature. The normalized β is plotted on the vertical axis. That is, this is not the expected value but is a ratio used to compare β under varying conditions.

2N3903, 2N3904

TYPICAL STATIC CHARACTERISTICS

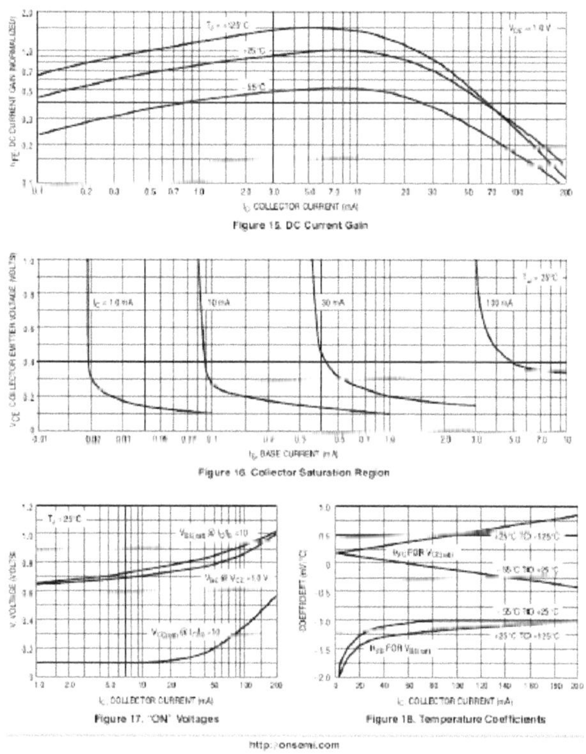

Figure 4.13c 2N3904 data sheet (cont).

For example, at room temperature and 10 mA, the normalized value is 1.0. The second page indicated a range of 100 to 300 for the 2N3904's β under these conditions. Let's say we measure one particular transistor to have a β of 200. If we were to operate this transistor at a lower current, say 0.2 mA, the β would drop under these conditions would be 0.7/1.0 • 200, or 140. The graph also shows that, generally speaking, β tends to increase with increasing temperature.

The middle graph plots the collector-emitter saturation voltage, or VCE(sat), for various current conditions. This is an important parameter when dealing with transistor switching circuits. We shall refer back to this graph a little later in this chapter.

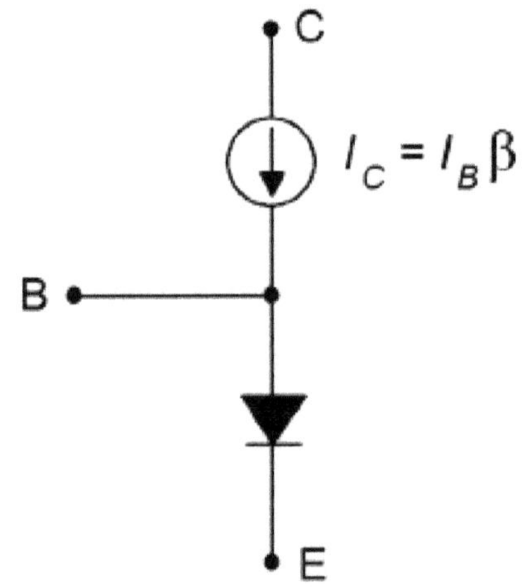

Figure 4.14 Ebers-Moll model of the NPNBJT.

4.5 EBERS-MOLL MODEL

The Ebers-Moll model is a widely used theoretical framework in semiconductor physics and electronics, specifically applied to bipolar junction transistors (BJTs). Named after its developers, John L. Moll and Bernard Ebers, this model provides a mathematical description of the behavior of BJTs under different operating conditions.

A good, functional model of the BJT is the simplified Ebers-Moll model shown in Figure 4.14. This utilizes an ideal diode to model the base-emitter junction and a current-controlled current source located at the collector-base. This model is sufficient to achieve good analysis results with a variety of DC and low frequency circuits. It is important to remember, though, that β varies not only from device to device, but also varies with changes in temperature, collector current and collector- emitter voltage.

We can put the Ebers-Moll model to use in basic DC biasing circuits. To properly bias the BJT we need to make the collector-base reverse-biased and the base-emitter forward-biased. In other words, *VC* > *VB* > *VE*. There are many ways to achieve this. One method places the emitter at ground, a modest DC source in the base-emitter loop, and a somewhat higher DC source at the collector. An example is shown in Figure 4.15.

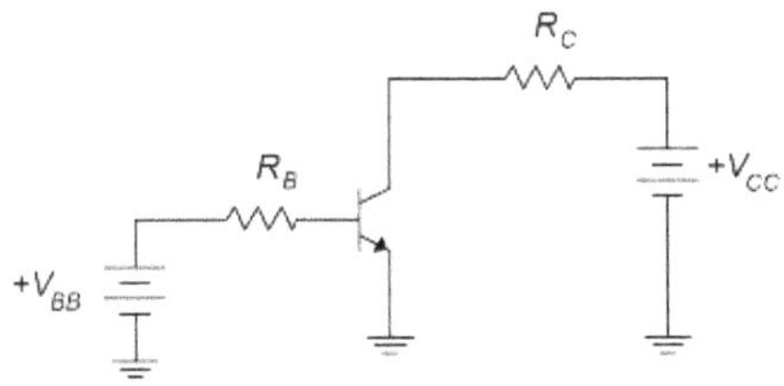

Figure 4.15 Simple base bias circuit.

The two resistors serve to limit the transistor's currents and voltages. Because the emitter is at ground, the common point, this circuit is classified as having a common emitter configuration. There are many possible common emitter circuits. We shall refer to this one specifically as base bias.

Now, let's replace the transistor with the Ebers-Moll model. The result, with added voltage polarities and current directions, is shown in Figure 4.16.

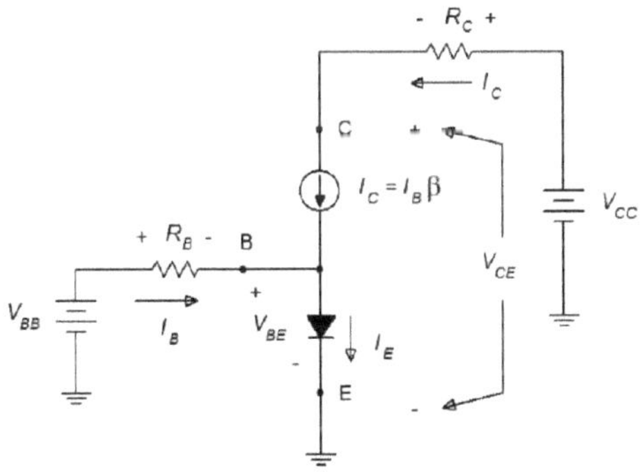

Figure 4.16 Base bias circuit with Ebers Moll model.

Given values for the resistors, power supplies and β, all of the currents and voltages may be determined using fundamental circuit analysis techniques. The basic idea is to create KVL equations for the two loops and then expand using Ohm's law. We begin with the base-emitter loop.

$$V_{BB} = V_{R_B} + V_{BE}$$
$$V_{BB} = I_B R_B + V_{BE}$$
$$I_B = \frac{V_{BB} - V_{BE}}{R_B} \tag{4.3}$$

And for the collector-emitter loop:

$$V_{CC} = V_{R_C} + V_{CE}$$
$$V_{CC} = I_C R_C + V_{CE}$$
$$V_{CE} = V_{CC} - I_C R_C \quad (4.4)$$

To obtain I_C, recall that $I_C = \beta I_B$.

Example 4.2

Determine the circulating currents and device voltages for the circuit of Figure 4.15 if VBB = 10 V, VCC = 15 V, RB = 200 kΩ, RC = 1 kΩ and β = 100.

Assume that the transistor is silicon.

First, find the base current. KVL dictates that the voltage across *RB* is 9.3 volts.

$$I_B = \frac{V_{BB} - V_{BE}}{R_B}$$
$$I_B = \frac{10\,\text{V} - 0.7\,\text{V}}{200\,\text{k}\Omega}$$
$$I_B = 46.5\,\mu\text{A}$$

Now find the collector current and follow with Ohm's law and KVL.

$$I_C = \beta I_B$$
$$I_C = 100 \times 46.5\,\mu\text{A}$$
$$I_C = 4.65\,\text{mA}$$

$$V_{CE} = V_{CC} - I_C R_C$$
$$V_{CE} = 15\,\text{V} - 4.65\,\text{mA} \times 1\,\text{k}\Omega$$
$$V_{CE} = 10.35\,\text{V}$$

For the sake of completeness, the voltage across *RC* is 4.65 volts, *VCB* is 9.65 volts and *IE* is 4.6965 mA.

The preceding example illustrates that the place to start the analysis is in the base- emitter loop instead of the collector-emitter loop. This is because in the base-emitter loop we have the forward-biased base-emitter junction which has a known potential (approximately 0.7 volts). In contrast, the collector-emitter voltage is an unknown as it includes the reverse-biased collector-base junction. That voltage will depend on other circuit elements, most notably the collector resistor and associated supply.

An improvement on the circuit of Example 4.2 would be to redesign it for a single power supply rather than two supplies. This is easy to do. All that is needed is to keep the base current unchanged. If that remains at its original value then the collector current won't change and consequently nothing in the collector-emitter loop will change either. Using the 15 volt source for *VBB* means that the voltage across *RB* will increase to 14.3 volts. Ohm's law then indicates that *RB* must be 14.3 volts divided by 46.5 μA, or 307.5 kΩ.

β Variation Issues

There is a major problem with the circuit of Figure 4.15, namely, it lacks stability of collector current and collector-emitter voltage. As we shall see in upcoming chapters, it is important to keep these parameters stable in order to ensure consistent performance for many kinds of circuits. As we noted from our inspection of the 2N3904 data sheet, the variation of β can be quite large at a given operating point. If we also add in the variance due to temperature and other factors, we may be looking at a 10:1 range. If we repeat Example 4.2 with a doubled β of 200, the base-emitter loop is unchanged but the collector current doubles to 9.3 mA. This increases the voltage drop across RC to 9.3 volts which then forces VCE to drop to 5.7 volts. Given a typical production run of transistors, this circuit might exhibit collector currents from less than 4 mA to more than 10 mA. In some applications this variation in current might be tolerable but not in all of them. For example, suppose an LED is placed in series with RC. Because the brightness of an LED depends on its current level, the brightness will now depend on the β of the specific BJT used. If this is one LED in a larger display made up of similar circuits, then the illumination will be uneven between them causing the entire display to appear off kilter.

In fact, if this circuit was built in the lab, it is quite likely that after turning on the power, you could watch IC slowly rise on your ammeter. This is because the BJT will begin to warm up as it dissipates power. As noted from the data sheet, β increases with increasing temperature. Because IB is a fixed value, any rise in β means that IC must also rise. This increased current will tend to cause a further rise in power dissipation and temperature which causes a further increase in β, and the process cycles. We have created an inadvertent thermal positive feedback loop. Left unchecked, devices could overheat and be destroyed. We will examine biasing circuits that achieve high stability in the next chapter.

There is another interesting aspect to this circuit. As noted, if we substituted the original BJT with another unit that had a higher β, the collector current would rise. What if we continued this to higher and higher β values? For example, if we increased β to 400 (admittedly, rather high) the new collector current would seem to jump up to 46.5 μA• 400, or 18.6 mA. There is a "small" problem with this value. Ohm's law indicates that this current would develop a drop of 18.6 volts across the 1 kΩ RC but that's impossible because VCC is only 15 volts. The only way that "works" is if somehow the BJT is transformed magically into a 3.6 volt battery. No amount of prayer or letters to Santa will make that happen17.

4.6 DC LOAD LINES

So how do we determine the range of possible values of collector current and collector-emitter voltage in any given DC BJT circuit? One answer is to employ the concept of the DC load line. In general, a load line is a plot of all possible coordinate pairs of IC and VCE for a transistor in a given circuit. Referring back to Figure 4.16, we pick up with Equation 4.4 and solve it for IC :

$$V_{CE} = V_{CC} - I_C R_C$$

$$I_C = \frac{1}{R_C}(V_{CC} - V_{CE})$$

$$I_C = -\frac{1}{R_C} V_{CE} + \frac{V_{CC}}{R_C} \qquad (4.5)$$

Equation 4.5 is a linear equation of the form $y = mx + b$. The y intercept (the value of *IC* when *VCE* = 0) is *VCC/RC*. This is the maximum collector current that can be achieved. At this point the transistor is saturated and this maximum is referred to as *IC(sat)*. The x intercept (the value of *VCE* when *IC* = 0) is *VCC*. This represents the largest possible voltage across the transistor's collector-emitter. At this point the current is cut off, and therefore this voltage is called *VCE(cutoff)*. Lastly, the slope of the line is -1/*RC*. A plot is shown in Figure 4.17.

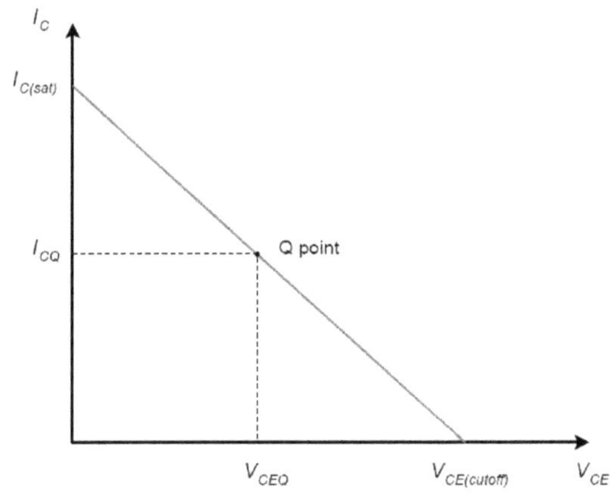

Figure 4.17 Generic DC load line.

To complete the graph, we also include the operating point for some specific transistor. This is called the quiescent point, or Q point, and the associated device current and voltage are called ICQ and VCEQ. All possible Q points lay on this line.

Referring back to Example 4.2 and using Equation 4.5, we can summarize the circuit as follows:

$I_{C(sat)}$ = 15 mA

$V_{CE(cutoff)}$ = 15 V

Q Point for β = 100:

I_C = 4.65 mA

V_{CE} = 10.35 V

Q Point for β = 200:

I_C = 9.3 mA

V_{CE} = 5.7 V

This is plotted in Figure 4.18.

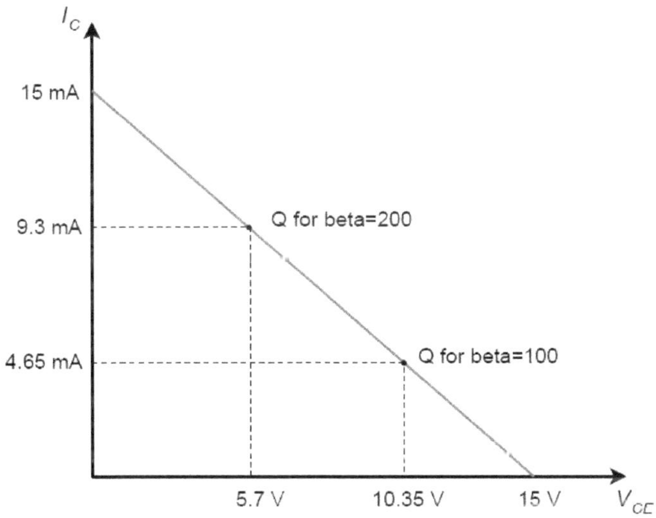

Figure 4.18 Load line for the variations on Example 4.2.

If we calculate a collector current that is greater than the saturation current, then we know that the actual current will be the saturation current maximum. For this circuit, any calculated value greater than 15 mA indicates that the transistor would produce only 15 mA (our earlier example using β = 400, for instance). In reality, the true value will be very slightly less. This is because the collector-emitter voltage does not go all the way down to zero volts when the device is saturated. Typically, $V_{CE(sat)}$ will be a tenth of a volt or so for small signal devices. Precise values can be determined from device graphs such as the middle graph of

Figure 4.13c, labeled "Collector Saturation Region". As an example, if IC = 10 mA and IB = 0.3 mA, then VCE(SAT) is approximately 0.15 V. It turns out that we can use saturation to our advantage in switching circuits, as we are about see.

4.7 BJT SWITCHING AND DRIVER APPLICATIONS

As mentioned, variation in β can cause changes in collector current. This can cause performance issues. For example, when driving an LED, this can lead to variance in brightness. But what if we purposely put the transistor into saturation? Saturation is a fixed value. It is inherently stable and β no longer matters. Effectively, when a BJT saturates, β is forced to drop to whatever value is needed to produce IC(sat). We just need to make sure that even the smallest β is large enough to cause saturation.

The Saturating Switch

A good example of this is the saturating LED driver circuit shown in Figure 4.19. To begin with, the whole point of the driver is to offload the current demand from the prior circuit. For example, we may wish to light an LED from the output of a logic gate or microcontroller chip. The problem is that those circuits might only be able to deliver, say, 5 mA when we might need well over 10 mA to achieve the desired brightness. The LED driver circuit is used to overcome this limitation.

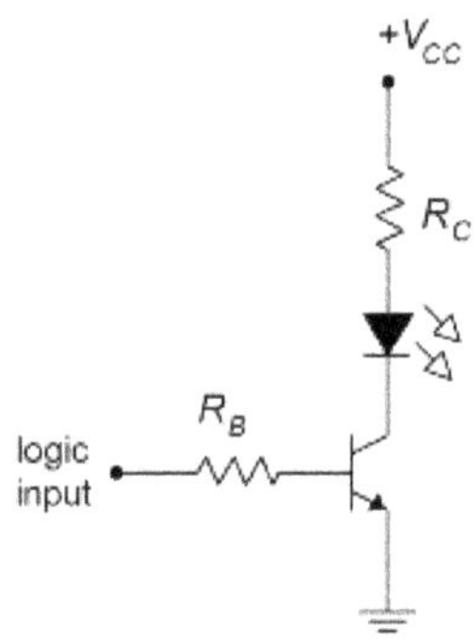

Figure 4.19 Saturating LED driver circuit(positive logic).

Note: The negative terminal of VCC is connected to ground (not shown).

With the driver, the logic circuit will only need to supply base current, not LED current. Here is how it works: If the logic input voltage is zero, there will be no base current. This means that there will be no collector current and therefore the LED will be off. At this point the BJT is in cutoff. In contrast, when the logic level goes high, all of the logic voltage drops across *RB*, with the exception of *VBE*. This creates *IB*. If properly designed, this current will be sufficient to put the BJT into saturation. The BJT acts as a switch, completing the circuit between the DC supply, the LED and the current limiting resistor, *RC*. For this to work reliably, we have to make sure that the ratio of saturation current to base current is much less than β. A value of 10 or so would guarantee hard saturation.

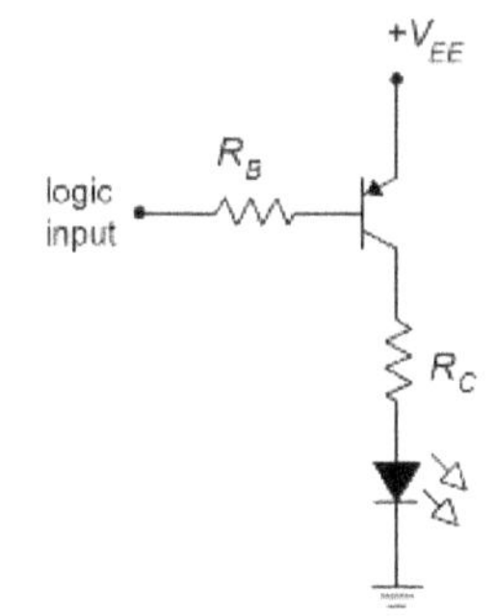

Figure 4.20 Saturating LED driver circuit (negative logic).

If we would like to invert the logic, that is, have a logic low turn on the LED and a logic high turn it off, we can achieve that with a PNP version of the circuit as shown in Figure 4.20.

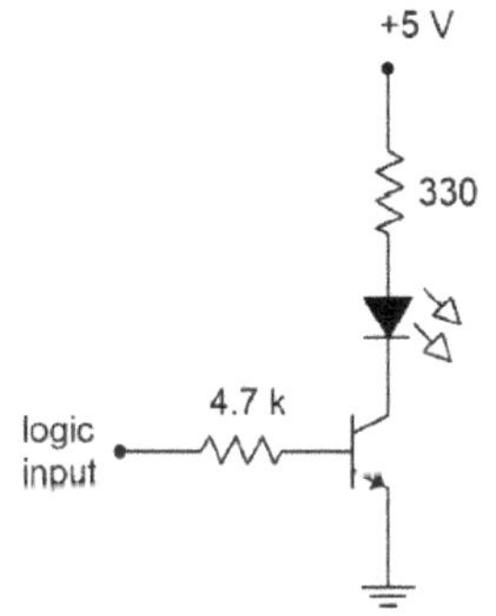

Figure 4.21 Circuit for Example 4.3.

Determine the LED "on" current for the circuit of Figure 4.21. Assume the logic "on" voltage is 5 volts, *VLED* = 1.8 volts and *VCE(sat)* = 0.

$$I_{C(sat)} = \frac{V_{CC} - V_{LED}}{R_C}$$

$$I_{C(sat)} = \frac{5\,\text{V} - 1.8\,\text{V}}{330\,\Omega}$$

$$I_{C(sat)} = 9.7\,\text{mA}$$

The ratio of these two currents is just over 10:1. This will guarantee hard saturation.

There are many different applications for saturating switches. Just about anywhere you can imagine a relay being used, you can consider a transistor switch.

The transistor switch has the advantages of small size, no moving parts to wear out and very fast switching speeds. Relays have the advantage for very high currents. Figure 4.22 shows an example of direct motor drive using a saturating BJT switch.

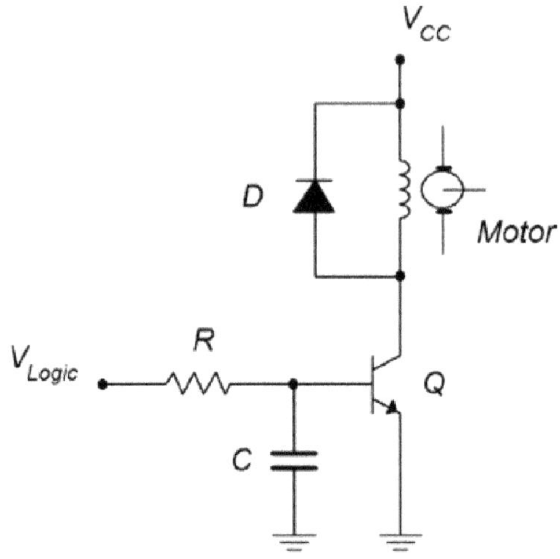

Figure 4.22 Direct DC motor drive.

This circuit is used to control the speed of a DC motor through a technique called *pulse width modulation.* The speed of the motor will depend on the average voltage applied to it. The trick here is that instead of applying a continuously variable voltage to the motor, we instead apply a series of pulses of varying width. These pulses are sufficient to saturate the BJT, causing it to behave as a switch. These pulses are so fast that the motor does not start and stop, but rather inertia keeps it going. Instead, the motor responds to the averaged value of these pulses. If the pulses are narrow and widely spaced, the average value will be low and the motor speed will be slow. If the pulses are wide and closely spaced, the average will high and the motor speed will be fast.

The resistor and capacitor at the base are used to shape the incoming pulse to improve performance. The diode across the motor winding is particularly important. It is referred to as a *snubbing diode*[18]. Without it, the switching transistor might experience large and damaging transient spikes. Here's why: Let's assume the BJT is on and conducting fully. This current is the same current flowing through the motor's armature, which is little more than a huge coil of wire. That means it exhibits a lot of inductance. When we turn off the transistor, we are attempting to turn off the armature current, but the current through an inductor cannot change instantaneously. The result is that the winding now generates a large flyback voltage (also called an "inductive kick") directly across the BJT. That is, the winding momentarily appears as a high voltage source of opposite polarity and, via KVL, this potential appears from collector to emitter. This could damage the BJT. The snubbing diode effectively short-circuits the winding when it reverses voltage polarity, preventing the large spike. The remainder of the time the diode is reverse-biased and effectively out of the circuit.

The Non-Saturating Driver

It is also possible to create a switch or driver that is non-saturating. An example of a non-saturating LED driver is shown in Figure 4.23.

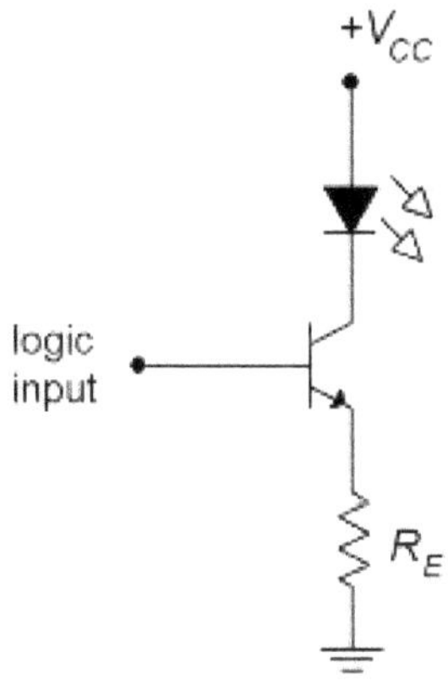

Figure 4.23 Non-saturating LED driver circuit (positive logic).

The advantage of this circuit is that it requires less current from the logic circuit. Unfortunately, it also exhibits higher transistor power dissipation and requires a DC source that is higher than the logic level. The operation is as follows: As in the saturating driver, if the logic level is zero, there is no rise in the base-emitter loop and the collector current will also be zero. With a high logic voltage, via KVL around the base-emitter loop, all of the logic input voltage drops across RE, with the exception of VBE. This creates IE which is virtually the same as IC (which is ILED). This circuit "programs" the emitter current via the resistor and logic voltage. Therefore it is fixed and stable. This process is sometimes referred to as bootstrapping. It might be said that the emitter voltage is "bootstrapped" to within 0.7 volts of the logic input level, keeping it stable19. In any case, if β varies, this will cause an inverse change in IB with no change in IC. A negative logic PNP version is also possible and left as an exercise.

Example 4.4

Determine the LED "on" current for the circuit of Figure 4.24. Assume the logic "on" voltage is 5 volts, *VLED* = 1.8 volts and β = 100.

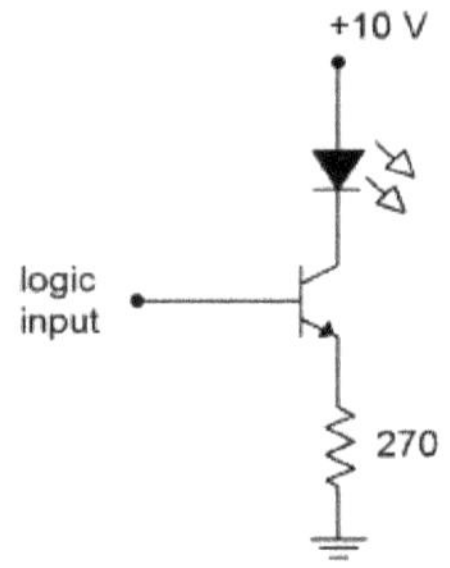

Figure 4.24 Circuit for Example 4.4.

We can find *IC* directly because *IC* ≈ *IE* . This will be the LED current.

$$I_C = \frac{V_{logic} - V_{BE}}{R_E}$$
$$I_C = \frac{5\text{V} - 0.7\text{V}}{270\Omega}$$
$$I_C = 15.9\,\text{mA}$$

Note that β was not used. All it tells us is that *IB* = 15.9 mA/100, or 159 μA. A higher β would simply lead to a lower base current.

For the sake of completeness, we might also note that

$$V_{CE} = V_{CC} - V_{LED} - V_{RE}$$
$$V_{CE} = 10\text{V} - 1.8\text{V} - 4.3\text{V}$$
$$V_{CE} = 3.9\text{V}$$

Clearly, if VCE is 3.9 volts, the transistor is not in saturation.

The Zener Follower

In the prior chapter we examined a method of regulating the output voltage of a filtered full-wave rectifier through the use of a Zener diode. The downfall of that specific circuit is that it was not particularly efficient because it drew a fair amount of current even when the demand for load current was light. Using the concept of locking one voltage to another, as in the non-saturating switch, we can create a nice improvement, the Zener Follower.

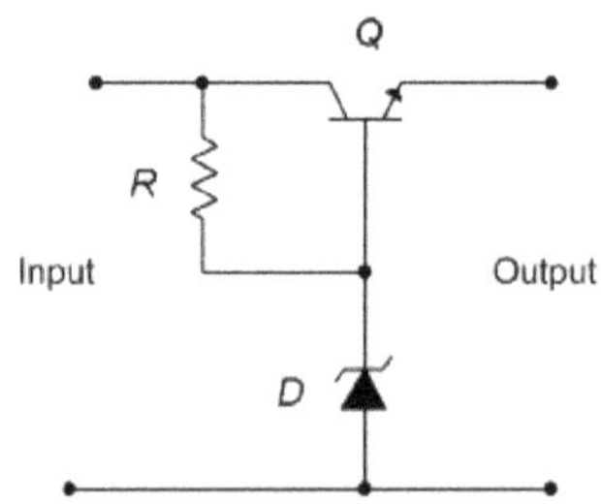

Figure 4.25 Zener Follower.

The primary thing to notice is that the Zener diode is reverse-biased via resistor R. That is, a current will flow down through R and into the Zener. The Zener presents a fixed potential, VZ. Consequently, the difference between the input voltage and VZ must drop across R, and by extension, VCB. Further, the final output voltage is the voltage at the BJT's emitter which must be VZ - VBE. As these are both fixed, stable potentials, the output must likewise be a fixed, stable voltage. Lastly, because VCE = VCB + VBE, it is apparent that any variation between the input voltage and the desired output (for example, due to ripple) must be dropped across the BJT.

The diode current is kept low in the Zener Follower and thus its power dissipation is also modest. Further, current draw from the input circuit is a direct reflection of load current demand. If the load current requirement is low, very little current will flow through the transistor, and ultimately, from the input circuit. This makes for a more efficient system.

CONCISE SUMMARY

- A bipolar junction transistor may be thought of as an extension of a simple diode or PN junction. Another layer of doped material is added, resulting in either an NPN or PNP configuration, both with two depletion regions. The two depletion regions create two hills in the energy diagram. The three terminals of the device are called the emitter, base (middle) and collector. BJTs are not normally constructed symmetrically and swapping the collector and emitter can result in unpredictable behavior.
- For proper operation, the base emitter junction is forward-biased while the collector- base junction is reverse-biased. This results in the emitter and collector currents being very nearly equal and much, much larger than the base current. The ratio of collector current to base current is called β (beta) while the ratio of collector current to emitter current is called α (alpha). β in particular is subject to wide variations and it can have a major impact on circuit parameters. A plot of collector current versus collector-emitter voltage reveals the three main regions of the BJT circuit: saturation, constant current and breakdown.
- The Ebers-Moll model consists of a diode from the base to emitter and a controlled current source from the collector to base. This simple model of the BJT can be used to solve a variety of transistor circuits, particularly when used in conjunction with a DC load line. The DC load line is a plot of all possible operating points for a given transistor circuit.
- Finally, it is possible to create switching and driver circuits using BJTs that produce stable output currents. These may utilize saturating or non-saturating configurations with NPN or PNP devices.

CHAPTER-5

BIASING IN A BJT CIRCUIT

5.1 INTRODUCTION: AN INTRODUCTION

Biasing in a bipolar junction transistor (BJT) circuit is a critical aspect of electronic design that involves applying external voltages to establish the desired operating point for the transistor. Proper biasing ensures that the transistor operates in its active region, allowing it to amplify signals effectively. There are three common biasing configurations: fixed bias, collector-emitter feedback bias, and emitter bias. Fixed bias involves applying a constant voltage to the base terminal, while collector-emitter feedback bias utilizes a portion of the output voltage to stabilize the operating point. Emitter bias, also known as self-bias or voltage-divider bias, employs a resistor network to set the base-emitter voltage. The choice of biasing configuration depends on the specific requirements of the circuit and the desired transistor characteristics. Effective biasing is crucial for optimizing the transistor's linearity, stability, and overall performance in electronic circuits such as amplifiers and oscillators.

As we saw in the preceding chapter, a bipolar junction transistor requires a forward-bias of the base-emitter junction and a reverse-bias of the collector-base junction in order to operate properly. One of the prime BJT parameters is the current gain, β. It can have a considerable impact on the operation of the circuit. Unfortunately, β also varies with changes in temperature, collector-emitter voltage, etc., and this can lead to circuit instability. In this chapter we shall investigate a variety of circuit topologies to bias the BJT, always with an eye toward stability.

5.2 THE NEED FOR BIASING

Why bias a transistor in the first place? After all, if the device exhibits current gain (i.e., β), why not just apply an AC signal at the base and obtain an amplified version of it at the collector? The first thing to remember is that current gain is an outgrowth of forward-reverse bias. Given that fact, and without an additional source of energy, amplification cannot be produced. Also, remember the magnitude of the energy hill required for forward-biasing the base-emitter. In order to achieve that, VBE needs to be around 0.7 volts. If we simply applied an AC signal to the base, we could only hope to forward-bias the base-emitter when that signal exceeded 0.7 volts. The entire negative half of the AC signal would be ignored along with everything positive that's below 0.7 volts. Seeing that the voltage generated from many input devices such as microphones and sensors may only be a few hundred millivolts, the entire signal could be ignored! The solution to these problems is to apply a DC bias to the transistor and then superimpose the AC signal on top of that. In other words, if the AC voltage is riding on a much larger DC voltage, then even the negative peak of the AC signal will be a net positive voltage, and we can maintain proper transistor function.

There are numerous ways to establish a proper polarity DC bias on a transistor. The trick is to find ways to make a stable bias, that is, to establish a Q point that doesn't move in spite of parameter changes such as changes in β. As we shall see in following chapters, an unstable Q point can have negative effects on the AC performance of an amplifier. For example, it could make the gain unstable, increase distortion or reduce output power. This lack of stability is a major problem with the base bias configuration examined in the prior chapter. What we would like is a circuit that will establish a collector current that does not shift even when β changes.

5.3 TWO-SUPPLY EMITTER BIAS

For proper functioning, the collector-base junction needs to be reverse-biased and the base-emitter junction needs to be forward-biased. For an NPN transistor that means that the collector must be at the highest potential, the base somewhat lower and the emitter at the lowest potential of the three. One way of doing this is to apply the usual positive supply to the collector, but instead of using a second potential at the base, the base is tied to ground through a resistor. The requisite forward-bias on the base-emitter is then achieved by connecting the emitter to a negative power supply. This circuit configuration is shown in Figure 5.1 using an NPN device. We shall refer to this as two-supply emitter bias.

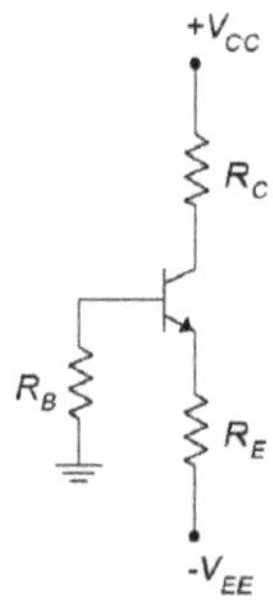

Figure 5.1 Two-supply emitter bias, NPN version.

We can derive an equation for the collector current by applying KVL to the base- emitter loop:

$$V_{EE} = V_{R_B} + V_{BE} + V_{R_E}$$
$$V_{EE} = I_B R_B + V_{BE} + I_E R_E$$

Recalling that $I_B = I_C/\beta$ and $I_E \approx I_C$,

$$V_{EE} = (I_C/\beta) R_B + V_{BE} + I_C R_E$$

Solving for I_C we arrive at

$$I_C = \frac{|V_{EE}| - V_{BE}}{R_E + R_B/\beta} \qquad (5.1)$$

The absolute value has been added to the emitter supply voltage so there is no confusion regarding the sign of this potential in the equation.

The thing to notice about equation 5.1 is that β only partly determines the collector current. In fact, if we can make $R_E >> R_B/\beta$, then the equation reduces to

$$I_C \approx \frac{|V_{EE}| - V_{BE}}{R_E} \qquad (5.2)$$

It is relatively easy to achieve the $R_E >> R_B/\beta$ stipulation. Given typical values of β, this will be the case if R_E is approximately equal to or larger than R_B. What we find in this instance is that almost all of the emitter supply drops across R_E to establish a stable I_C with β playing virtually no role. If β changes, the result will be an inverse change in I_B with I_C remaining largely unchanged.

Now that we have the collector current, any other current or voltage in the circuit may be derived by applying Ohm's law, KVL and the like. For example, to find V_C, the voltage from the collector to ground,

$$V_C = V_{CC} - V_{R_C}$$
$$V_C = V_{CC} - I_C R_C$$

And to find the transistor's collector-emitter voltage, V_{CE},

$$V_{CE} = V_{CC} + |V_{EE}| - V_{R_C} - V_{R_E}$$
$$V_{CE} = V_{CC} + |V_{EE}| - I_C R_C - I_C R_E$$
$$V_{CE} = V_{CC} + |V_{EE}| - I_C (R_C + R_E) \qquad (5.3)$$

Note that *VCE* can also be found via *VCE* = *VC* – *VE*. Dropping voltages along the base- emitter loop yields

$$V_E = -V_{R_B} - V_{BE}$$
$$V_E = -I_B R_B - V_{BE}$$

Also, it is to our advantage to develop the DC load line for this configuration. The load line can serve as a "sanity check" for our computations. To find the endpoints, *IC(sat)* is the maximum current and will occur when *VCE* = 0. If we imagine the current rising as *VCE* collapses, eventually all of the available supply voltage will have dropped across *RC* and *RE*. Thus

$$I_{C(sat)} = \frac{V_{CC} + |V_{EE}|}{R_C + R_E} \tag{5.4}$$

Similarly, *VCE(cutoff)* occurs when *IC* = 0. That means that there will be no potentials across *RC* and *RE*. Therefore, *VCE* "absorbs" the entire available source voltage.

$$V_{CE(cutoff)} = V_{CC} + |V_{EE}| \tag{5.5}$$

Do not attempt to memorize all of the myriad equations presented. There are simply too many variations on the theme and it will only get worse when other biasing configurations are introduced. Instead, remember how to find the collector current and then get in the habit of applying Ohm's law and KVL to derive whatever else you may need.

At this point a comprehensive example is called for.

Example 5.1

Assuming β = 100, plot the Q point (*IC* and *VCE*) on the load line for the circuit of Figure 5.2.

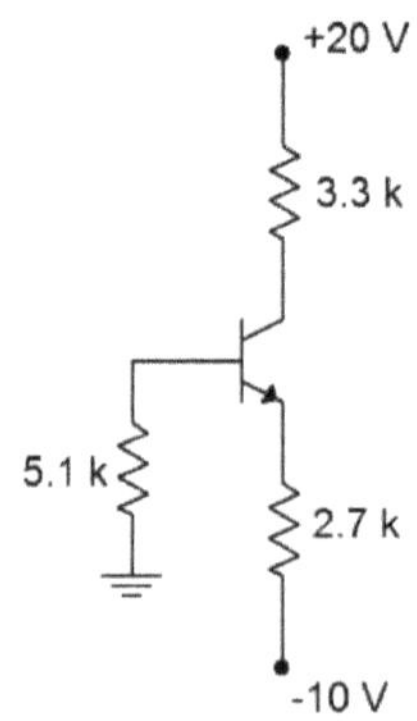

Figure 5.2 Circuit for Example 5.1.

Using Equation 5.1:

$$I_C = \frac{|V_{EE}| - V_{BE}}{R_E + R_B/\beta}$$
$$I_C = \frac{10 - 0.7}{2.7\,\text{k}\Omega + 5.1\,\text{k}\Omega/100}$$
$$I_C = 3.38\,\text{mA}$$

Noting the relative sizes of *RE* and *RB*, the approximation should be close.

$$I_C = \frac{|V_{EE}| - V_{BE}}{R_E}$$
$$I_C = \frac{10 - 0.7}{2.7\,\text{k}\Omega}$$
$$I_C = 3.44\,\text{mA}$$

To find *VCE* we can use the equation derived above (Equation 5.3).

$$V_{CE} = V_{CC} + |V_{EE}| - I_C(R_C + R_E)$$
$$V_{CE} = 20\,\text{V} + 10\,\text{V} - 3.38\,\text{mA}(3.3\,\text{k}\Omega + 2.7\,\text{k}\Omega)$$
$$V_{CE} = 9.72\,\text{V}$$

Now calculate the load line endpoints:

$$I_{C(sat)} = \frac{V_{CC} + |V_{EE}|}{R_C + R_E}$$
$$I_{C(sat)} = \frac{20\,\text{V} + 10\,\text{V}}{3.3\,\text{K}\Omega + 2.7\,\text{K}\Omega}$$
$$I_{C(sat)} = 5\,\text{mA}$$

$$V_{CE(cutoff)} = V_{CC} + |V_{EE}|$$
$$V_{CE(cutoff)} = 20\,\text{V} + -10\,\text{V}$$
$$V_{CE(cutoff)} = 30\,\text{V}$$

V CE (cutoff) = V CC +|V EE | V CE (cutoff) = 20 V+-10 V V CE (cutoff) = 30 V

The load line for the circuit in Example 5.1 is shown in Figure 5.3.

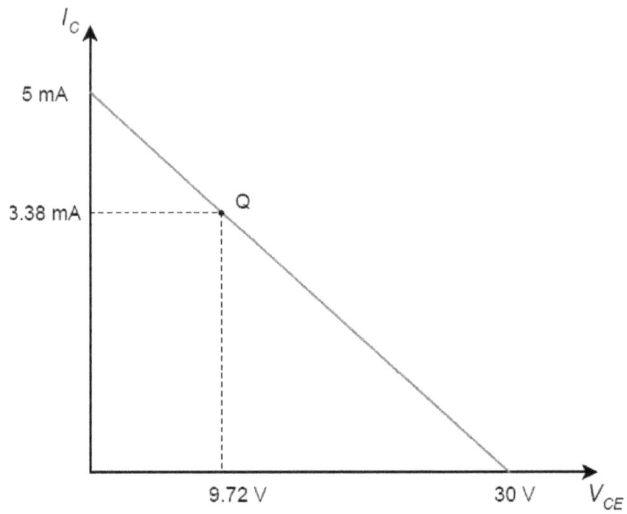

Figure 5.3 DC load line for the circuit of Figure 5.2.

Note the proportions between voltage and current for the Q point. The voltage is a little less than one-third of the maximum while the current is a little more than two-thirds of its maximum.

Verification off Stability

The claim was made that two-supply emitter bias circuits like the one Figure 5.1 potentially have a stable Q point. If we were to plot a second Q point for a large change in β, it should hardly move, thus indicating very high stability. For example, doubling β to 200 results in *IC* = 3.41 mA and *VCE* = 9.53 V. The new Q point has edged just slightly closer to saturation, producing about a 1% change in current for a 100% change in β. Clearly, this configuration can produce very small changes in the Q point in spite of very large changes in β.

Computer Simulation

The two-supply emitter bias circuit of Figure 5.4 is simulated using the DC Bias function. A quick estimation shows that we expect about 2 mA of collector current (9.3 V/4.7 kΩ) and a collector voltage of about 8 volts (15 V - 2 mA · 3.6 kΩ). We also expect a small negative potential at the base -IBRB). Given typical β values for the 2N3904 (200-ish at this current, refer back to the data sheet), we expect a base current of around 10 to 15 μA, leaving us with a VB of approximately -0.1 volts. The emitter voltage would be about 0.7 volts less than that, perhaps -0.8 volts or so.

In short, for a properly designed circuit of this type we expect VB to be pretty close to 0 V and VE to be about -0.7 volts.

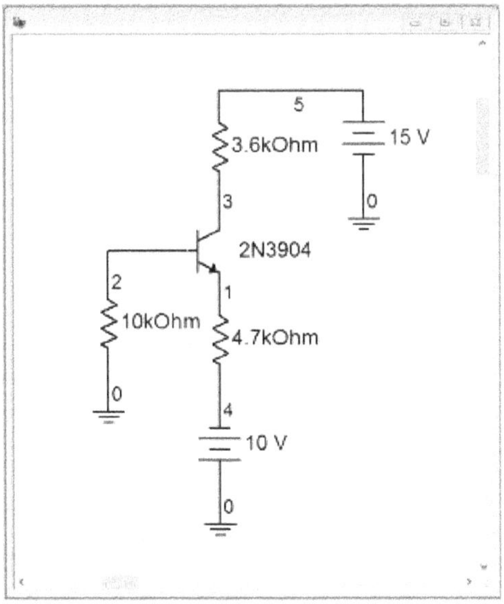

Figure 5.4 Schematic for two-supply emitter bias simulation.

The results are shown in Figure 5.5. The node voltages agree with our estimations. Node 3 is the collector voltage, very close to the estimation. The results for the base voltage (node 2) and the emitter voltage (node 1) are also in line with the estimates.

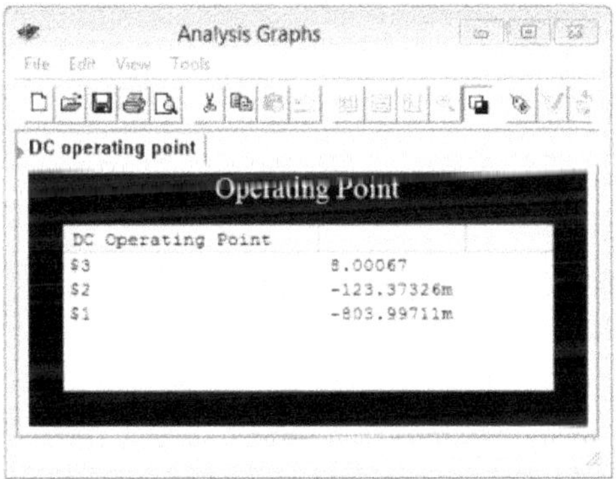

Figure 5.5 Simulation results for two- supply emitter bias circuit.

PNP Two-Supply Emitter Bias

While it is possible to create a PNP version of bias circuits by simply swapping out the device and then changing the signs of the power supplies, it is common to "flip" the entire circuit from top to bottom so that the emitter winds up on top and the collector on the bottom. One advantage of this is that, in a multi-transistor circuit schematic, all of the DC bias currents "run down the page", that is, the collector currents flow from the top of the page to the bottom of the page. Figure 5.6 shows a PNP two-supply emitter bias circuit.

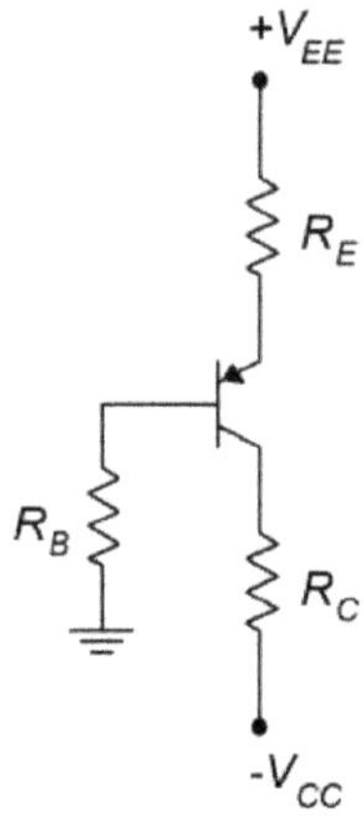

Figure 5.6 PNP two-supply emitter bias circuit.

All of the device current equations and component voltage equations derived for the NPN version will hold for the PNP version. The differences to remember are that the voltage polarities will be reversed (what was positive in the NPN is negative in the PNP) and the current directions will be reversed (e.g., conventional current flows into the NPN's collector but out of the PNP's collector. For example, in the NPN we expect the base current to flow into the base terminal. This creates a small negative voltage at the base and a somewhat more negative voltage (by 0.7 V) at the emitter. In the PNP, the base current flows out of the base. This creates a small positive voltage at the base and results in the emitter being slightly more positive (by 0.7 V). This is perhaps best illustrated with an example.

Example 5.2

Assuming $\beta = 100$, determine the Q point and load line endpoints of the circuit of Figure 5.7.

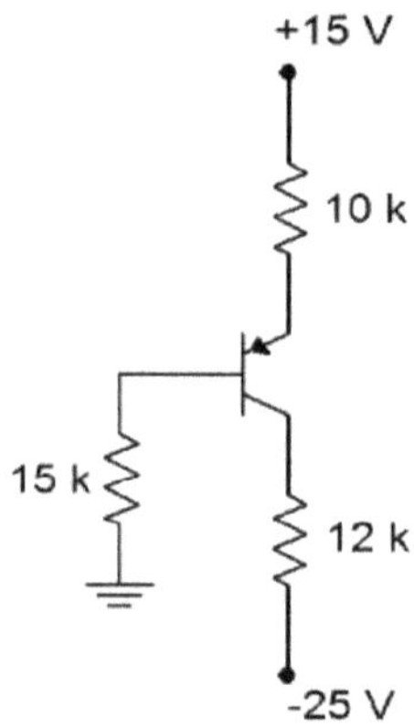

Figure 5.7 Circuit for Example 5.2.

First, note that this is a PNP drawn upside down so the emitter is at the top. Using Equation 5.1:

$$I_C = \frac{|V_{EE}| - V_{BE}}{R_E + R_B/\beta}$$
$$I_C = \frac{15 - 0.7}{10\,\text{k}\Omega + 15\,\text{k}\Omega/100}$$
$$I_C = 1.409\,\text{mA}$$

As a cross check, noting the relative sizes of RE and RB, the approximation should be close.

$$I_C = \frac{|V_{EE}| - V_{BE}}{R_E}$$
$$I_C = \frac{15 - 0.7}{10\,\text{k}\Omega}$$
$$I_C = 1.43\,\text{mA}$$

To find VCE we can use Equation 5.3 with a slight modification.

$$V_{CE} = V_{EE} + |V_{CC}| - I_C(R_C + R_E)$$
$$V_{CE} = 15\,\text{V} + 25\,\text{V} - 1.409\,\text{mA}(12\,\text{k}\Omega + 10\,\text{k}\Omega)$$
$$V_{CE} = 9\,\text{V}$$

We complete the picture by determining the endpoints of the load line.

$$I_{C(sat)} = \frac{V_{EE} + |V_{CC}|}{R_C + R_E}$$
$$I_{C(sat)} = \frac{10\,\text{V} + 25\,\text{V}}{12\,\text{K}\Omega + 10\,\text{K}\Omega}$$
$$I_{C(sat)} = 1.818\,\text{mA}$$

$$V_{CE(cutoff)} = V_{EE} + |V_{CC}|$$
$$V_{CE(cutoff)} = 15\,\text{V} + 25\,\text{V}$$
$$V_{CE(cutoff)} = 40\,\text{V}$$

The Q point is about $3/4^{ths}$ of the maximum current and $1/4^{th}$ of the maximum voltage.

5.4 VOLTAGE DIVIDER BIAS

Another configuration that can provide high bias stability is *voltage divider bias.* Instead of using a negative supply off of the emitter resistor, like two-supply emitter bias, this configuration returns the emitter resistor to ground and raises the base voltage. So as to avoid issues with a second power supply, this base voltage is derived from the collector power supply via a voltage divider. The bias template is shown in Figure 5.8.

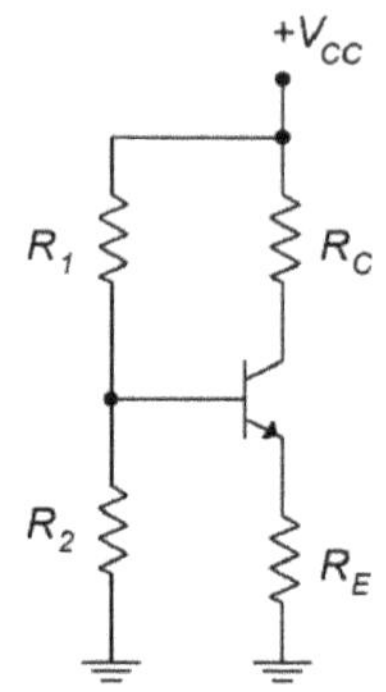

Figure 5.8 Voltage divider bias.

Let's derive the equations for the load line. First, let's consider the saturation and cutoff endpoints. For saturation, assume *VCE* goes to 0. What resistances are left to limit the current?

$$I_{C(sat)} = \frac{V_{CC}}{R_C + R_E} \tag{5.6}$$

VCE(cutoff) occurs when *IC* = 0 and that means that there will be no potentials across *RC* and *RE*. Therefore, *VCE* takes on the entire available source voltage.

$$V_{CE(cutoff)} = V_{CC} \tag{5.7}$$

The key to finding the Q point (and pretty much any other current or voltage in the circuit) is to find IC. To simplify the process, Thevenize the voltage divider as shown in Figure 5.9.

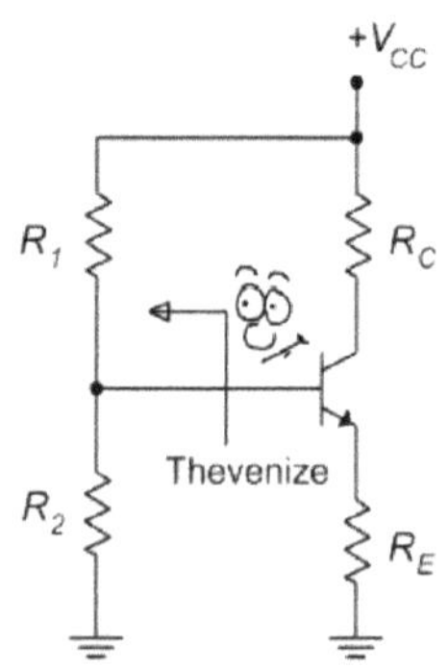

(a)

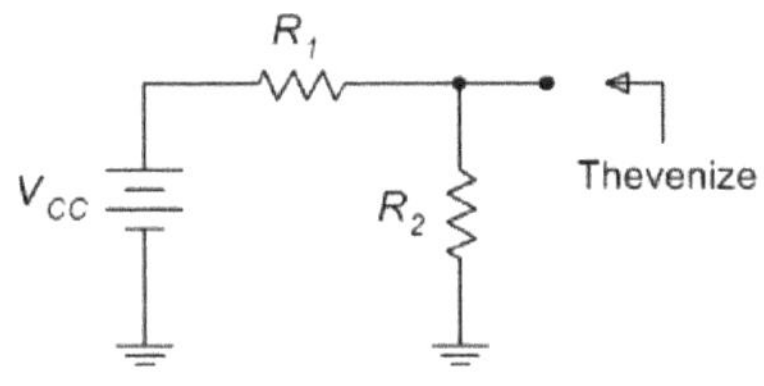

(b)

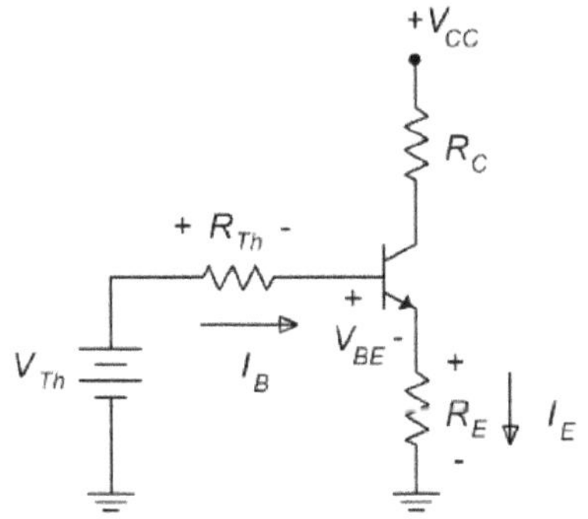

(c)

Figure 5.9 Thevenizing the voltage divider.

By inspection of Figure 5.9b,

$$V_{TH} = V_{CC}\frac{R_2}{R_1+R_2}$$

$$R_{TH} = R_1 \parallel R_2 = \frac{R_1 R_2}{R_1+R_2}$$

Now we can derive an equation for the collector current by applying KVL to the base-emitter loop of Figure 5.9c:

$$V_{TH} = V_{R_{TH}} + V_{BE} + V_{R_E}$$

$$V_{TH} = I_B R_{TH} + V_{BE} + I_E R_E$$

Recalling that $I_B = I_C/\beta$ and $I_E \approx I_C$,

$$V_{TH} = (I_C/\beta) R_{TH} + V_{BE} + I_C R_E$$

Solving for I_C we arrive at

$$I_C = \frac{V_{TH} - V_{BE}}{R_E + R_{TH}/\beta} \qquad (5.8)$$

Can we find a quick approximation for *IC* as well? If we assume that the voltage divider of *R1* and *R2* is lightly loaded, in other words, that the divider current is much, much less than the base current, finding *IC* is easy. The divider voltage yields the base voltage. We then subtract the 0.7 volt drop on the base-emitter and what's left drops across *RE*. From there it's one short application of Ohm's law to get *IE*, which is approximately equal to *IC*. But how do we know if the divider is lightly loaded in the first place without going through the Thevenin equivalent? Looking at Equation 5.8, as long as *RE* >> *RTH*/β, we can ignore the second term in the denominator, leaving us with our quick approximation.

Given typical values for β, as long as *R2* is not much larger than *RE*, the approximation will be reasonably accurate.

Once *IC* is obtained we can find the transistor's collector-emitter voltage, *VCE*,

$$V_{CE} = V_{CC} - V_{R_C} - V_{R_E}$$
$$V_{CE} = V_{CC} - I_C R_C - I_C R_E$$
$$V_{CE} = V_{CC} - I_C (R_C + R_E) \quad (5.9)$$

Time for yet another thrilling illustrative example.

Example 5.3

Assuming β = 200, plot the Q point (*IC* and *VCE*) on the load line for the circuit of Figure 5.10. Also determine the value of *VB*.

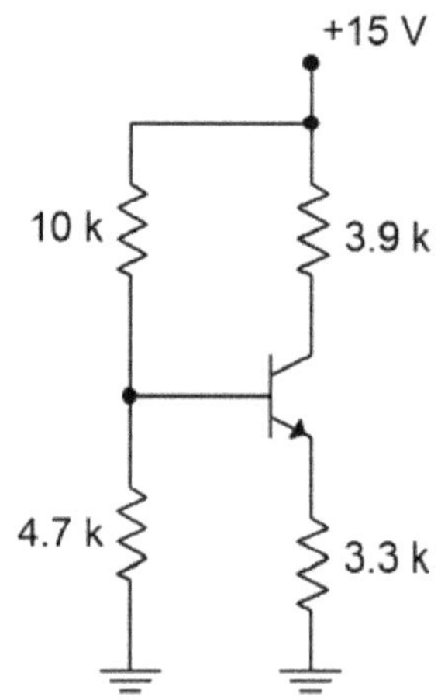

Figure 5.10 Circuit for Example 5.3.

Calculate the load line endpoints so we know the maximums.

$$I_{C(sat)} = \frac{V_{CC}}{R_C + R_E}$$

$$I_{C(sat)} = \frac{15\,\text{V}}{3.9\,\text{K}\Omega + 3.3\,\text{K}\Omega}$$

$$I_{C(sat)} = 2.08\,\text{mA}$$

$$V_{CE(cutoff)} = V_{CC}$$

$$V_{CE(cutoff)} = 15\,\text{V}$$

As far as finding *VB* is concerned, a decent approximation would be the value of *VTH* because we have determined that the divider is lightly loaded. In a more general sense, we could also find the drop across *RE* and then add *VBE*. The approximation yields 4.8 volts and the more accurate method yields

$$V_{TH} = V_{CC} \frac{R_2}{R_1 + R_2}$$

$$V_{TH} = 15\,\text{V} \frac{4.7\,\text{k}\Omega}{10\,\text{k}\Omega + 4.7\,\text{k}\Omega}$$

$$V_{TH} = 4.8\,\text{V}$$

$$R_{TH} = R_1 \| R_2 = \frac{R_1 R_2}{R_1 + R_2}$$

$$R_{TH} = \frac{10\,\text{k}\Omega \times 4.7\,\text{k}\Omega}{10\,\text{k}\Omega + 4.7\,\text{k}\Omega}$$

$$R_{TH} = 3.2\,\text{k}\Omega$$

Using Equation 5.8:

$$I_C = \frac{V_{TH} - V_{BE}}{R_E + R_{TH}/\beta}$$

$$I_C = \frac{4.8\,\text{V} - 0.7\,\text{V}}{3.3\,\text{k}\Omega + 3.2\,\text{k}\Omega/200}$$

$$I_C = 1.236\,\text{mA}$$

Noting the relative sizes of R_E and R_2, the approximation should be fairly accurate.

$$I_C = \frac{V_{TH} - V_{BE}}{R_E}$$
$$I_C = \frac{4.8\,\text{V} - 0.7\,\text{V}}{3.3\,\text{k}\Omega}$$
$$I_C = 1.242\,\text{mA}$$

To find V_{CE} we can use Equation 5.9.

$$V_{CE} = V_{CC} - I_C(R_C + R_E)$$
$$V_{CE} = 15\,\text{V} - 1.236\,\text{mA}(3.9\,\text{k}\Omega + 3.3\,\text{k}\Omega)$$
$$V_{CE} = 6.1\,\text{V}$$

As far as finding V_B is concerned, a decent approximation would be the value of V_{TH} because we have determined that the divider is lightly loaded. In a more general sense, we could also find the drop across RE and then add V_{BE}. The approximation yields 4.8 volts and the more accurate method yields

$$V_B = V_{BE} + I_C R_E$$
$$V_B = 0.7\,\text{V} + 1.236\,\text{mA} \times 3.3\,\text{k}\Omega$$
$$V_B = 4.78\,\text{V}$$

The load line for the circuit in Example 5.3 is shown in Figure 5.11.

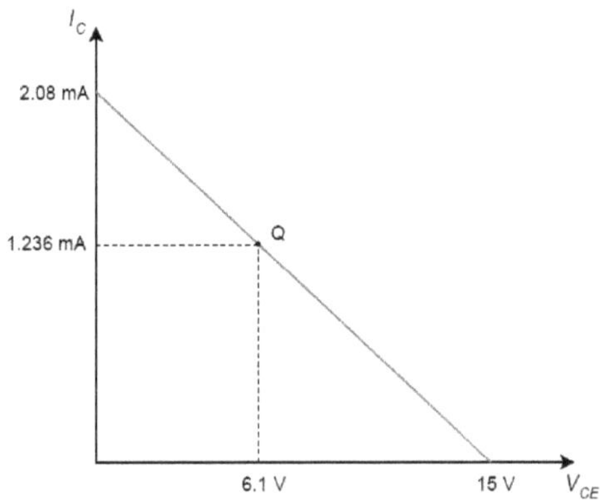

Figure 5.11 DC load line for the circuit of Figure 5.10.

Once again the proportions between voltage and current for the Q point appear to be proper when compared against the endpoints.

Verification off Stability

How much does the Q point move if β were to get cut half? Recalculating with a β of 100 yields IC = 1.23 mA and VCE = 6.14 V. This represents a shift in both current and voltage of less than 1%. This will, of course, cause a near doubling of IB but this will be hardly noticed here as the divider current is so much larger; approximately 15V/ (10 k + 4.7 k) or 1 mA versus about 1.23 mA/100 or 12.3 μA.

PNP Voltage Divider Bias

To create the PNP version of the voltage divider bias, we replace the NPN with a PNP and then change the sign of the power supply. As mentioned with the two-supply emitter bias, these circuits are usually flipped top to bottom resulting in the flow of DC current going down the page. All of the currents and component voltages are unchanged except that their directions and polarities are reversed. The current equations and so forth remain valid. Something a little odd-looking happens with the voltage divider bias, though: we end up with ground being the most positive potential and a negative supply at the bottom of the schematic. It works, but it's an issue if we're using a traditional positive supply elsewhere in the circuit. After all, why have two supplies where one will do? It turns out that we can make a positive supply version fairly easily. All we need to do is add the magnitude of the negative source voltage to the ground and power connections. This progression is shown in Figure 5.12.

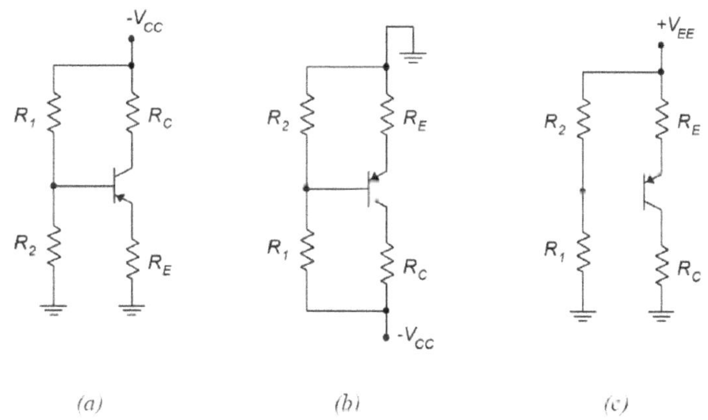

Figure 5.12 Progression of PNP voltage divider bias circuit (a) Direct conversion from NPN (b) Top-to-bottom flip (c) DC supply offset added to (d) achieve a positive supply.

There is nothing magic about this procedure. In essence, all we've really done is renamed the reference point. All of the individual component voltages remain unchanged. For example, looking at Figure 5.12c versus 5.12b, it is still the case that the top connection to RE is more positive than the bottom connection to RC by the voltage VCC (although we did rename the supply to VEE to be consistent with where it's connected). What has happened is that all ground-referenced (i.e., single subscript) voltages have changed. For example, VB in Figures 5.12a and 5.12b is the voltage across R2. In contrast, VB in Figure 5.12c is the voltage across R1. That makes sense. If we move the reference then any voltage that is measured against the reference will change.

When analyzing the PNP voltage divider, we could simply parrot the collector current formula developed for the NPN, but there are other techniques. Two methods are illustrated in the following example.

Example 5.4

Assuming β = 200, determine the Q point (*IC* and *VCE*) for the circuit of Figure 5.13. Also determine the values of *VC* and *VB*.

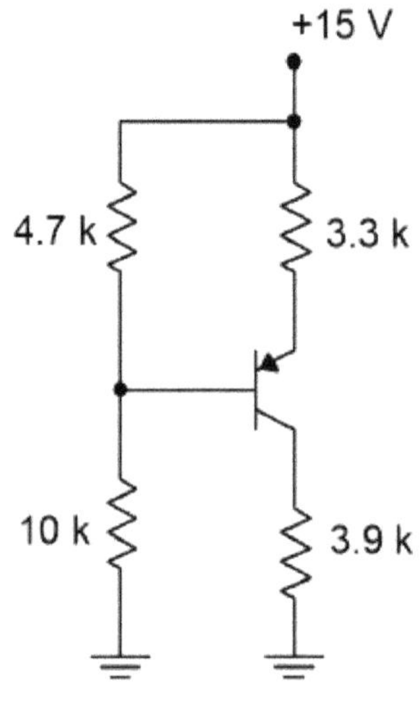

Figure 5.13 Circuit for Example 5.4.

First off, R2 (now on top) is around the same size as RE so the approximation method should be accurate and we can assume the divider is lightly loaded.

Method One

We will focus on the base-emitter loop as usual because VBE is a known potential. Our immediate goal is to find the voltage across RE so that we can use Ohm's law to find IC. First we note that the voltage drop across R2 is equal to the combined drops across RE and VBE. The drop across R2 is found via the voltage divider rule.

$$V_{R_2} = V_{EE}\frac{R_2}{R_1+R_2}$$
$$V_{R_2} = 15\,\text{V}\frac{4.7\,\text{k}\Omega}{10\,\text{k}\Omega+4.7\,\text{k}\Omega}$$
$$V_{R_2} = 4.8\,\text{V}$$

And

$$V_{R_E} = V_{R_2} - V_{BE}$$
$$V_{R_E} = 4.8\,\text{V} - 0.7\,\text{V}$$
$$V_{R_E} = 4.1\,\text{V}$$

Therefore

$$I_C = \frac{V_{R_E}}{R_E}$$
$$I_C = \frac{4.1\,\text{V}}{3.3\,\text{k}\Omega}$$
$$I_C = 1.24\,\text{mA}$$

Method Two

Here we will determine all voltages with respect to ground.

$$V_B = V_{EE}\frac{R_1}{R_1+R_2}$$
$$V_B = 15\,\text{V}\frac{10\,\text{k}\Omega}{10\,\text{k}\Omega+4.7\,\text{k}\Omega}$$
$$V_B = 10.2\,\text{V}$$

The voltage from base to emitter has a - to + polarity, meaning it is a rise of0.7 volts. Therefore

$$V_E = V_B + V_{BE}$$
$$V_E = 10.2\,\text{V} + 0.7\,\text{V}$$
$$V_E = 10.9\,\text{V}$$

The voltage across *RE* is the difference between *VEE* and *VE*.

$$V_{R_E} = V_{EE} - V_E$$
$$V_{R_E} = 15\,\text{V} - 10.9\,\text{V}$$
$$V_{R_E} = 4.1\,\text{V}$$

This is the same value we arrived at using method one, so the collector current must be the same at 1.24 mA.

To find *VCE* we also have options. One path is to use a slightly modified Equation 5.9.

$$V_{CE} = -(V_{EE} - I_C(R_C + R_E))$$
$$V_{CE} = -15\,\text{V} + 1.24\,\text{mA}\,(3.9\,\text{k}\Omega + 3.3\,\text{k}\Omega)$$
$$V_{CE} = -6.07\,\text{V}$$

The collector is negative relative to the emitter, hence the negative sign. To avoid this, we could just swap the leads and refer to *VEC* instead.

Alternately, we could find *VCE* by determining *VC* and then subtracting *VE* from it.

$$V_C = I_C R_C$$
$$V_C = 1.24\,\text{mA} \times 3.9\,\text{k}\Omega$$
$$V_C = 4.84\,\text{V}$$

$$V_{CE} = V_C - V_E$$
$$V_{CE} = 4.84\,\text{V} - 10.9\,\text{V}$$
$$V_{CE} = -6.06\,\text{V}$$

We see a very slight difference here due to carried rounding errors.

It is instructive to compare the results of Example 5.4 back to Example 5.3. These circuits are otherwise identical except for the fact that one is NPN and the other is PNP. We find the same results for device currents (IC) and component voltage magnitudes (VCE or the voltage across RE); only the signs and directions are reversed. On the other hand, we find that ground referenced potentials such as VB, VC and VE are decidedly different between the two circuits.

Feedback Biasing

While two-supply emitter bias and voltage divider bias can produce very high stability, there are other bias configurations available. Their stability tends not to be quite as good, but they are superior to simple base bias. They also tend to use fewer components than their high stability cousins. As a group, we refer to these as *feedback biasing* configurations. They use the concept of *negative feedback*. This is a technique where a change in the output can be reflected back to the input in such a way that it tends to partially offset the output change.

Collector Feedback Bias

With a simple move of RB in the basic base bias configuration, we arrive at collector feedback bias. The NPN template is shown in Figure 5.14. Compared to base bias, all that has changed is that RB is connected to the lower part of RC rather than to the power supply. That small change can have a noticeable effect on stability.

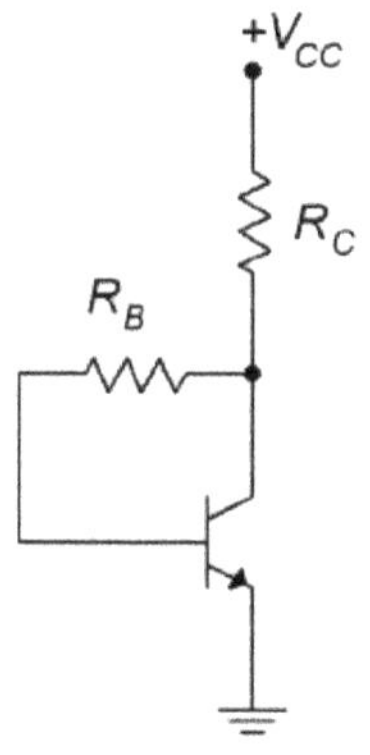

Figure 5.14 Collector feedback bias.

To understand how feedback works, assume that a current is flowing from the supply, through *RC*, into the collector and finally, out of the emitter to ground. Via KVL, *VCE* = *VC* = *VCC* - *IC*·*RC*. Now suppose for some reason, a temperature change perhaps, β increases. This should cause an increase in *IC*. An increase in *IC*, though, would cause an increase in the drop across *RC* due to Ohm's law. This, in turn, would force *VC* to drop. Here is the key: *VC* is also equal to the drop across *RB* plus the voltage *VBE*. The base-emitter potential is fixed at approximately 0.7 volts so any decrease in *VC* is reflected as a decrease in voltage across *RB*. By Ohm's law, that means that *IB* must decrease by a similar proportion. This decrease tends to offset the initial tendency of the collector current to increase.

To derive an equation for the collector current, we can use KVL.

$$V_{CC} = V_{R_C} + V_{R_B} + V_{BE}$$

$$V_{CC} = I_E R_C + I_B R_B + V_{BE}$$

$$V_{CC} = I_C R_C + \frac{I_C}{\beta} R_B + V_{BE}$$

$$I_C = \frac{V_{CC} - V_{BE}}{R_C + R_B/\beta} \tag{5.10}$$

This equation is very similar to the current derivations for the two-supply emitter bias (Eq 5.1) and voltage divider bias (Eq 5.8). Again, if we can set RC >> RB/β then IC will be relatively immune from Q point shifts due to β. The problem here is that it's not nearly so easy to meet that stipulation in this circuit. Consequently, collector feedback tends to have only modest stability.

Concerning the cutoff and saturation endpoints on the DC load line, once again, cutoff is determined by the DC power supply while saturation is determined by the amount of resistance in the collector-emitter to limit said power supply's current.

$$I_{C(sat)} = \frac{V_{CC}}{R_C} \tag{5.11}$$

$$V_{CE(cutoff)} = V_{CC} \tag{5.12}$$

Example 5.5

Assuming β = 100, determine the Q point (IC and VCE) for the circuit of Figure 5.15. How much does the Q point change if β is halved?

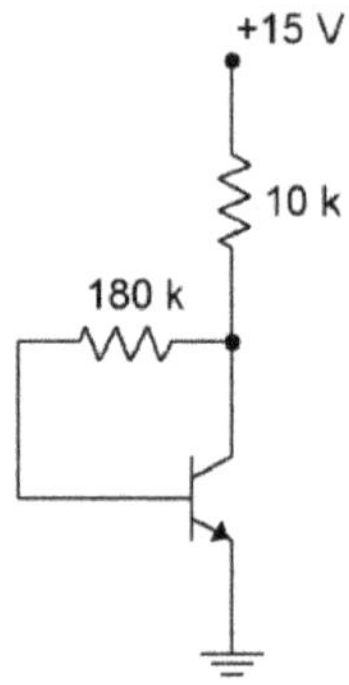

Figure 5.15 Circuit for Example 5.5.

Using Equation 5.10

$$I_C = \frac{V_{CC} - V_{BE}}{R_C + R_B/\beta}$$
$$I_C = \frac{15\,\text{V} - 0.7\,\text{V}}{10\,\text{k}\,\Omega + 180\,\text{k}\,\Omega/100}$$
$$I_C = 1.21\,\text{mA}$$

Using KVL

$$V_{CE} = V_{CC} - V_{R_C}$$
$$V_{CE} = V_{CC} - I_C R_C$$
$$V_{CE} = 15\text{V} - 1.21\,\text{mA} \times 10\,\text{k}\,\Omega$$
$$V_{CE} = 2.9\,\text{V}$$

If β is halved to 50

$$I_C = \frac{V_{CC} - V_{BE}}{R_C + R_B/\beta}$$
$$I_C = \frac{15\text{V} - 0.7\text{V}}{10\,\text{k}\,\Omega + 180\,\text{k}\,\Omega/50}$$
$$I_C = 1.05\,\text{mA}$$

$$V_{CE} = V_{CC} - I_C R_C$$
$$V_{CE} = 15\text{V} - 1.05\,\text{mA} \times 10\,\text{k}\,\Omega$$
$$V_{CE} = 4.5\,\text{V}$$

For a 2:1 drop in β we see about a 13% reduction in *IC* with a somewhat larger change in *VCE*. This circuit is clearly not as stable as the two-supply emitter bias or the voltage divider bias but it is superior to base bias.

The PNP version of the collector feedback bias configuration should come as no surprise. The template is shown in Figure 5.16. Here, we use the same technique of power supply shifting that was used with the PNP voltage divider in order to wind up with a positive power supply. As with the PNP voltage divider, because we have changed the reference point, all ground referenced voltages will be different from their NPN counterparts. All currents and component voltages will have the same magnitudes but with opposite directions and polarities.

Emitter Feedback Bias

The emitter feedback bias uses the same overall idea as the collector feedback circuit, namely, that changes at the output will be reflected back to the input and thus help mitigate the initial change. While collector feedback focuses on collector current establishing VC via RC, emitter feedback uses the fact that emitter current establishes VE via RE. In both cases, these voltages are used to change the voltage across RB, which results in a change in IB that opposes the original collector current change.

A basic emitter feedback bias circuit is shown in Figure 5.17. We shall use KVL to develop an equation for collector current.

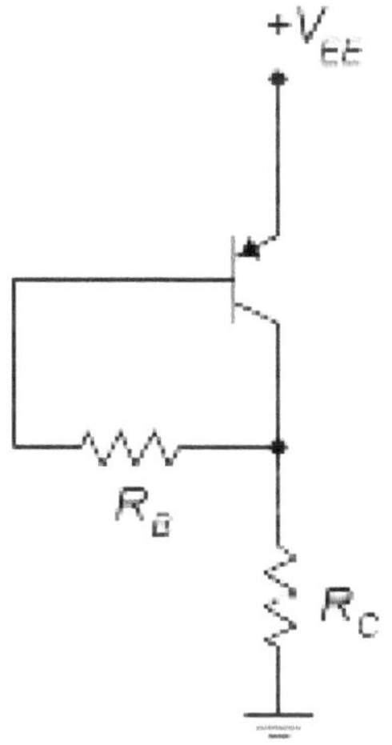

Figure 5.16 PNP Collector feedback bias.

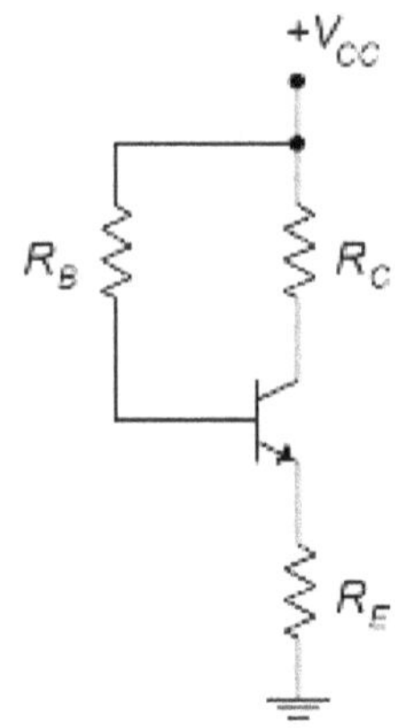

Figure 5.17 Emitter feedback bias.

$$V_{CC} = V_{R_B} + V_{BE} + V_{R_E}$$
$$V_{CC} = I_B R_B + V_{BE} + I_E R_E$$
$$V_{CC} = \frac{I_C}{\beta} R_B + I_C R_E + V_{BE}$$
$$I_C = \frac{V_{CC} - V_{BE}}{R_E + R_B/\beta} \quad (5.13)$$

If we can set RE >> RB/β then the Q point will be stable in spite of changes in β. The problem here is the same as was the case in collector feedback, namely that this stipulation is not easy to achieve. Consequently, the emitter feedback configuration tends to have only modest stability. In any event, once the collector current is known, VCE can be found using the techniques illustrated with the voltage divider configuration. The endpoints for the DC load line are found in the usual manner.

$$I_{C(sat)} = \frac{V_{CC}}{R_C + R_E} \quad (5.14)$$
$$V_{CE(cutoff)} = V_{CC} \quad (5.15)$$

Example5.6

Assuming β = 100, determine the Q point (*IC* and *VCE*) for the circuit of Figure 5.18.

Using Equation 5.13

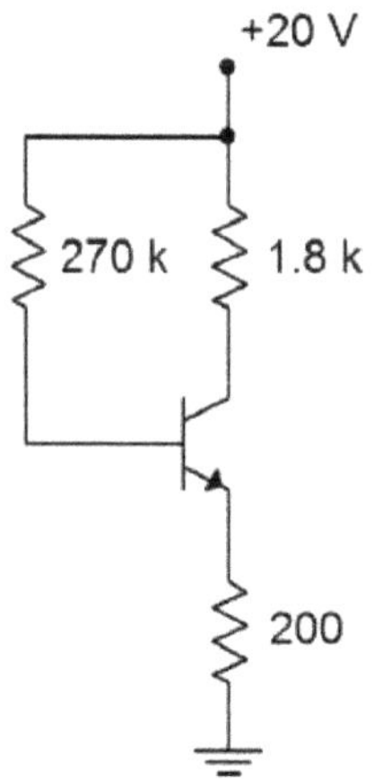

Figure 5.18 Circuit for Example 5.6.

$$I_C = \frac{V_{CC} - V_{BE}}{R_E + R_B/\beta}$$

$$I_C = \frac{20\,\text{V} - 0.7\,\text{V}}{200\,\Omega + 270\,\text{k}\,\Omega/100}$$

$$I_C = 6.66\,\text{mA}$$

Using KVL

$$V_{CE} = V_{CC} - V_{R_C} - V_{R_E}$$

$$V_{CE} = V_{CC} - I_C(R_C + R_E)$$

$$V_{CE} = 20\,\text{V} - 6.66\,\text{mA}\,(1.8\,\text{k}\Omega + 200\,\Omega)$$

$$V_{CE} = 6.68\,\text{V}$$

To complete the load line, we find

$$I_{C(sat)} = 10\,\text{mA}$$

$$V_{CE(cutoff)} = 20\,\text{V}$$

Dropping β to 50 will result in

$$I_C = 3.45\,\text{mA}$$

$$V_{CE} = 13.1\,\text{V}$$

We see less than a 2:1 change in IC and VCE but the stability is not dramatic.

Combination Feedback Bias

The final feedback bias configuration combines both collector feedback and emitter feedback to arrive at the circuit depicted in Figure 5.19. We shall bestow upon it the highly original name of combination feedback bias.

This circuit applies feedback to RB from both ends, so to speak, so it tends to have slightly better stability than either collector feedback or emitter feedback bias. Of course, it is now only one resistor shy from the voltage divider circuit which is considerably more stable.

The equations for the load line are listed below. The derivations are left as an exercise.

$$I_C = \frac{V_{CC} - V_{BE}}{R_C + R_E + R_B/\beta} \tag{5.16}$$

$$V_{CE} = V_{CC} - I_C(R_C + R_E) \tag{5.17}$$

$$I_{C(sat)} = \frac{V_{CC}}{R_C + R_E} \tag{5.18}$$

$$V_{CE(cutoff)} = V_{CC} \tag{5.19}$$

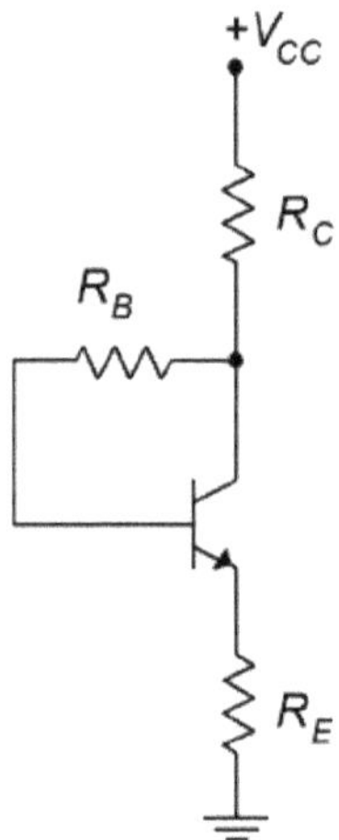

Figure 5.19 Combination feedback bias. Example 5.7

Example 5.7

Assuming β = 125, determine the Q point (IC and VCE) for the circuit of Figure 5.20.

Note that this is the upside down PNP version. Using Equation 5.16

$$I_C = \frac{V_{CC} - V_{BE}}{R_C + R_E + R_B/\beta}$$

$$I_C = \frac{18\,\text{V} - 0.7\,\text{V}}{7.5\,\text{k}\Omega + 330\,\Omega + 100\,\text{k}\Omega/125}$$

$$I_C = 2\,\text{mA}$$

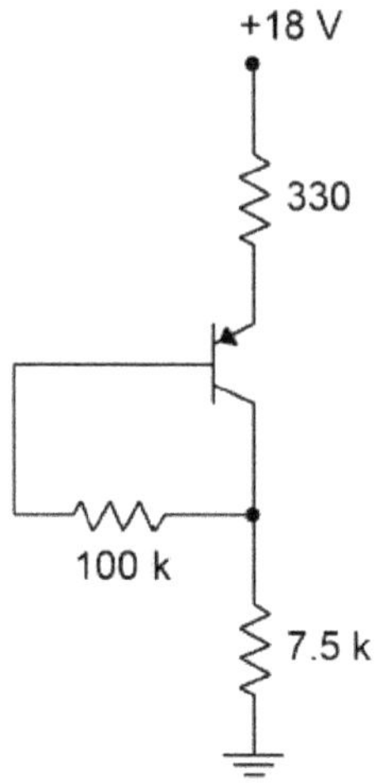

Figure 5.20 Circuit for Example 5.7.

Using KVL

$$V_{CE} = V_{CC} - V_{R_C} - V_{R_E}$$
$$V_{CE} = V_{CC} - I_C(R_C + R_E)$$
$$V_{CE} = 18\,V - 2\,mA(7.5\,k\Omega + 330\,\Omega)$$
$$V_{CE} = 2.34\,V$$

CONCISE SUMMARY

- DC biasing is required in order to maintain the proper junction potentials and operation of the BJT. Several different circuit configurations are available to establish a DC bias on both NPN and PNP transistors. These circuits vary in complexity and their ability to maintain a constant operating point, or Q point, in the face of variations of β.
- The two-supply emitter bias topology offers very high Q point stability. It achieves this through the use of two powers supplies; one connected through a resistor to the emitter and a second unit connected through a resistor to the collector. It is unique in that the supplies are bipolar; one being positive and the other being negative.
- The voltage divider bias circuit offers similar stability performance to the two- supply emitter bias circuit. It uses a single supply and a resistive voltage divider to establish a second, lower potential at the base terminal.
- The three feedback bias configurations offer only modest enhancements in stability but use the least amount of parts. They all rely on a single DC power source.

- A DC load line is a plot of all possible collector current and corresponding collector- emitter voltage operating points. No matter what the β for a circuit happens to be, the transistor's operating point must lie on this line. It is a valuable DC analysis tool.

CHAPTER-6

CONCEPTS OF AMPLIFIER

6.1 INTRODUCTION: AN OVERVIEW

Amplifiers are fundamental electronic devices that enhance the strength of electrical signals, making them a cornerstone in various applications. The core concept of an amplifier involves taking a weak input signal and increasing its magnitude, usually in terms of voltage, current, or power. This process allows the amplifier to boost signals for transmission, processing, or driving loads such as speakers. Amplifiers come in various types, including operational amplifiers, power amplifiers, and radio-frequency amplifiers, each tailored to specific tasks. The concept of signal amplification finds numerous uses in the field of electronics. This includes applications such as boosting the signal level from a sensor or driving loads like loudspeakers or antennas. Reduced to its most simple terms, amplification is just multiplication. The ideal amplifier multiplies the amplitude of the input signal by a constant. It should not change the frequency of the signal, alter its shape, add noise or in any other way warp or distort the signal.

Amplifiers can be designed to be voltage sensing or current sensing and can be modeled as either controlled voltage sources or controlled current sources. As a functional block, we are primarily interested in describing an amplifier in terms of its amplification factor, input impedance and output impedance. The amplification factor is also referred to as the gain and may be expressed in terms of voltage gain, current gain or power gain, depending on the application. Other items of interest include the maximum output level or compliance, useful frequency range, noise and distortion characteristics.

6.2 AMPLIFIER MODEL

The ideal amplifier does nothing except increase the amplitude of the input signal. The factor of increase is defined as the ratio of the output signal to the input signal. It is a unit-less quantity. For example, if the input signal has a power of 10 millwatts and the circuit boosts the signal up to 50 milliwatts, we say it has a power gain of 50 milliwatts /10 milliwatts, or 5. Similarly, if the input signal is 2 volts and the output signal is 16 volts, we say it has a voltage gain of 16 volts/2 volts, or 8. Historically, power gain is denoted as G. For voltage gain and current gain we use Av and Ai, where A stands for Amplification factor. Some amplifiers invert the signal from input to output. Basically, they flip the wave shape upside down. For a simple sine wave this is equivalent to shifting the phase of the signal by 180°, and for a sine wave input the amplifier produces a -sine output. To reflect this effect, the amplification factor is denoted as negative. For example, an Av of -10 indicates an amplification factor of 10 with a signal inversion.

The size and complexity of an amplifier circuit can vary considerably, ranging from a single transistor to dozens of transistors. To ease system design it is helpful to use simplified functional models. Typically, these models use a resistor to represent the impedance seen looking into the amplifier along with a controlled source and its associated internal resistance. An example is shown in Figure 6.1.

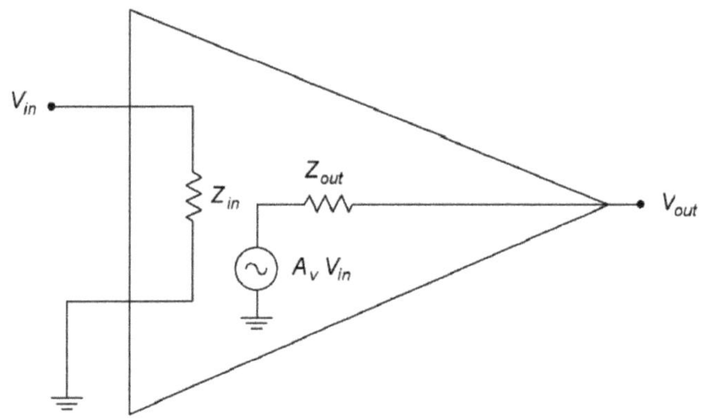

Figure 6.1 A simple voltage amplifier model.

This is a model of a voltage amplifier (note that Vin and Vout are specified along with a controlled voltage source within the model). The controlled voltage source and its series Zout is the Thevenin equivalent of the output when viewed from the load (i.e., the Vout pin). Likewise, Zin is the equivalent impedance seen by the driving source.

Because a voltage amplifier is designed to maximize voltage transfer, the input impedance tends to be high to minimize loading (think of a voltmeter). Similarly, the output impedance would tend to be low (think of an ideal voltage source). In contrast, a circuit designed for maximum current transfer would tend to have a low Zin and a high Zout. Precisely how the circuit creates the signal boost is not a concern of this model, we only care that it does.

Loading Effects

Once the model is established it is relatively easy to recognize and compute loading effects. Loading effects are signal losses caused by interactions between the amplifier's impedances and those of the circuits and loads connected to it. A generic model including loading effects is shown in Figure 6.2.

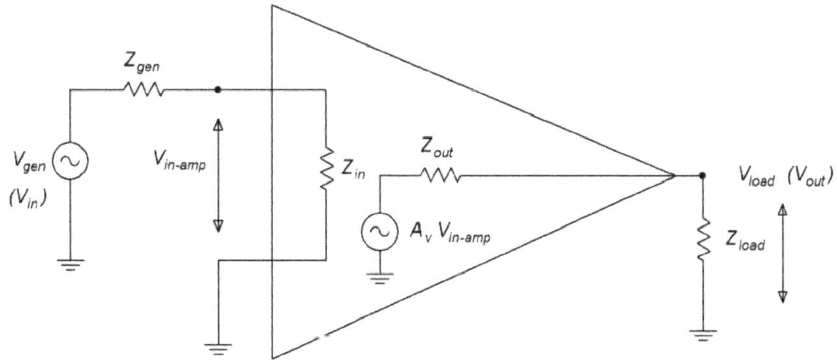

Figure 6.2 Voltage amplifier model with loading effects.

A cursory examination of Figure 6.2 shows that there is a voltage divider between Zgen and Zin along with a second divider between Zout and Zload. Each of these dividers causes signal loss, that is, they reduce the final output voltage. The input voltage to the amplifier is reduced as follows

$$V_{in-amp} = \frac{Z_{in}}{Z_{in} + Z_{gen}} \times V_{gen}$$

Example 6.1

A voltage amplifier has the following specifications: Av=20, Zin=10 kΩ, Zout=200 Ω. It is driven by a 30 millivolt source with a 600 Ω internal impedance and drives a 1 kΩ load. Determine the load voltage. The voltage that appears at the amplifier's input is

The load voltage is reduced as follows

$$V_{load} = \frac{Z_{load}}{Z_{load} + Z_{out}} \times A_v \times V_{in-amp}$$

Therefore the combined gain is

$$A_{system} = \frac{V_{load}}{V_{gen}} = \frac{Z_{in}}{Z_{in} + Z_{gen}} \times A_v \times \frac{Z_{load}}{Z_{load} + Z_{out}}$$

This is multiplied by the voltage gain of 20 and then reduced by the output divider.

To minimize these losses we'd like $Z_{in} >> Z_{gen}$ and $Z_{load} >> Z_{out}$.

Without the loading effects the output signal would be simply 30 millivolts times the voltage gain of 20, or 600 millivolts. Further, note that if the source is replaced with a typical laboratory grade function generator exhibiting an internal impedance of 50 Ω and the load is removed, being replaced by an oscilloscope exhibiting a typical 1 MΩ input impedance, the loading effects would be minimal and we would measure just a few millivolts shy of the ideal 600 millivolts.

Example 6.1

A voltage amplifier has the following specifications: Av=20, Zin=10 kΩ, Zout=200 Ω. It is driven by a 30 millivolt source with a 600 Ω internal impedance and drives a 1 kΩ load. Determine the load voltage.

The voltage that appears at the amplifier's input is

$$V_{in-amp} = \frac{Z_{in}}{Z_{in}+Z_{gen}} \times V_{gen}$$

$$V_{in-amp} = \frac{10\,\mathrm{k}\Omega}{10\,\mathrm{k}\Omega+600\,\Omega} \times 30\,\mathrm{mV}$$

$$V_{in-amp} = 28.3\,\mathrm{mV}$$

This is multiplied by the voltage gain of 20 and then reduced by the output divider.

$$V_{in-amp} = \frac{Z_{load}}{Z_{load}+Z_{out}} \times A_v \times V_{in-amp}$$

$$V_{in-amp} = \frac{1\,\mathrm{k}\Omega}{1\,\mathrm{k}\Omega+200\,\Omega} \times 20 \times 28.3\,\mathrm{mV}$$

$$V_{in-amp} = 471.7\,\mathrm{mV}$$

6.3 COMPLIANCE AND DISTORTION

At some point the idealization that the output signal is merely the input signal times the gain fails. All amplifiers have a limit on just how large the output signal can be. This is set by the DC power supply and the amplifier design. The maximum output signal (typically, the maximum output voltage) is referred to as the compliance. Any attempt to produce an output signal that swings beyond the compliance will result in waveform distortion. In the simplest case, the output signal is strictly and abruptly limited to the compliance level and any portion of the output waveform that would otherwise lay above that will be removed. It is as if some form of electronic scissors clipped off the top of the waveform. Hence, this is often referred to as clipping. An example of clipping is illustrated in Figure 6.3. The ideal output waveform is shown in brown and the clipped waveform is shown in blue. The clipping is so severe here that the clipped waveform now looks less like a sine wave and more like a square wave. This is extreme waveform distortion and has important consequences.

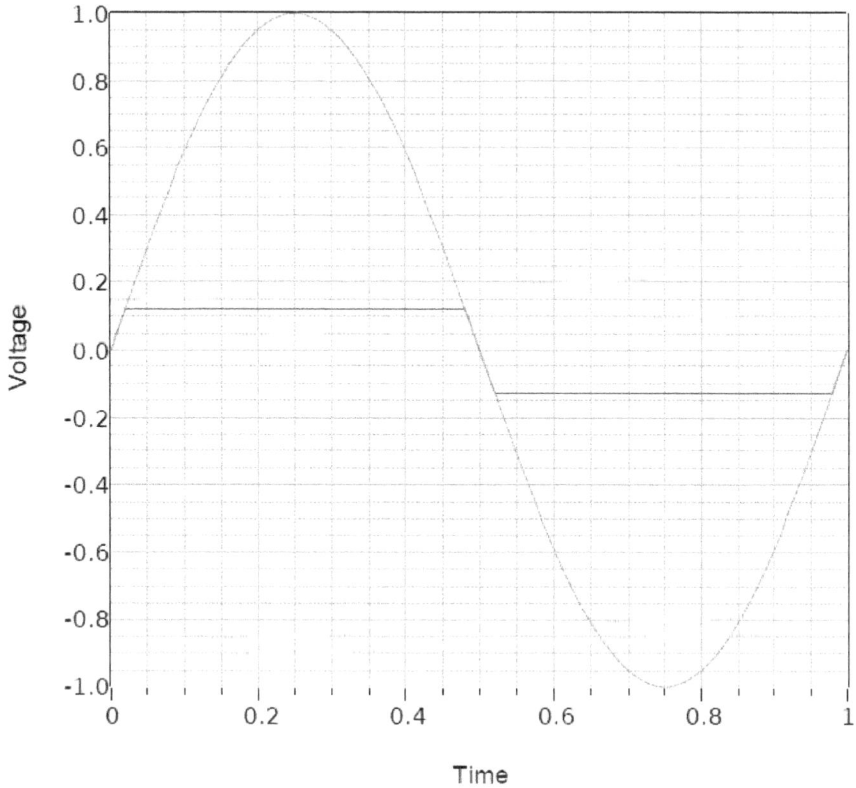

Figure 6.3 Clipped waveform.

Whenever a signal is altered in the time domain, there will be an alteration of its frequency content. Depending on what we started with and the manner in which the wave shape is altered, new frequency components may be added to the signal and the levels of existing components may be changed or even deleted. The extreme clipping that created a near-square wave added a large number of new frequency components to the signal.

To see how the wave shape and frequency content are connected, consider the waveform depicted in Figure 6.4. We start with a simple sine wave shown in green. We refer to this as the fundamental or base frequency. We then add a harmonic. A harmonic is another sine wave that is an integer multiple of the fundamental frequency. It may be larger or smaller in amplitude and the phase may be shifted. In this example we have a single sine at three times the fundamental frequency (blue). When we add this harmonic to the fundamental we arrive at a new waveform shown in red. This new waveform looks something like a square wave but with a "lumpy" top and bottom. If we add more harmonics, these variations will begin to smooth out, as shown in Figure 6.5. This waveform appears to be fairly close to a square wave and not too distant from our earlier clipped waveform. Based on this, we can conclude that the clipped sine wave has new frequency components added to it. We can also conclude that if our amplifier clips a more complicated waveform such as a snippet of music, that process will add new harmonics as well. Further, it is likely that these harmonics will be audible and can change our perception of the music, perhaps subtly but maybe drastically.

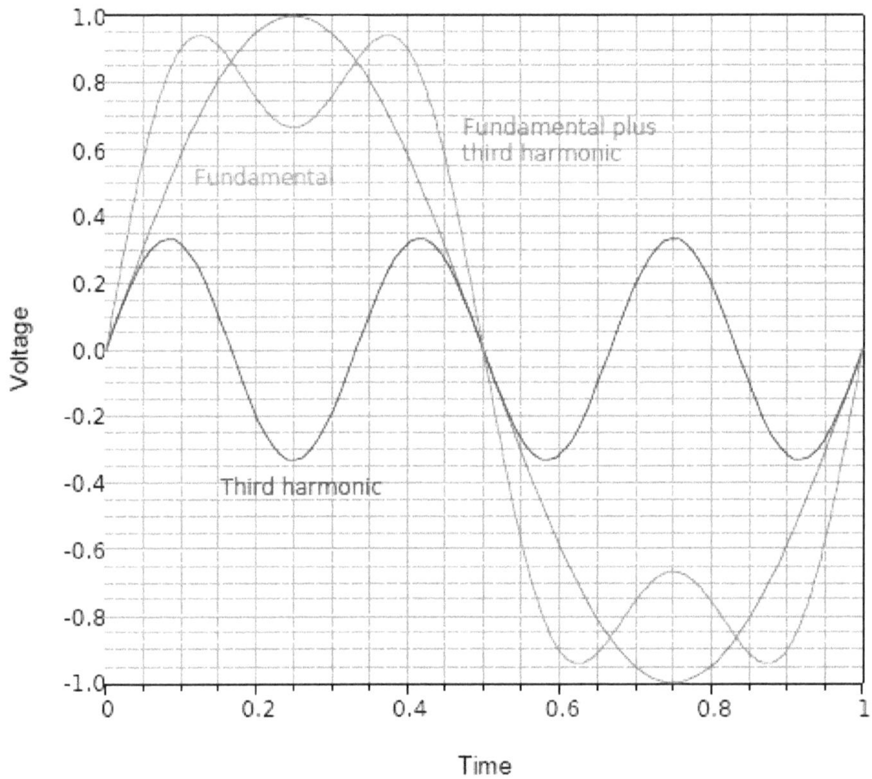

Figure 6.4 Sine wave with third harmonic.

Along with clipping, amplifiers can exhibit more subtle forms of distortion due to internal nonlinearity. For example, it is possible for the gain to vary slightly as the signal swings from low to high or from negative to positive. An example is shown in Figure 6.6 with the distorted wave shown in red.

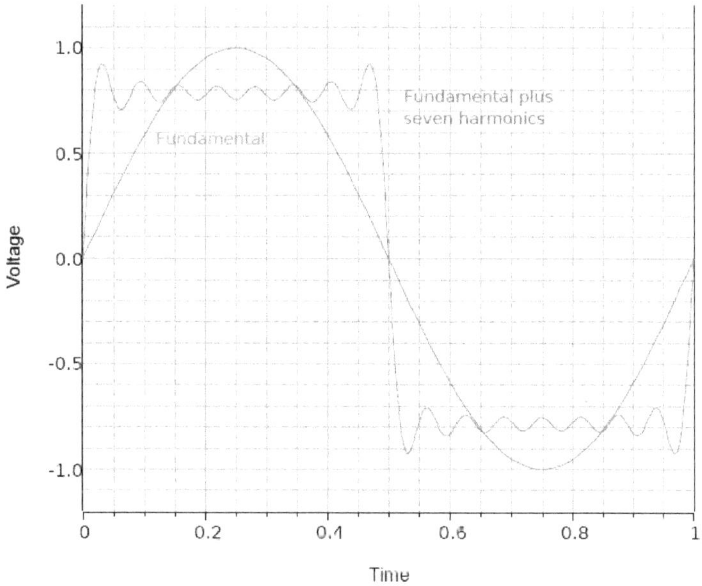

Figure 6.5 Sine wave with seven harmonics, approaching a square wave.

At first glance it may appear as though the wave is merely offset negatively. This is not the case. If we shift the wave vertically, as in Figure 6.7, it becomes apparent that the wave is truly distorted and is no longer a pure sine wave.

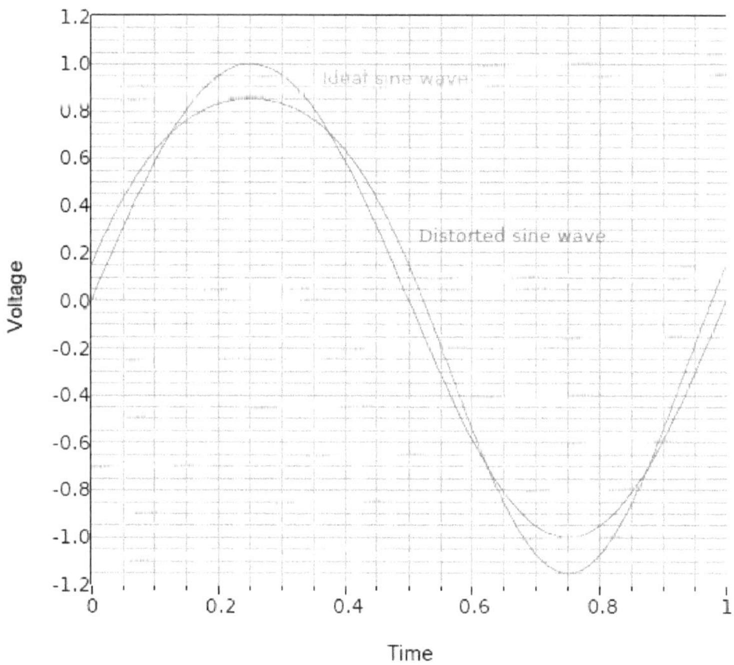

Figure 6.6 Sine wave with distortion.

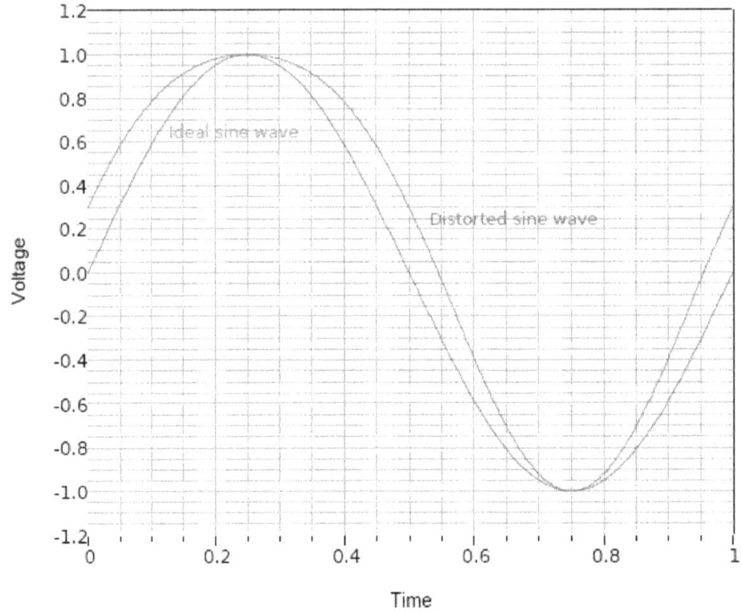

Figure 6.7 Sine wave with distortion, level shifted.

Unlike the clipped wave, the distorted wave in Figure 6.7 exhibits an asymmetry; the negative portion does not appear to be a mirror image of the positive portion. In other words, this wave lacks half-wave symmetry. Waves that exhibit half-wave symmetry contain only odd harmonic distortion (harmonics that are odd integer multiples of the fundamental). In contrast, waves that lack half-symmetry have at least one even harmonic. Here is how to test for half-wave symmetry. First, consider the saw tooth wave shown in Figure 6.8.

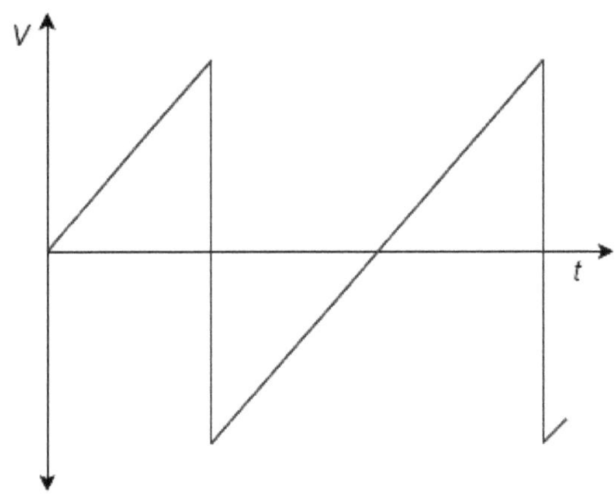

Figure 6.8 Half-wave symmetry test: Saw tooth wave.

Rotate the negative portion of the wave around the time axis as shown in Figure 6.9.

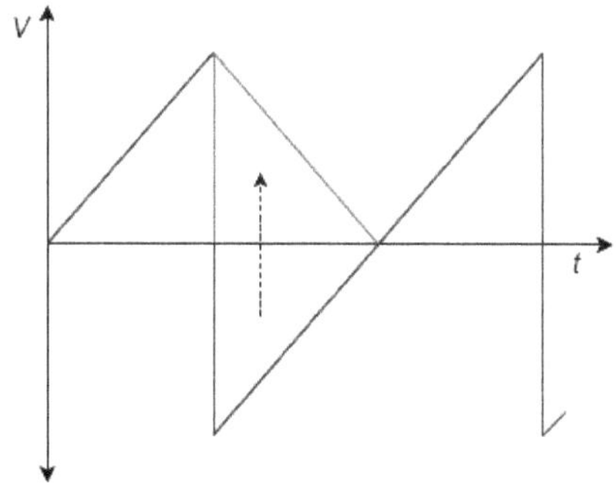

Figure 6.9 Half-wave symmetry test: Saw tooth wave, negative portion rotated.

Finally, slide the negative portion over the positive portion and see if they're identical, as in Figure 6.10.

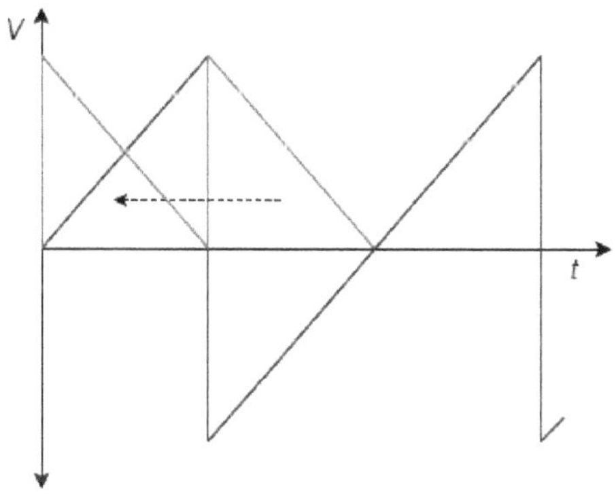

Figure 6.10 Half-wave symmetry test: Saw tooth wave, negative portion rotated and slid.

If the two halves are identical then the wave has half-wave symmetry. The saw tooth wave does not exhibit half-wave symmetry, therefore it must contain at least one even harmonic.

An amplifier's linearity is often quantified through a Total Harmonic Distortion, or THD, measurement. The measurement is carried out by applying a very pure, low distortion sine wave to the amplifier. This is the fundamental. At the output of the amplifier, a very selective filter is used to remove the fundamental. This leaves behind just the added distortion harmonics.20 These harmonics are then treated as a lumped value and presented as a percentage of the total signal. On an oscilloscope, it is relatively easy for a person to discern THD levels in the double digits. On the other hand, it is very difficult, if not impossible, for an individual to discern THD levels much below 1% by eye. Of course, what matters is what we can hear, not how the waveform looks. To put this in perspective, many high fidelity audio amplifiers exhibit THD levels below 0.1% while an over-driven guitar amplifier might be running over 20%. THD is not the final word on distortion though. It has its limits.

For example, all of the distortion products are lumped together. It says nothing about which harmonics are particularly strong or their distribution. It also doesn't say much about what happens when multiple frequencies interact. One method of trying to quantify that is to apply two sine waves at different frequencies to the amplifier simultaneously. The result is called an Intermediation Distortion rating, or IMD. This is also expressed as a percentage.

6.4 FREQUENCY RESPONSE AND NOISE

Like compliance and distortion, two other practical limits on amplifier performance are its frequency response and output noise. First, let's discuss frequency response.

Although we describe an amplifier as having a specific gain or amplification factor, this is true only for a certain range of frequencies. All amplifiers are limited in terms of the range of frequencies over which they can operate. If we examine an amplifier's performance at extreme frequencies, the gain may be much less than the nominal value. In fact, if we go far enough, the gain may even be fractional, meaning that the "amplifier" is actually reducing the signal level.

The region where the nominal gain is accurate is referred to as the mid-band. This range is defined by one or two corner or break frequencies. The lower limit is referred to as f1 while the upper limit is referred to as f2. At these frequencies, the output level has dropped to half the power exhibited by a mid-band frequency of the same input level. A gain versus frequency response plot that encapsulates this concept is shown in Figure 6.11.

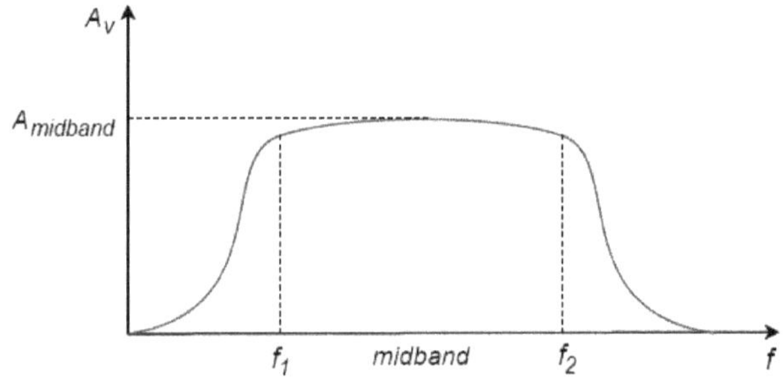

Figure 6.11 A generic gain versus frequency plot.

In this representative plot it is apparent that only input signals whose frequencies lay between f1 and f2 will receive full amplification. As the input frequency moves to either side of the middle band, the gain begins to drop off. The drop-off increases as the signal frequency moves farther and farther away. Eventually the gain will fall to practically zero and virtually no trace of the input signal will appear at the output.

Precise values of the corner frequencies will depend on the application. For example, a high fidelity audio amplifier will most likely have an f1 below 20 Hz and an f2 above 20 kHz21 while an amplifier used for telephone systems might range from 300 Hz to 4 kHz22. In contrast, a radio frequency amplifier may be operating at frequencies orders of magnitude higher than these.

Without exception, all amplifiers have an upper limit frequency, f2, but not all of them have a lower frequency limit, f1. Amplifiers without a lower limit can amplify signals with frequencies all the way down to DC. They are referred to as direct coupled or DC amplifiers. The lower frequency limit is usually caused by in-line coupling capacitors, and in some cases, transformers. Among other uses, these components are added to purposely block DC. There are good reasons to do this, as we shall see in upcoming work, however, it is possible to design amplifiers without them. The resulting amplifier will then have no limit on how low of a frequency it can amplify.

The upper limit frequency is another story. While components are often added to tailor the upper frequency response of an amplifier, even if no tailoring was desired the amplifier would still have an upper limit frequency. This would be due to small and unavoidable capacitances and inductances that exist in the circuit, for example stray wiring capacitance. Ultimately, the corresponding reactance will cause a signal level reduction that worsens as frequency increases. As you might guess, these reactance's also cause varying phase shifts between the input signal and the output signal.

Amplifier performance is also limited by its internal noise. Noise is an undesired signal that appears at the output of an amplifier. Unlike distortion, noise is usually not correlated with the input signal level. Generally, noise is broad-band, meaning that it contains a very wide range of frequencies. As such, it does not have a discernible pitch. Examples in nature include the sound of leaves rustling in the wind or the sound of a waterfall. Noise is best thought of as a truly random signal. As such, it cannot be accurately predicted and therefore there is no easy way to remove it once it has been added to a desired signal. There are many potential sources of noise in an amplifier. They range from process issues in semiconductors to thermal effects in resistive elements. In general, noise gets worse as temperature, resistance and frequency range increase. Noise is unavoidable in absolute terms but ultimately what we care about is whether or not it is low enough for a given application. In other words, is the noise level significantly lower than the signal level, to the point where it is no longer a problem? This is quantified by simply creating a ratio between the nominal output signal level and the output noise level. This ratio is given the very creative name *signal-to-noise ratio*, or *S/N* for short. All other factors being equal, the higher the *S/N*, the better.

6.5 MILLER'S THEOREM

Some inverting voltage amplifier designs employ an impedance bridged between the input and output, as shown in Figure 6.12.

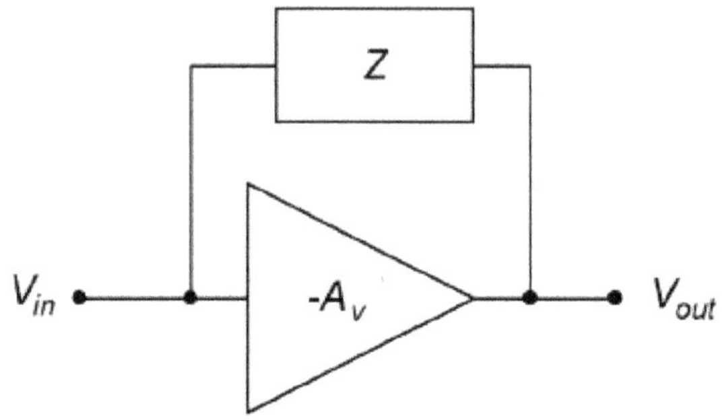

Figure 6.12 An input/output bridged impedance.

There are different reasons for doing this, a prime example involving shaping the frequency response of the amplifier. To simplify the analysis of such a configuration, we may employ *Miller's Theorem*, named after American engineer John Milton Miller.

The goal here is to determine equivalent impedances that lie in parallel with the input and output of the amplifier. These equivalents simply become part of the input and output networks around the amplifier.

First, let's consider the equivalent impedance at the input, *Zin-miller*. By definition, this is the impedance in parallel with the input of the amplifier that would draw the same amount of current as the original bridging Miller impedance. The current through the Miller impedance is simply the voltage across it divided by the Miller impedance, Z. The voltage across it is the difference between the input and output voltages.

$$i_{in-miller} = \frac{V_{in} - V_{out}}{Z}$$

$$i_{in-miller} = \frac{V_{in} - A_v V_{in}}{Z}, \text{ the gain is negative so}$$

$$i_{in-miller} = \frac{V_{in}(|A_v| + 1)}{Z}$$

Dividing this current into the input voltage yields the equivalent impedance.

$$Z_{in-miller} = \frac{Z}{|A_v| + 1}$$

A similar derivation yields the output equivalent.

$$Z_{out-miller} = \frac{Z|A_v|}{|A_v|+1}$$

The general rule to remember is that the Miller equivalent presents equivalent impedances that are less than the original bridging impedance. In the case of the input section, the reduction effect is very large at higher gains.

The Miller equivalent circuit for a general impedance Z is shown in Figure 6.13.

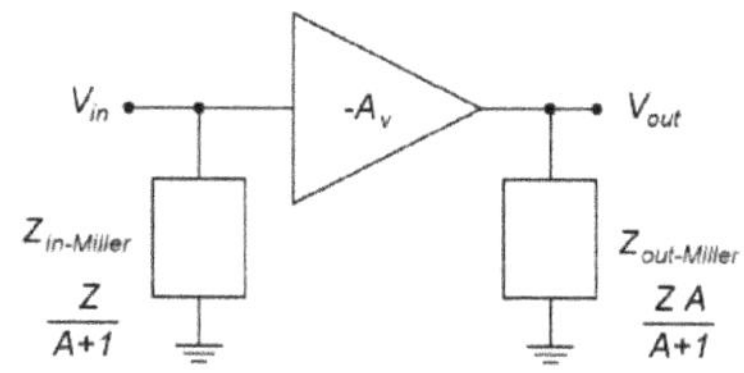

Figure 6.13 Miller general equivalent.

The general rule to remember is that the Miller equivalent presents equivalent impedances that are less than the original bridging impedance. In the case of the input section, the reduction effect is very large at higher gains.

Two typical cases for the impedance Z are a resistance, R, and a capacitance, C. For a pure resistance, we can perform a direct substitution for Z. The original and Miller equivalents are shown in Figure 6.14.

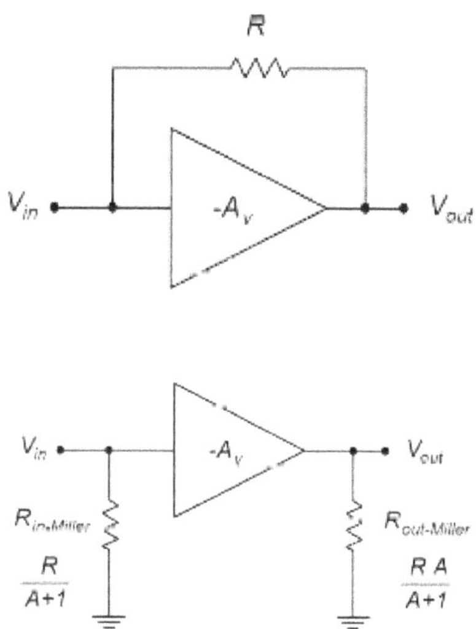

Figure 6.14 Original and Miller equivalent for a resistor.

For a capacitor, the situation is similar, however, we will substitute Xc for Z and recall that C=1/(2πfXc).

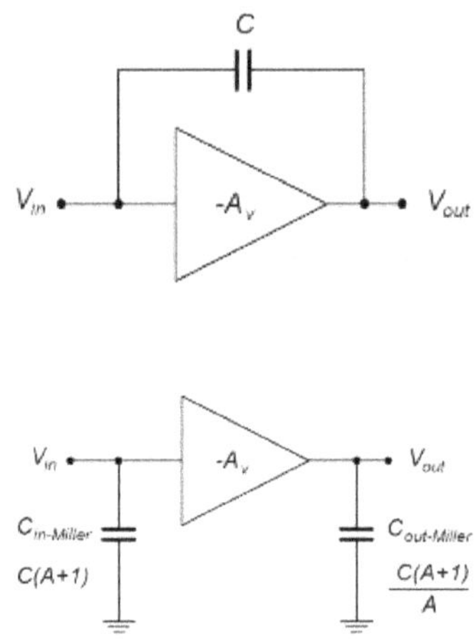

Figure 6.15 Original and Miller equivalent for a capacitor.

Note that for the capacitor, there is a multiplicative effect. That is, the effect of the original bridging capacitor on a high gain amplifier is equivalent to a much larger input shunt capacitor.

CONCISE SUMMARY

- Complex amplifier circuits can be modeled with a functional block. The ideal model includes the input and output impedances along with a controlled source. This source would exhibit a signal gain or amplification factor. Usually, G stands for power gain while Av and Ai represent voltage and current amplification, respectively. This amplification factor may be negative which indicates that the amplifier inverts the phase of the input, that is, the waveform is flipped upside down. The impedances allow calculation of loading effects while the gain determines the size of the output signal.

- If the input signal is too large, the output signal may be limited in amplitude or clipped. The maximum output amplitude is referred to as the compliance. Clipping is a gross form of distortion but more subtle forms exist as well. In general, distortion creates new frequency components. If these new components are integer multiples of the original input frequency, which they are typically, they are referred to as harmonics. One method of quantifying distortion performance is to sum all of the harmonics and compare that to the original signal. This is called THD or total harmonic distortion. Along with distortion, the amplifier might also add undesirable noise to the output signal. Noise is a random signal that contains many different frequencies. Typically, this is measured via a signal-to-noise ratio, or S/N, at the output.

- An amplifier also operates over a given range of frequencies, from a lower limit, f1, to a high limit, f2. Some amplifiers are able to amplify down to 0 Hz (DC) and effectively do not have an f1 but all amplifiers do have an upper limit.
- Finally, Miller's Theorem is an analysis technique that allows an impedance that bridges from the input of an inverting voltage amplifier to its output to be split into equivalent input and output parallel impedances. These impedances will be smaller than the original bridging impedance and are a function of the gain of the amplifier.

CHAPTER-7

BIPOLAR JUNCTION TRANSISTORS (BJTS) SMALL SIGNAL AMPLIFIERS

7.1 INTRODUCTION: AN OVERVIEW

Small signal amplifiers using Bipolar Junction Transistors (BJTs) are foundational components in electronic circuits designed for signal amplification. These amplifiers operate within the linear region of the transistor's characteristics, making them suitable for amplifying small signals without distortion. BJT small-signal amplifiers play a crucial role in electronic systems such as audio amplifiers, communication circuits, and instrumentation. By carefully biasing the transistor in its active region, these amplifiers provide controlled amplification of weak input signals, making them more robust and suitable for subsequent stages of signal processing. Understanding the principles of small-signal analysis, biasing techniques, and the various amplifier configurations is essential for engineers designing electronic systems where signal fidelity is paramount. These amplifiers form the building blocks of complex electronic devices, contributing to the functionality and performance of a broad range of applications in modern technology.

In the prior chapter we discussed the general operational characteristics of amplifiers including voltage gain, input and output impedance, compliance, distortion and so forth. In this chapter we shall focus on the analysis of small signal amplifiers, specifically, their voltage gain and input/output impedances. As we will be performing a *small signal analysis*, we will not be concerned with compliance, maximum load power, device dissipation or the like. There is no specific definition of small signal versus large signal but for our purposes we shall define small signal as output signals that are well below the clipping limit and with power dissipation of no more than a few hundred milliwatts for either the load or transistor.

There are two popular techniques used to analyze BJT amplifier circuits. One is through the use of *hybrid parameters*. There are four different hybrid parameters. We have already seen one of them, the forward current gain, *hfe*. We simply call it β. The other three are *hie*, the input impedance; *hoe*, the output admittance; and *hre*, the reverse voltage gain. The second letter of the subscript (the "*e*" in *hfe*) indicates it is for the common emitter configuration (that is, input applied to the base, output taken at the collector and the emitter at the common ground).

The second approach uses r' parameters (pronounced "r prime"). The r' approach is sufficient for all of our analyses and, given an understanding of Ohm's law, KVL and KCL, produces straightforward equations for circuit gain, input impedance and the like. As a consequence, we shall focus on the r' system.

7.2 SIMPLIFIED AC MODEL OF THE BJT

Just as we created a DC model to ease the analysis of DC bias circuits, we shall make use of an AC BJT model for our AC analyses. In fact, our AC model is based on the DC model. The collector-base region is still represented with a current-controlled current source although it's AC instead of DC: iC = βiB. The base-emitter junction is a bit trickier. Although a simple 0.7 volt junction worked fine for DC, we now have to consider the AC resistance of the diode.

To find the dynamic resistance of the junction, first recall that the AC signal is riding on the DC bias current, as plotted in Figure 7.1.23 We can imagine that the AC signal is causing this point to trace back and forth along the curve. Of course, as this is a small signal analysis, this sweep will be very small, perhaps only a few percent of the quiescent current and can be approximated as a straight line segment. The slope of this line segment represents its conductance.24 The reciprocal of conductance is resistance; therefore, the reciprocal of the slope represents the resistance of the device. Consequently, we can approximate the dynamic resistance of the device as the reciprocal of the slope of the line tangent to the operating point (that is, the reciprocal of the slope of the line tangent to the quiescent bias current IC).

In reality, this slope is changing slightly as the signal swings back and forth along the base-emitter I-V curve. As the signal swings positive and goes above the quiescent point the slope is a little steeper producing a slight reduction in dynamic resistance. In contrast, as it swings negative, going below the quiescent point, the slope becomes a bit more shallow and produces a slightly higher resistance. As a result, we are effectively computing an average value for the dynamic resistance by assuming this is a straight line segment. The variance in this resistance will be a source of asymmetrical distortion in the amplifier of the type shown in Chapter 6, Figure 6.6. We shall see more on this later.

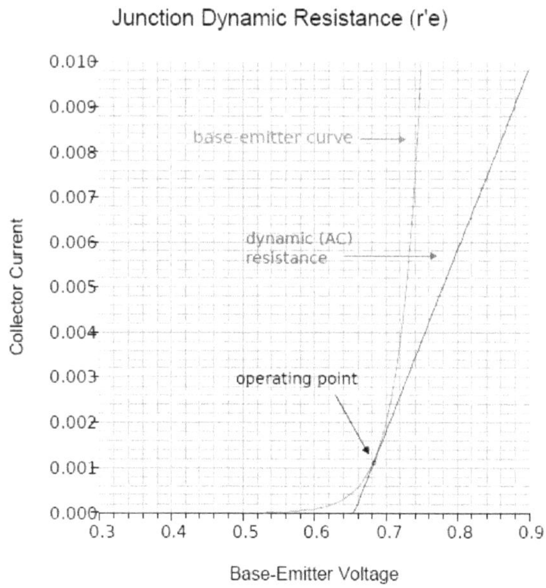

Figure 7.1 Base-emitter junction plot and dynamic resistance.

In order to derive an equation for the dynamic resistance, we begin with the Shockley equation from Chapter 2, Equation 2.1, slightly modified to reflect the terminal names of a BJT.

$$I_C = I_S\left(e^{\frac{V_{BE}\,q}{n\,k\,T}} - 1\right)$$

Where

I_C is the junction (collector) current,

I_S is the reverse saturation current,

V_{BE} is the voltage across the junction (base-emitter),

q is the charge on an electron, 1.6E−19 coulombs,

n is the quality factor (typically between 1 and 2),

k is the Boltzmann constant, 1.38E−23 joules/kelvin,

T is the temperature in kelvin.

At 300 kelvin (about 80°F), q/kT is approximately 38.6, thus for any reasonable value of V_{BE} the "−1" term is small enough to ignore. Also, we shall take n as 1.

The equation then reduces to

$$I_C = I_S e^{38.6 V_{BE}} \qquad (7.1)$$

To find the slope we take the first derivative of Equation 7.1 with respect to VBE.

$$\frac{dI_C}{dV_{BE}} = 38.6\, I_S e^{38.6 V_{BE}} \qquad (7.2)$$

Substituting Equation 7.1 into Equation 7.2 yields

$$\frac{dI_C}{dV_{BE}} = 38.6\, I_C$$

We call this r_{e}. This value is slightly low as it doesn't include bulk resistance so a good approximation is

$$r'_e = \frac{26\ \text{mV}}{I_C} \qquad (7.3)$$

It is important to note that IS in Equation 7.2 varies with temperature. Therefore r'e varies with temperature as well, decreasing with increasing temperature. This carries important ramifications with the thermal stability of higher power amplifiers as we shall see in subsequent work.

One of the most important things to remember here is that the DC collector current sets up the resistance of the AC model. In other words, the stability of the AC circuit will depend in part on the stability of the DC bias (hence our emphasis on stable bias circuits in Chapter 5).

We now have our AC model, as shown in Figure 7.2. This is a simplified model in that it does not include junction capacitance effects, lead inductance and the like. It is appropriate, therefore, as a low to mid-band frequency model. To summarize, the AC collector current, iC, is determined by the AC input current, iB; which is in turn a function of the size of the applied input signal. In contrast, r'e is set by the DC bias current, IC. The AC input can produce small variations in r'e which are manifested as waveform distortion.

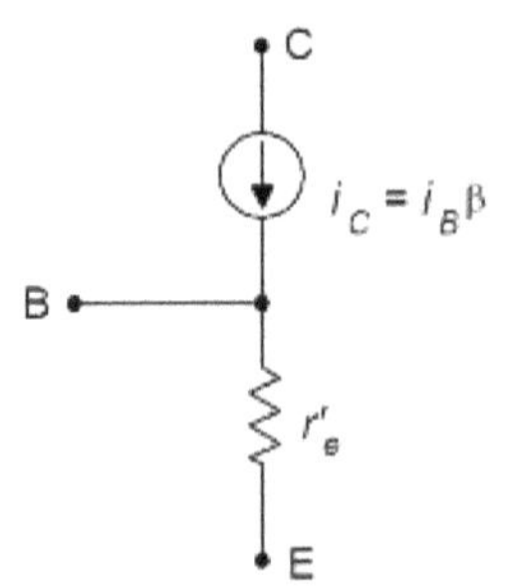

Figure 7.2 Simplified AC model of BJT.

Given the model, there are three ways to configure the transistor as an amplifier:

Common Emitter: The input is applied to the base and the output is taken at the collector. The emitter terminal is at the common or ground point. This configuration exhibits both voltage gain and current gain. It also inverts the phase of the signal.

Common Collector: The input is applied to the base and the output is taken at the emitter. The collector terminal is at the common or ground point. This configuration offers a voltage gain of about unity but does exhibit current gain. It maintains the phase of the input signal. It is also referred to as an emitter follower or voltage follower.

Common Base: The input is applied to the emitter and the output is taken at the collector. The base terminal is at the common or ground point. This configuration exhibits voltage gain but the current gain is unity at best. It also maintains the phase of the input signal.

We shall examine each of these topologies in turn. Each of these can be made using a variety of DC bias techniques. For example, a two-supply emitter bias or voltage divider bias could be used for any of the three AC topologies, and further, they could utilize either an NPN or PNP transistor.

7.3 THE COMMON EMITTER AMPLIFIER

The common emitter configuration finds wide use as a general purpose voltage amplifier. We begin with a basic DC biasing circuit and then add a few other components. For example, refer to Figure 7.3.

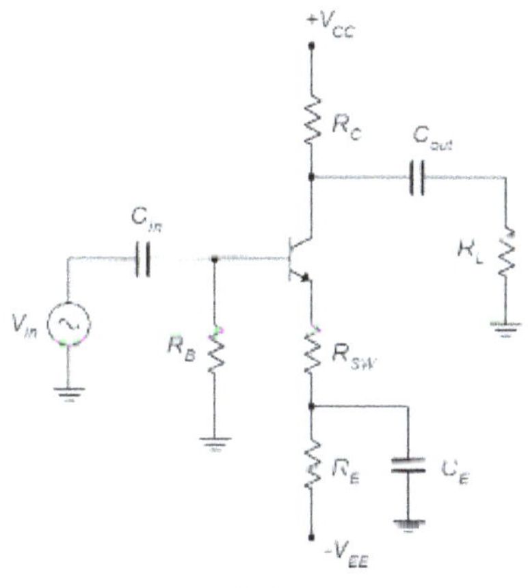

Figure 7.3 Common emitter amplifier using two-supply emitter bias.

This amplifier is based on a two-supply emitter bias circuit. The notable changes are the inclusion of an input signal voltage, Vin, and a load, RL. So that these components do not alter the bias, we isolate the input and load through the use of coupling capacitors Cin and Cout. These capacitors will act as opens to DC creating the desired isolation. As for the AC signal, the capacitances will be chosen such that their reactance's will be much smaller than the surrounding resistors at the frequency of the input. Consequently, the capacitors will appear as shorts and allow the AC signal to pass through the amplifier.

The final alteration involves the emitter resistor. The single resistor of the bias network is replaced by a pair of resistors, *RE* and *RSW*, along with a *bypass capacitor*, *CE*. For DC, the capacitor is open and the effective emitter bias resistance is *RE* + *RSW*. For AC, the capacitor will behave ideally as a short so the AC emitter resistance will fall to just *RSW*. This resistor is called a *swamping* or *emitter degeneration* resistor. It is used primarily to help control the voltage gain of the amplifier.

We can use our AC transistor model along with the Superposition Theorem to arrive at an equivalent AC circuit of the amplifier, as shown in Figure 7.4.

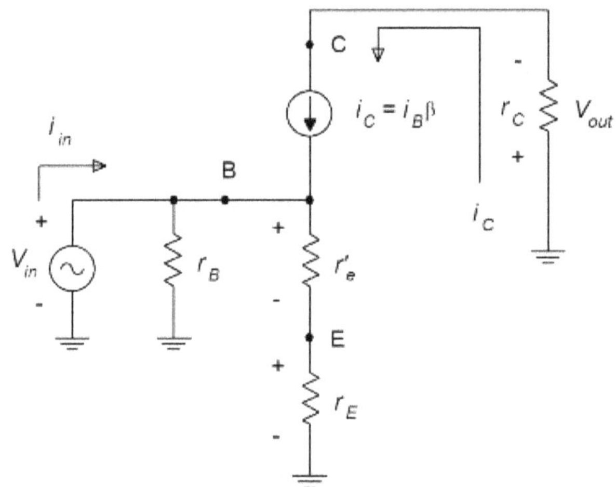

Figure 7.4 AC equivalent of common emitter amplifier.

First, we have shorted all of the capacitors. Second, we have replaced the DC sources with their ideal internal resistance (a short) which places those points at AC ground. Third, we swapped out the transistor for the model. Lastly, we have combined and/or renamed resistances where needed. Because this is an AC circuit, we use the convention of lower case *r* for resistance to avoid confusion with the DC resistance (which are upper case). Thus, *rE* is the AC resistance from the emitter to AC ground. This corresponds to *RSW* in the original schematic. Similarly, *rC* represents the total resistance seen from the collector to AC ground. In the original schematic this corresponds to *RC* in parallel with *RL*. If this circuit was unloaded, then *rC* would just be equal to *RC*. Finally, *rB* corresponds to *RB* but in a voltage divider bias it would be equal to *R1* in parallel with *R2*.

Voltage Gain

Voltage gain, *Av*, is defined as the ratio of *vout* to *vin*. Using Ohm's law, we find

$$A_v = \frac{v_{out}}{v_{in}} = \frac{v_C}{v_B}$$

$$A_v = \frac{-i_C r_C}{i_C(r'_e + r_E)}$$

$$A_v = -\frac{r_C}{r'_e + r_E} \qquad (7.4)$$

First, the negative sign indicates that this amplifier inverts the waveform, top to bottom. For a sine wave, this is equivalent to shifting the phase 180°. In some applications this can be a major issue, in others, not so much. If it is an issue, it can be resolved by using a second inverting gain amplifier in sequence with the first (inverting the inversion).

The second thing we see is that the gain is little more than a ratio of collector to emitter resistances. This is where splitting the emitter resistor into two parts comes in. In the equation, *rE* is the swamping resistor *RSW*. The larger the swamping resistor, the lower the gain. The maximum gain will be achieved when *RSW* = 0. That is, when the emitter is completely bypassed. The down side of this is that the gain will now depend entirely on *r'e*. This will increase the distortion. The reason is because *RSW*, being so much larger, effectively "swamps out" the variation in *r'e* and reduces distortion. The larger *RSW* is relative to *r'e*, the greater the reduction in distortion, but with the cost of reduced gain. This is why a swamping resistor is also called an emitter degeneration resistor: it degrades the voltage gain.

Input Impedance

Input impedance, *Zin*, is defined as the ratio of *vin* to *iin*. In Figure 7.2 this is equal to *rB* in parallel with the impedance looking into the base terminal, *Zin(base)*. Using Ohm's law we find

$$Z_{in(base)} = \frac{v_B}{i_B}$$

$$Z_{in(base)} = \frac{i_C(r'_e + r_E)}{i_B}$$

$$Z_{in(base)} = \frac{i_C(r'_e + r_E)}{i_C/\beta}$$

$$Z_{in(base)} = \beta(r'_e + r_E) \qquad (7.5)$$

Therefore

$$Z_{in} = r_B \,||\, Z_{in(base)} \qquad (7.6)$$

We see that both the swamping resistor and β play a role in setting the input impedance. Larger values of *RSW* and β produce larger input impedances.

In sum, we find that while swamping decreases voltage gain, it reduces distortion and increases input impedance, the latter two generally desirable for a voltage amplifier. A non-swamped amplifier will have the largest gain but will suffer from the worst distortion and a low input impedance. This is a classic "quality versus quantity" trade-off: a large low quality gain versus a modest high quality gain[25].

Output Impedance

Output impedance, *Zout*, is defined as the internal impedance of the equivalent source that drives the load. If we position ourselves at the load and look back into the amplifier shown in Figure 7.1, *Cout* is shorted ideally and *VCC* is at AC ground. This leaves us with *RC* in parallel with the transistor. The transistor is modeled as a current source and its ideal internal resistance would approach infinity. In reality, the effective value, *r'C*, is likely in the region of 100 kΩ or so, depending on bias current. This parallel combination comprises the output impedance of the current source. We model this circuit as a voltage amplifier so to be proper, we'd convert the current source with parallel internal resistance to a voltage source with series internal resistance. Those resistance values are identical, though, and we arrive at

$$Z_{out} = r'_C \,||\, R_C$$

In many circuits, *RC* is considerably smaller than *r'C*, therefore

$$Z_{out} \approx R_C \qquad (7.7)$$

Example 7.1

Determine the input and output impedances of the amplifier shown in Figure7.5. Also compute the voltage gain. Assume β = 150.

First, the easy bit. We can determine the output impedance by inspection. It is approximately equal to *RC*, or 22 kΩ.

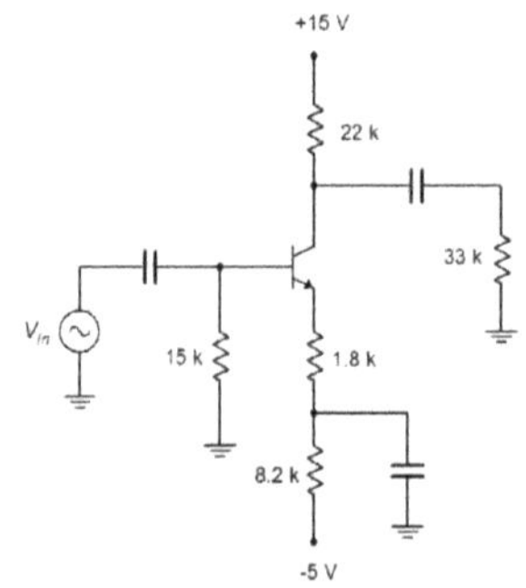

Figure 7.5 Schematic for Example 7.1.

In order to find *Zin* and *Av*, we will need to determine *r'e*. To obtain *r'e* we need to find *IC*. Using KVL around the base-emitter loop, if we approximate the DC base voltage to be near zero, then all of the emitter supply drops across the DC emitter resistance, with the exception of *VBE*.

$$I_C = \frac{|V_{EE}| - V_{BE}}{R_E + R_{SW}}$$
$$I_C = \frac{5\text{ V} - 0.7\text{ V}}{8.2\text{ k}\Omega + 1.8\text{ k}\Omega}$$
$$I_C = 0.43\text{ mA}$$

$$r'_e = \frac{26\text{mV}}{I_C}$$
$$r'_e = \frac{26\text{ mV}}{0.43\text{ mA}}$$
$$r'_e = 60.5\,\Omega$$

$$Z_{in-base} = \beta(r'_e + r_E)$$
$$Z_{in-base} = 150(60.5\,\Omega + 1.8\text{k}\,\Omega)$$
$$Z_{in-base} = 279\text{k}\Omega$$

This value in parallel with the base biasing resistor creates the input impedance.

$$Z_{in} = R_B \,||\, Z_{in(base)}$$
$$Z_{in} = 15\text{k}\,\Omega \,||\, 279\text{k}\,\Omega$$
$$Z_{in} = 14.2\text{k}\,\Omega$$

Now for the voltage gain. We have two options. The first option is to find the loaded gain directly, meaning we include the load resistor as part of *rC*. This would be the fastest method for this particular problem. The second option is to find the unloaded gain (i.e., with the load resistor removed) and then include the voltage divider effect caused by the load interacting with the output impedance. This would be the preferred method if we were swapping out different load values. We'll do both just for funsies.[26]

First, method one.

$$r_C = R_C \,||\, R_L$$
$$r_C = 22\text{k}\Omega \,||\, 33\text{k}\Omega$$
$$r_C = 13.2\text{k}\,\Omega$$

$$A_v = -\frac{r_C}{r'_e + r_E}$$
$$A_v = -\frac{13.2\text{k}\Omega}{60.5\,\Omega + 1.8\text{k}\Omega}$$
$$A_v = -7.1$$

And now method two; first the unloaded gain, then the divider effect and finally, the composite gain.

$$A_{v(unloaded)} = -\frac{r_C}{r'_e + r_E}$$

$$A_{v(unloaded)} = -\frac{22\,\text{k}\Omega}{60.5\,\Omega + 1.8\,\text{k}\Omega}$$

$$A_{v(unloaded)} = -11.82$$

$$A_{divider} = \frac{R_L}{R_L + R_C}$$

$$A_{divider} = \frac{33\,\text{k}\Omega}{33\,\text{k}\Omega + 22\,\text{k}\Omega}$$

$$A_{divider} = 0.6$$

$$A_v = A_{v(unloaded)} \times A_{divider}$$

$$A_v = -11.82 \times 0.6$$

$$A_v = -7.1$$

We shall repeat the prior example using the same circuit but with one change: the emitter resistor will be completely bypassed. This will show the effect that swamping has on voltage gain and input impedance.

Example 7.2

Determine the voltage gain and input impedance of the amplifier shown in Figure 7.6. Assume β = 150.

The DC equivalent of this circuit is identical to that of the circuit shown in Figure 7.5. In both cases, the DC emitter resistance is 10 kΩ. Therefore, *IC* and *r'e* are unchanged. The bypass capacitor shorts this entire value for the AC equivalent because there is no swamping resistor. Consequently, *rE* = 0. We can simply use 0 for *rE* in the equations previously derived.

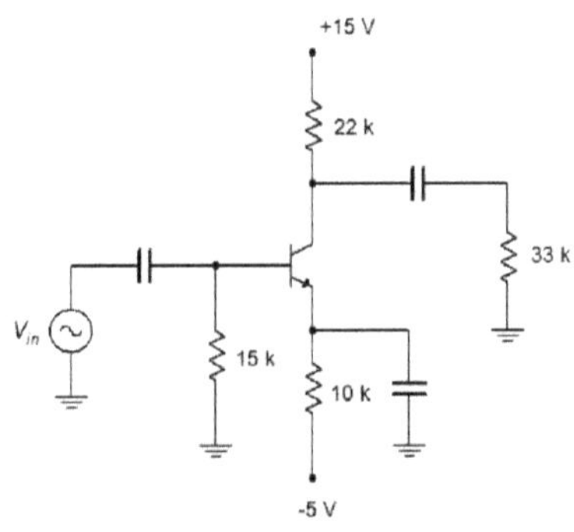

Figure 7.6 Schematic for Example 7.2.

We begin with the input impedance.

$$Z_{in(base)} = \beta(r'_e + r_E)$$
$$Z_{in(base)} = 150(60.5\,\Omega + 0)$$
$$Z_{in(base)} = 9075\,\Omega$$

This value is in parallel with the voltage divider biasing resistors, creating the input impedance.

$$Z_{in} = R_B \parallel Z_{in(base)}$$
$$Z_{in} = 15\,k\Omega \parallel 9075\,\Omega$$
$$Z_{in} = 5654\,\Omega$$

$$A_v = -\frac{r_C}{r'_e + r_E}$$
$$A_v = -\frac{13.2\,k\Omega}{60.5\,\Omega + 0}$$
$$A_v = -218.2$$

The end result is an input impedance less than half of the swamped case and a voltage gain over 30 times greater. What these calculations do not show is the increase in distortion that will be created by this change. More on that in a moment.

Let's consider something slightly different: a voltage divider bias PNP amplifier.

Example 7.3

Determine the input impedance and voltage gain for the circuit shown in Figure 7.7. Also determine *vload* if *vin* = 20 mV peak. Assume β = 100.

We need to first determine *r'e* which means that we need to find the collector current. If we assume a lightly loaded divider, the base voltage will be approximately 15 volts and the emitter will be 0.7 volts higher, or 15.7 volts. This leaves 20 volts - 15.7 volts, or 4.3 volts, across the DC equivalent emitter resistance. That's 4.1 k Ω + 200 Ω, or 4.3 kΩ , yielding 1 mA for *IC*. This will produce *r'e* = 26 Ω.

$$Z_{in} = R_1 \parallel R_2 \parallel Z_{in(base)}$$
$$Z_{in} = 15\,k\Omega \parallel 5\,k\Omega \parallel 22.6\,k\Omega$$
$$Z_{in} = 3.22\,k\Omega$$

$$A_v = -\frac{r_C}{r'_e + r_E}$$
$$A_v = -\frac{7.5\,k\Omega \parallel 10\,k\Omega}{26\,\Omega + 200\,\Omega}$$
$$A_v = -19$$

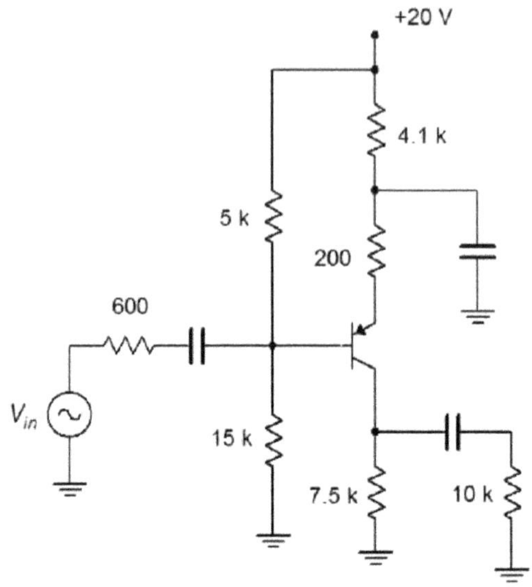

Figure 7.7 Schematic for Example 7.3.

We also need to include the effect of the 600 Ω source impedance. This will create a voltage divider with the input impedance.

$$A_{divider} = \frac{Z_{in}}{Z_{in} + Z_{source}}$$

$$A_{divider} = \frac{3.22\,\text{k}\Omega}{3.22\,\text{k}\Omega + 600\,\Omega}$$

$$A_{divider} = 0.843$$

$$A_{v(system)} = A_v \times A_{divider}$$

$$A_{v(system)} = -19 \times 0.843$$

$$A_{v(system)} = -16$$

Finally, we get to the load voltage.

$$V_{load} = A_{v(system)} \times V_{in}$$

$$V_{load} = -16 \times 20\,\text{mV}$$

$$V_{load} = 320\,\text{mV peak, inverted}$$

If we were to inspect the circuit of Figure 7.7 using a direct coupled oscilloscope, we would see the superposition of the AC and DC components. In other words, we'd see the AC signal riding on a DC offset. In some cases, the AC signal would be too small to notice compared to the DC portion. In proper scale it might be no thicker than the trace itself. In order to measure it accurately, we'd have to AC couple the oscilloscope.

The voltages at the source and load would be just AC as the coupling capacitors serve to block DC. At the base we'd have 15 volts DC with an AC signal riding on top of it. The AC would be the 20 mV input times the input impedance/source impedance divider of 0.843, or 16.86 mV. Recalling that *IC* is 1 mA, the DC drop across *RC* must be 7.5 volts. This is, of course, *VC*. Therefore, at the collector we'd see an inverted 320 mV signal riding on 7.5 volts DC.

Computer Simulation

In order to get some insight into the swamping-versus-distortion issue, we shall take a look at a more involved circuit simulation. This will echo Examples 7.1 and 7.2 in that we will simulate two circuits with the same DC equivalents. The only circuit change will be that one version will have a fully bypassed emitter while the other version will utilize a swamping resistor. In order to keep the comparison fair, we will increase the input signal voltage of the lower gain swamped amplifier so that both versions have a similar load voltage. In this way we guarantee that they are both using a similar percentage of the junction curve.

The unswamped circuit is shown in Figure 7.8. This utilizes a straightforward two- supply emitter bias.

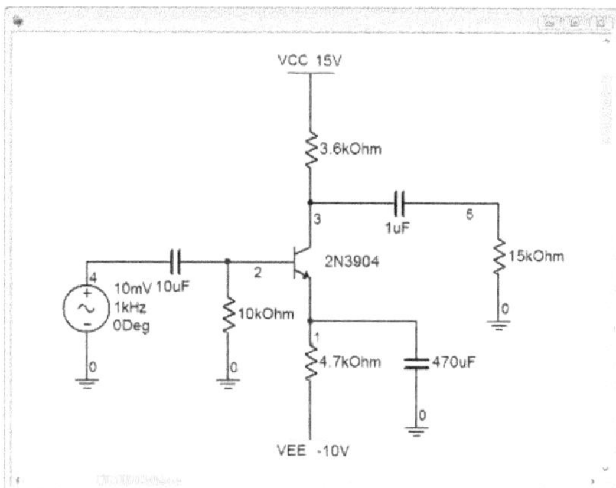

Figure 7.8 Unswamped CE amplifier in simulator.

A quick "back-of-an-envelope" estimate gives IC ≈ 2 mA, yielding r'e ≈ 13 Ω. The load will be around 3 kΩ which gives a gain in the low 200s. Thus, we expect the load voltage to be around 2 volts.

The transient analysis graph is depicted in Figure 7.9. Several traces are shown.

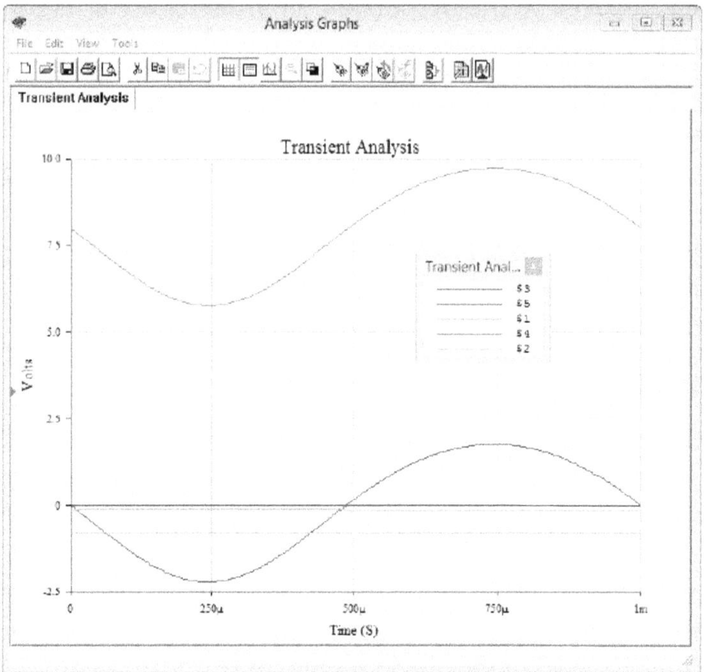

Figure 7.9 Unswamped CE amplifier, Transient Analysis.

At this scale, the AC signal at the input (node 4, purple) and the base (node 2, aqua) cannot be seen. As expected, we see a small negative DC value at the base and at the emitter, around -0.7 VDC. The DC offset at the collector is around 8 volts, as expected. Finally, the load voltage (node 5, blue) is sitting right around 2 volts.

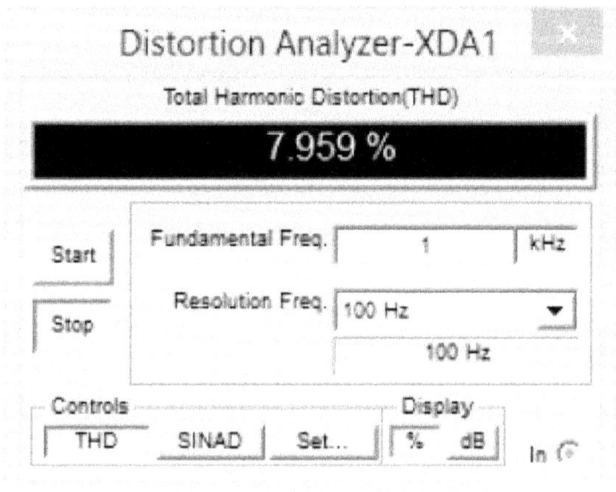

Figure 7.10 Non-swamped CE amplifier, THD Analysis.

What might not be visible immediately in the load voltage plot is some waveform asymmetry distortion. This can be quantified through a THD simulation, the output of which is shown in Figure 7.10. THD is nearly 8%. Not so good.

For the second pass, the circuit is modified to include a swamping resistor, as illustrated in Figure 7.11. The original 4.7 kΩ emitter resistor has been split into a4.5 kΩ and a 200 Ω swamping resistor. The bias in this circuit is identical to the first, therefore r'e is unchanged. This will lower our expected gain to around 13, decreasing by a factor of 15. The input signal is raised by a factor of 15 to compensate so that our load voltage will still be around 2 volts.

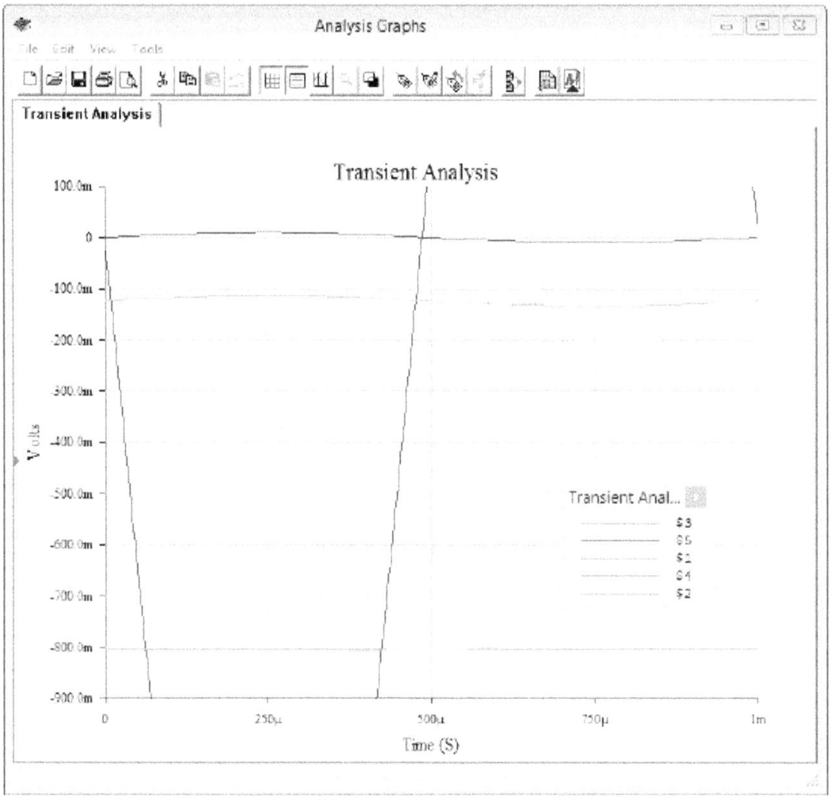

Figure 7.12 Swamped CE amplifier, Transient Analysis.

Once again, we run a transient analysis. The results are shown in Figure 7.12. In this case we have done something a little different. By zooming in, we can now confirm the signal inversion. The input signal is the purple trace at node 4. We can also see this signal at the base, riding on the small negative DC bias voltage (aqua trace, node 2). The DC offset is about -0.1 volts. Looking at the emitter we see the expected 0.7 volt DC base-emitter drop below this, or about -0.8 volts DC. Notice that there is no AC signal at the emitter whatsoever. This is expected as the emitter bypass capacitor forces this point to an AC ground.

The load voltage is the blue trace, node 5. While much of it is not visible at this zoom level, clearly it is an inverted waveform when compared to the input signal.

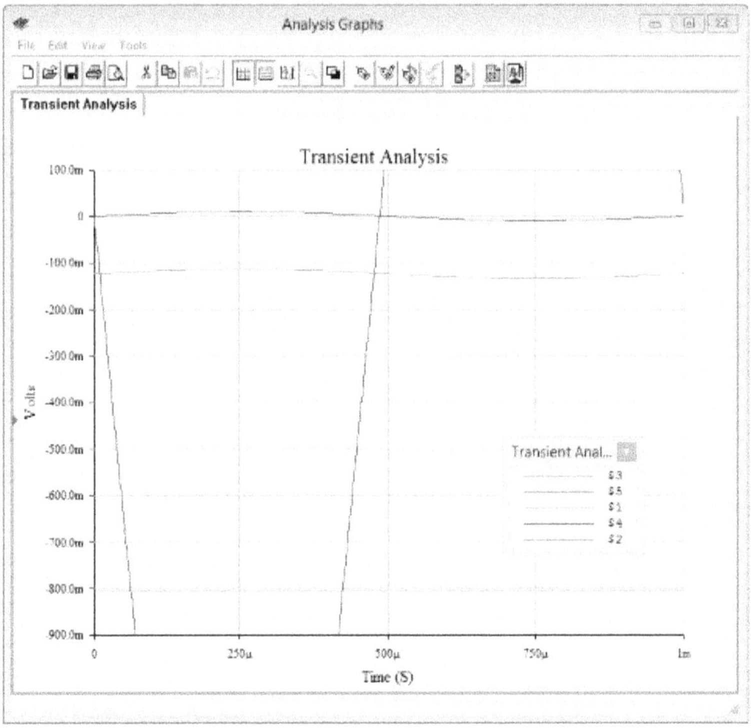

Figure 7.12: Swamped CE amplifier, Transient Analysis.

But what about the load voltage distortion? A THD simulation is performed on the swamped amplifier with the results shown in Figure 7.13. The THD is now under0.6 %, a considerable improvement, even if not audiophile quality. Interestingly, as a ratio, the reduction in distortion is roughly equal to the reduction in gain. The more you give up, the more you get.

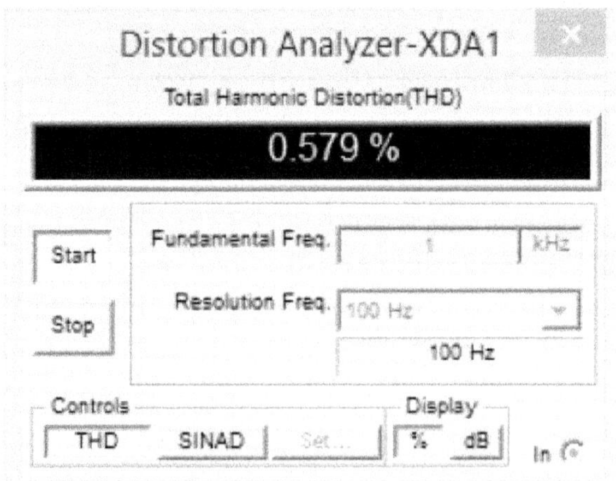

Figure 7.13 Swamped CE amplifier, THD Analysis.

Finally, the change in signal quality can be seen readily by plotting both load voltages concurrently, as shown in Figure 7.14. The non-swamped output (in blue) exhibits telltale asymmetry. Notice that the positive peak does not quite reach 2 volts but the negative peak exceeds -2 volts. The positive peak is also broadened and flattened, while the negative peak is sharper. In contrast, the swamped output (in red) has virtually identical positive and negative peak values with no apparent shape changes on them. Compare this simulation to the waveform distortion discussion from Chapter 6. In particular, compare Figure 7.14 to Figure 6.6.

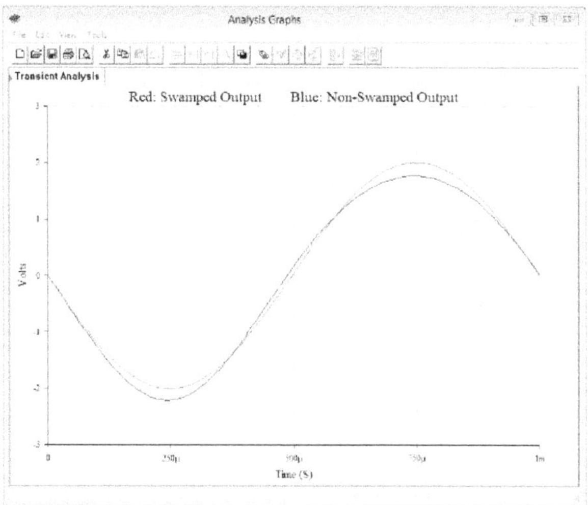

Figure 7.14 Swamped versus non-swamped CE amplifiers, Transient Analysis.

Power Supply Bypass and Decoupling

In the prior analyses we have assumed ideal behavior from the DC power sources. First, we assumed that they present a perfect AC ground and second, that they exhibit no ripple or noise. In reality, this may not be the case and non-ideal behavior may lead to a number of problems that diminish the quality of the amplified output signal, including hum and oscillations.

To battle the first issue, power supply bypass capacitors may be used. These capacitors are usually modest in size, perhaps 1 μF or so, although they can be much larger, particularly with high output power amplifiers. Power supply bypass capacitors are located physically close to the active devices. This location minimizes the resistive and inductive effects of power supply circuit board traces and wiring that could result in the power supply not being a good AC ground.

The second issue involves the noise and ripple from the power supply finding its way into the input signal and becoming part of the output signal. A classic example of this is amplifiers that use voltage divider biasing such as the one shown in Figure 7.7. Not only does the divider create the needed DC potential at the base terminal, but it also couples in any noise or ripple that might be riding on the DC voltage. This is particularly nasty because this undesirable signal is being applied to the base where it will get amplified.

The obvious solution to this problem is to create a very high quality, regulated DC supply, but this is not always practical given cost constraints. A relatively simple solution is to decouple the undesirable AC components through an RC network as shown in Figure 7.15.

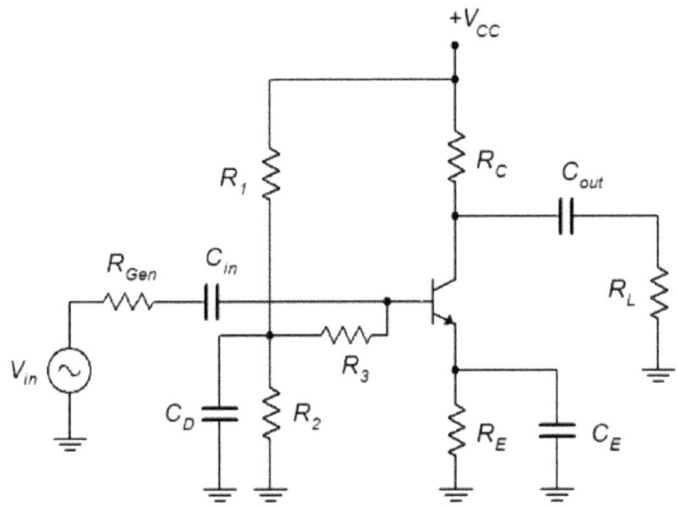

Figure 7.15 Decoupled voltage divider.

The capacitor CD is used to create a low impedance at the divider junction, thus shunting any noise or ripple to ground. Unfortunately, this would also short out the input signal so R3 is added to prevent this. R3 is in parallel with Zin(base) to create the input impedance.

7.4 THE COMMON COLLECTOR AMPLIFIER

The common collector amplifier is often referred to as an *emitter follower*, or more generically, as a *voltage follower*. The key characteristics of a voltage follower are a high input impedance, a low output impedance and a non-inverting voltage gain of approximately one. The name comes from the fact that output voltage *follows* the input, that is, it's at the same voltage level and is in phase with the input. While this configuration does not produce voltage gain, it does produce current gain, and therefore, power gain. It's primary purpose is to reduce impedance loading effects, for example, to match a high impedance source to a low impedance load. Consequently, they are used as high-Z input buffer stages or as drivers for low impedance loads such as loudspeakers.

A common collector amplifier using two-supply emitter bias is shown in Figure7.16. The input is coupled into the base like the common emitter amplifier, however, the output signal is taken at the emitter instead of at the collector. Because the collector is at the AC common, there is no need for a collector resistor.

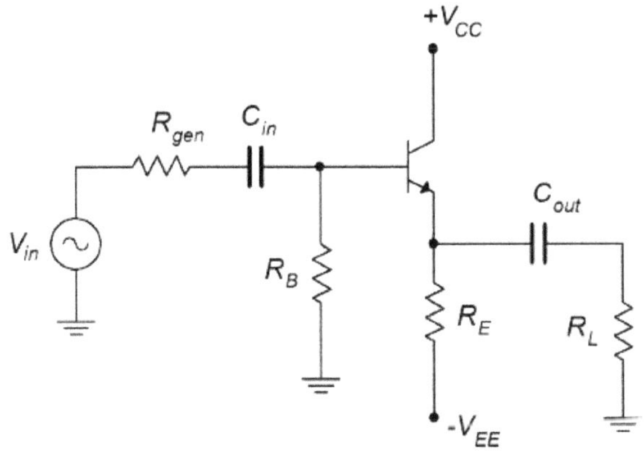

Figure 7.16 Common collector amplifier.

Perhaps the best way to think about the follower is not that it gives a voltage gain of one, but that it will prevent signal loss. The analysis follows, using Figure 7.17.

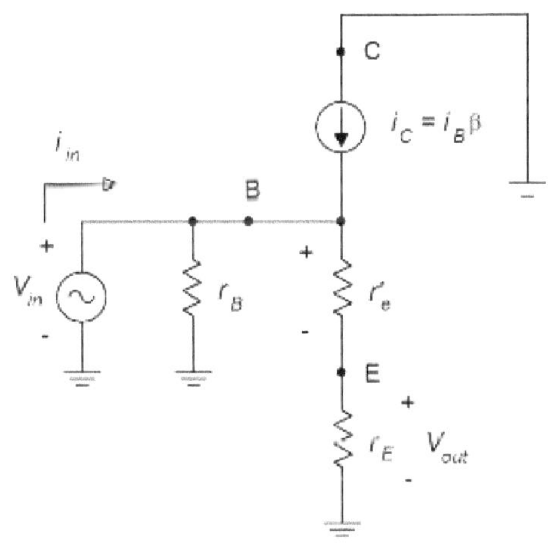

Figure 7.17 AC equivalent of common collector amplifier.

First, the AC emitter resistance, *rE*, is either the emitter bias resistor, *RE*, or the parallel combination of *RE* and the load resistance, *RL*. We'll use the former in order to determine the unloaded gain and the latter to determine the loaded gain, similar to what we did with the common emitter amplifier concerning *RC* and *RL*. The AC base resistance, *rB*, typically boils down to the base biasing resistor just as we saw with the common emitter amplifier (*RB* in a two-supply emitter bias or *R1* || *R2* for a voltage divider bias).

Voltage Gain

The derivation for the emitter follower's voltage gain equation is similar to that shown for the common emitter amplifier. We begin with the basic definition of voltage gain and then expand using Ohm's law.

$$A_v = \frac{v_{out}}{v_{in}} = \frac{v_E}{v_B}$$

$$A_v = \frac{i_C r_E}{i_C(r'_e + r_E)}$$

$$A_v = \frac{r_E}{r'_e + r_E} \qquad (7.8)$$

This equation is very similar to that of Equation 7.4. Here we see that the output signal is in phase with the input and that if *rE* >> *r'e*, the gain approaches unity. Signal distortion tends to be low in followers because a gain of one is a desired goal.

Input Impedance

The derivation for Zin and Zin(base) are unchanged compared to the common emitter configuration. The formulas are repeated below for convenience.

$$Z_{in(base)} = \beta(r'_e + r_E)$$

$$Z_{in} = r_B \,||\, Z_{in(base)}$$

Output Impedance

The derivation for common collector output impedance varies considerably from that of the common emitter. We shall use Figure 7.18 for the analysis.

First, note that this diagram splits the AC emitter resistance into its two components, RL and the biasing resistor RE. This is because we want to find the effective resistance of the source that drives the load, so logically we can't include the load in that value. We begin by looking back into the emitter from the perspective of the load. We see the emitter bias resistor in parallel with whatever the impedance is looking back into the emitter terminal.

$$Z_{out} = R_E \,||\, Z_{out(emitter)} \qquad (7.9)$$

Zout(emitter) is equal to r'e in series with the equivalent resistance of the network above it and to the left. The internal resistance of the current source is high enough to ignore so we're left with the equivalent resistance looking back off the base. We'll call this ZB (equivalent). At first glance this might appear to be the parallel combination of rgen and rB, but this ignores the effect of the collector current source. What we really want is the effective resistance as seen from the perspective of r'e, not as seen from the base terminal.

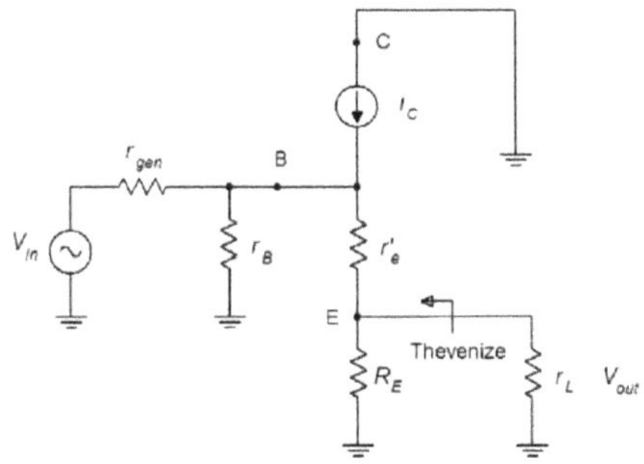

Figure 7.18 Common collector output impedance analysis.

$$Z_{out(emitter)} = r'_e + Z_{B(equivalent)} \tag{7.10}$$

$$Z_{B(equivalent)} = \frac{v_B}{i_C}$$

$$Z_{B(equivalent)} = \frac{i_B(r_B \parallel r_{gen})}{\beta i_B}$$

$$Z_{B(equivalent)} = \frac{r_B \parallel r_{gen}}{\beta} \tag{7.11}$$

In many instances the emitter bias resistor is large enough to ignore.

$$Z_{out} = R_E \parallel \left(r'_e + \frac{r_B \parallel r_{gen}}{\beta} \right) \tag{7.12}$$

Example 7.4

For the follower shown in Figure 7.19, determine the input impedance, output impedance and load voltage. Assume β = 100 and *Vin* = 100 mV.

First, find *IC* in order to find *r'e*. Assuming an unloaded divider, *VB* will equal half of the DC supply, or 10 volts. We lose 0.7 volts across the base-emitter junction leaving 9.3 volts across the 10 kΩ. This results in a collector current of 930 μA and an *r'e* of 28 Ω.

To find *Zin*

$$Z_{in(base)} = \beta(r'_e + r_E)$$
$$Z_{in(base)} = 100(28\,\Omega + 10\,k\Omega \parallel 500\,\Omega)$$
$$Z_{in(base)} = 50.4\,k\Omega$$

$$Z_{in} = R_1 \parallel R_2 \parallel Z_{in(base)}$$
$$Z_{in} = 22\,k\Omega \parallel 22\,k\Omega \parallel 50.4\,k\Omega$$
$$Z_{in} = 9.03\,k\Omega$$

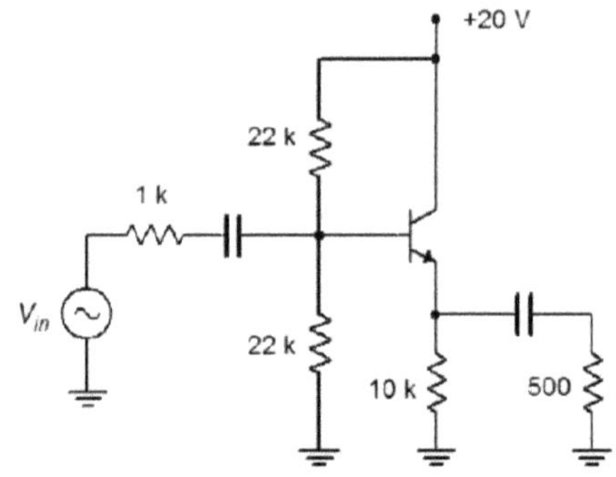

Figure 7.19 Schematic for Example 7.4.

This value is not particularly high when compared with the rather large source resistance of 1 kΩ. There will be some signal loss here due to the voltage divider effect between the two impedances. And now for Zout

$$Z_{out} = R_E \parallel \left(r'_e + \frac{r_B \parallel r_{gen}}{\beta} \right)$$

$$Z_{out} = 10\,\text{k}\Omega \parallel \left(28\,\Omega + \frac{22\,\text{k}\Omega \parallel 22\,\text{k}\Omega \parallel 1\,\text{k}\Omega}{100} \right)$$

$$Z_{out} = 37\,\Omega$$

This value is much, much lower than anything we saw with the common emitter amplifiers. Therefore this circuit can drive much lower impedance loads with minimal signal loss. The loaded gain from base to emitter is

$$A_v = \frac{r_E}{r'_e + r_E}$$

$$A_v = \frac{500\,\Omega \parallel 10\,\text{k}\Omega}{28\,\Omega + 500\,\Omega \parallel 10\,\text{k}\Omega}$$

$$A_v = 0.9444$$

As mentioned, we need to include the effect of the 1 kΩ source impedance. This will create a voltage divider with the input impedance.

$$A_{divider} = \frac{Z_{in}}{Z_{in} + Z_{source}}$$

$$A_{divider} = \frac{9.03\,\text{k}\Omega}{9.03\text{k}\,\Omega + 1\,\text{k}\Omega}$$

$$A_{divider} = 0.9$$

$$A_{v(system)} = A_v \times A_{divider}$$

$$A_{v(system)} = 0.9444 \times 0.9$$

$$A_{v(system)} = 0.85$$

Finally, we get to the load voltage.

$$V_{load} = A_{v(system)} \times V_{in}$$
$$V_{load} = 0.85 \times 100\,\text{mV}$$
$$V_{load} = 85\,\text{mV}$$

At this point the question might be, "Why did we go to the trouble of building this circuit when we lost 15% of the input signal?" Well, consider what would have happened without the circuit. If we had connected the source directly to the load, the resulting 1 kΩ/500 Ω voltage divider would have dropped the load voltage to 33 mV. This circuit prevented that loss.

A High Impedance Source: The Guitar Pickup

In Example 7.4, the source had an internal impedance of 1 kΩ, much higher than we would see with, say, a laboratory function generator (probably 50 Ω). Things could be much worse. Consider the electric guitar pickup. The job of a pickup is to transform the vibrations of the guitar strings into an electrical signal so that it can be amplified. It is commonly thought that a pickup is some form of microphone but this is not true27.

A guitar pickup is little more than a magnet surrounded by numerous turns of fine wire, as shown in Figure 7.20. This particular pickup is for a bass guitar but the construction is similar for all types of guitars and basses.

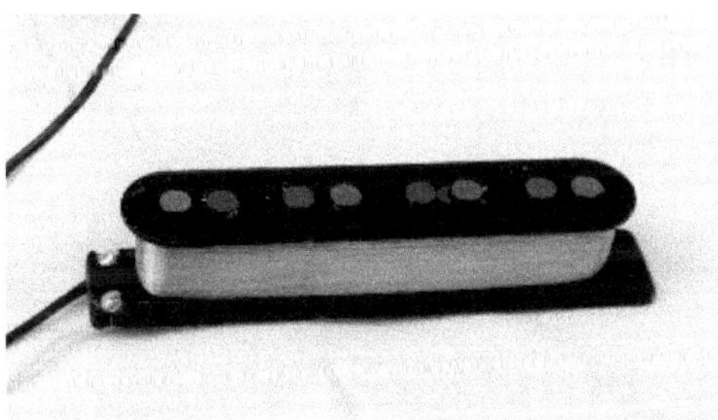

Figure 7.20 Electric bass pickup (cover removed).

Here is it how works: The magnet creates a field around the guitar strings. Because the strings are steel, their reluctance is much less than the surrounding air, therefore, they distort or bend the magnetic field. When a string is plucked, the field moves back and forth along with it. As the field moves, the flux lines cut across the coil's wire and this action induces a current in the conductor in accordance with Faraday's Law of Induction. This current is then fed to the amplifier.

A typical guitar pickup consists of perhaps 5000 turns of very fine wire, 42 AWG being typical. 42 gauge copper wire has a resistance of around 1.6 Ω per foot so the DC resistance of the coil can be over 5 kΩ. Further, that many turns of wire around a magnet can produce a very large inductance, perhaps several henries, that is in series with this resistance. There is also distributed capacitance and cable capacitance in parallel that could be upwards of 1 nF. The result is a complex impedance with resonance effects, regions of which can be tens of kΩ in magnitude. What makes this more challenging is that because the impedance is a function of frequency, the voltage divider effect with the amplifier's input impedance also becomes a function of frequency. For example, the increasing impedance due to XL will result in increasing attenuation with frequency. This is akin to turning down the treble on the amplifier. Generally, not a good result. How do we limit this effect? Simple. We make a circuit with a very, very high input impedance. How do we do that? Well, there are several ways, including the use of field effect transistors and operational amplifiers, but we can also obtain high input impedances through the use of a dual BJT configuration called the Darlington pair.

The Darlington Pair

The Darlington pair was invented by Sidney Darlington, an American engineer. The configuration leads to a compound device with a very high β. Used properly, this can lead to amplifier circuits with very high input impedance. A Darlington pair is shown in Figure 7.21.

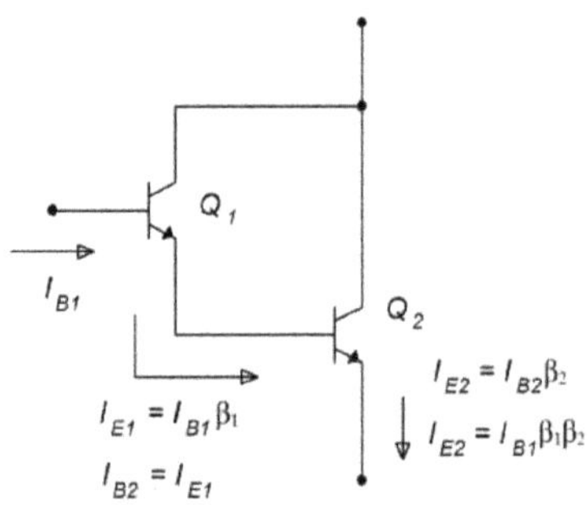

Figure 7.21 Darlington pair.

The operation is as follows. The base current of the first transistor, Q1, is multiplied by the β of Q1 resulting in Q1's emitter current. This current is fed into the base of the second transistor, Q2, where it is multiplied by the β of Q2 resulting in Q2's emitter current. If we treat the pair as a single device, then the effective β of the pair is β1β2. Given typical values for β, the compound value can be in the vicinity of 5000 to 10,000. The functional downside to this arrangement is that VBE is now doubled to 1.4 volts (for silicon) and the effective r'e of the pair is doubled as well. These issues are minor when compared to the advantage of the huge current gain that can be obtained.

The bottom line when using a Darlington pair is to treat it like an ordinary transistor except that it has a very large β and both VBE and r'e are doubled compared to the ordinary values.

Example 7.5

Determine the output voltage for the follower shown in Figure 7.22. Assume the input is 100 mV peak and the β for the Darlington pair is 10,000.

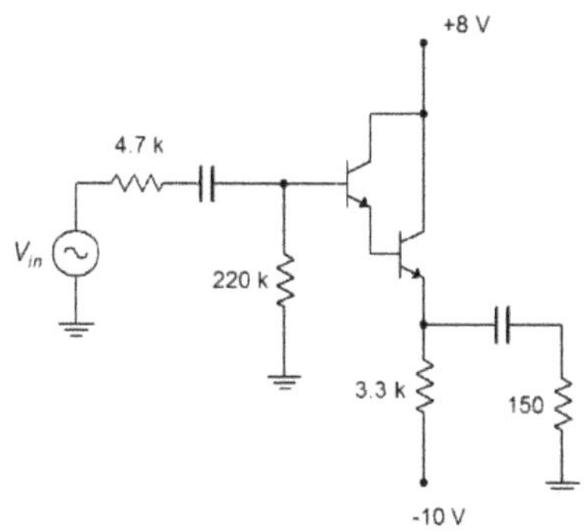

Figure 7.22 Schematic for Example 7.5.

The first thing that might look a little odd, at least compared to previous bias circuits, is that the base biasing resistor is so much larger than the emitter biasing resistor. Normally, this would lead to an unstable Q point but that's not a problem here. Because β is so large, RB can be much larger than normal and we'll still achieve good stability. In fact, we can still use the approximation that the base is at DC ground. This being true, the analysis proceeds as follows

$$I_C = \frac{|V_{EE}| - V_{BE}}{R_E}$$

$$I_C = \frac{10\,\text{V} - 1.4\,\text{V}}{3.3\,\text{k}\Omega}$$

$$I_C = 2.61\,\text{mA}$$

$$r'_e = 2 \times \frac{26\text{mV}}{I_C}$$

$$r'_e = \frac{52\,\text{mV}}{2.61\,\text{mA}}$$

$$r'_e = 20\,\Omega$$

$$Z_{in(base)} = \beta(r'_e + r_E)$$

$$Z_{in(base)} = 10{,}000(20\,\Omega + 3.3\,\text{k}\Omega \,||\, 150\,\Omega)$$

$$Z_{in(base)} = 1.63\,\text{M}\Omega$$

This value is in parallel with the base biasing resistor, creating the input impedance.

$$Z_{in} = R_B \parallel Z_{in(base)}$$
$$Z_{in} = 220\,\text{k}\Omega \parallel 1.63\,\text{M}\Omega$$
$$Z_{in} = 194\,\text{k}\Omega$$

This is much higher than we have seen in previous circuits. The loaded gain from base to emitter is

$$A_v = \frac{r_E}{r'_e + r_E}$$
$$A_v = \frac{150\,\Omega \parallel 3.3\,\text{k}\Omega}{20\,\Omega + 150\,\Omega \parallel 3.3\,\text{k}\Omega}$$
$$A_v = 0.88$$

Now to include the effect of the 4.7 kΩ source impedance. This will create a voltage divider with the input impedance, minimal as it turns out.

$$A_{divider} = \frac{Z_{in}}{Z_{in} + Z_{source}}$$
$$A_{divider} = \frac{194\,\text{k}\Omega}{194\,\text{k}\Omega + 4.7\,\text{k}\Omega}$$
$$A_{divider} = 0.976$$

$$A_{v(system)} = A_v \times A_{divider}$$
$$A_{v(system)} = 0.88 \times 0.976$$
$$A_{v(system)} = 0.86$$

The load voltage is

$$V_{load} = A_{v(system)} \times V_{in}$$
$$V_{load} = 0.86 \times 100\,\text{mV}$$
$$V_{load} = 86\,\text{mV}$$

If we had connected the source directly to the load, the 4.7 kΩ/150 Ω divider would have squashed the applied signal into a shadow of its former size, leaving us with just 3 mV.

Computer Simulation

To verify the results of Example 7.5, we'll run a transient analysis. The input schematic is shown in Figure 7.23.

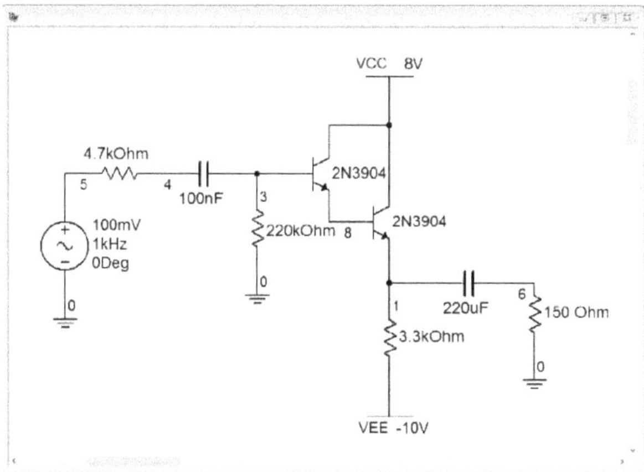

Figure 7.23 Simulation schematic for Darlington pair follower.

Of interest here will be the voltages at the source, base and load. As the input impedance/source impedance divider was 0.976, we expect 97.6 mV at node 4. At the output, node 6, we expect to see our final computed value of 86 mV. The output plot of the simulation is shown in Figure 7.24. The simulation concurs.

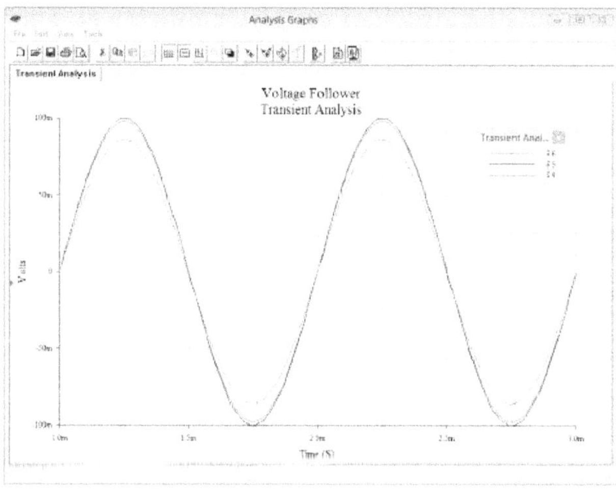

Figure 7.24 Transient analysis for Darlington pair follower.

The Phase Splitter

A phase splitter is a combination of a common emitter amplifier and a common collector follower using a single transistor. The purpose of the circuit is to produce two versions of the input signal: a buffered version identical to the input and an inverted version, both waves having the same amplitude. The circuit is used for differential line driver systems. This scheme helps to minimize outside noise and interference picked up by communications cables. There are other ways to create phase splitters, including using differential amplifiers or op amps, but this BJT-based version is a minimalist solution28. The basic circuit is shown in Figure 7.25.

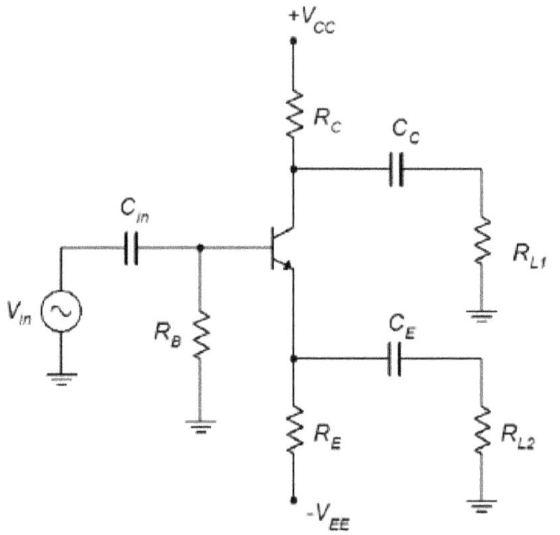

Figure 7.25 A simple phase splitter.

For proper operation, the circuit is largely symmetrical. That is, *RL1* = *RL2*, *RE* = *RC*, and *CC* = *CE*. That being the case, the AC collector and emitter resistances will be equal (*rC* = *rE*). If we then look at the basic gain equations, we find that both loads will receive the same gain magnitude (just under unity), although *RL1* will see the signal inverted.

$$A_v = -\frac{r_C}{r'_e + r_E} \quad \text{Common emitter amplifier}$$

$$A_v = \frac{r_E}{r'_e + r_E} \quad \text{Common collector follower}$$

7.5 THE COMMON BASE AMPLIFIER

The third and final prototype is the common base amplifier. In this configuration the input signal is applied to the emitter and the output is taken from the collector. The base terminal is at the common ground point. An example, using two-supply emitter bias, is shown in Figure 7.26. Note that because neither the input nor output is connected to the base, there is no need for a base resistor. Consequently, the base terminal is connected directly to ground.

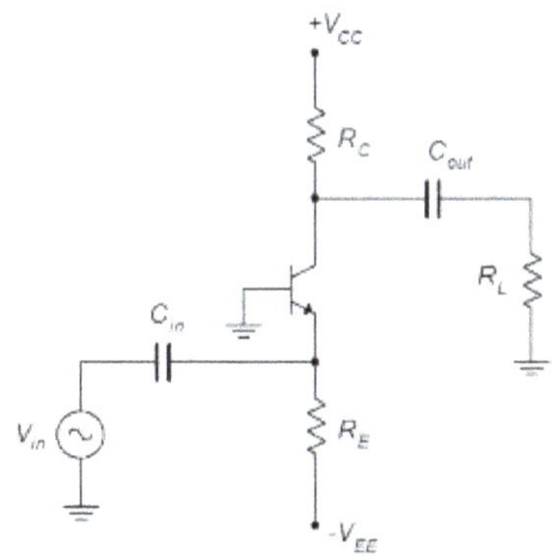

Figure 7.26 Common base amplifiers.

Some people find that redrawing the schematic horizontally helps to visualize the signal flow. This version is shown in Figure 7.27.

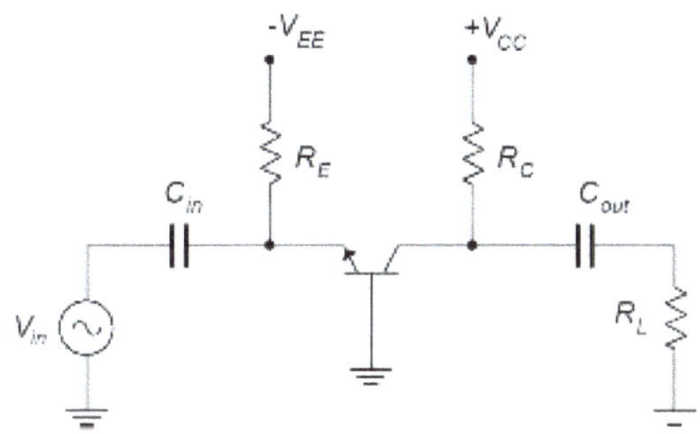

Figure 7.27 Common base amplifier redrawn.

One nice thing about the horizontal version is that when we make the AC equivalent, it becomes obvious that *RE* is in parallel with the input and *RC* is in parallel with the load. For the AC analysis we shall modify Figure 7.27 by substituting the BJT model for transistor, shorting the capacitors and taking the DC sources to AC ground. The result is shown in Figure 7.28.

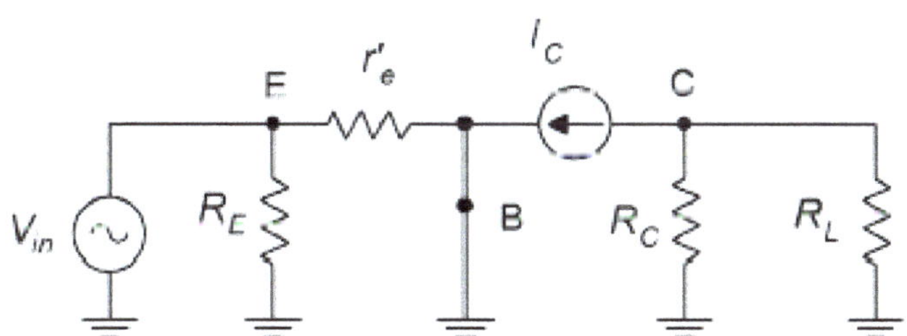

Figure 7.28 Common base amplifier with BJT model.

Voltage Gain

We begin with the basic definition of voltage gain and then expand using Ohm's law.

$$A_v = \frac{v_{out}}{v_{in}} = \frac{v_C}{v_E}$$

$$A_v = \frac{i_C r_C}{i_E r'_e}$$

$$A_v = \frac{r_C}{r'_e} \qquad (7.13)$$

This equation is very similar to that of a non-swamped common emitter amplifier except that it does not invert the input signal. Therefore, gain potential is fairly high.

Input Impedance

The derivation for *Zin* is obtained via direct inspection of the schematic.

$$Z_{in} = R_E \,||\, r'_e \qquad (7.14)$$

r'e normally dominates and thus we see that the common base configuration tends to have a low input impedance. For audio frequencies this can be an issue but it is less of a problem at higher frequencies as, generally speaking, system impedances need to be lower to avoid complications with capacitive effects.

Output Impedance

The derivation for *Zout* unchanged compared to the common emitter configuration. The formula is repeated below for convenience.

$$Z_{out} \approx R_C$$

Example 7.6

For the amplifier shown in Figure 7.29, determine the voltage gain and input impedance.

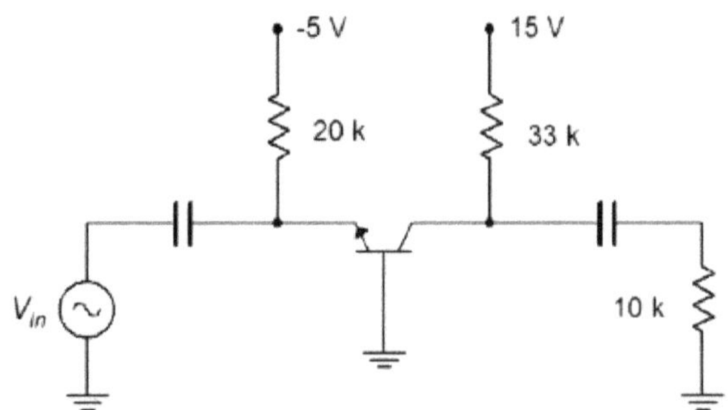

Figure 7.29 Schematic for Example 7.6.

$$I_C = \frac{|V_{EE}| - V_{BE}}{R_E}$$
$$I_C = \frac{5\,\text{V} - 0.7\,\text{V}}{20\,\text{k}\Omega}$$
$$I_C = 0.215\,\text{mA}$$

$$r'_e = \frac{26\,\text{mV}}{I_C}$$
$$r'_e = \frac{26\,\text{mV}}{0.215\,\text{mA}}$$
$$r'_e = 121\,\Omega$$

$$Z_{in} = r'_e \parallel r_E$$
$$Z_{in} = 121\,\Omega \parallel 20\,\text{k}\Omega$$
$$Z_{in} = 120\,\Omega$$

$$A_v = \frac{r_C}{r'_e}$$
$$A_v = \frac{33\,\text{k}\Omega \parallel 10\,\text{k}\Omega}{120\,\Omega}$$
$$A_v = 64$$

7.6 MULTI-STAGE AMPLIFIERS

In order to achieve a higher gain than we can obtain from a single stage, it is possible to cascade two or more stages. Different biasing types might be used along with a mix of AC configurations such as a common collector follower for the first stage that drives a common emitter voltage amplifier. A mix of NPN and PNP devices may also be present.

In general terms, each stage serves as the load for the preceding stage. That is, the Zin of one stage is the RL of the previous stage. The gains of the individual stages are then multiplied together to arrive at the system gain. The system input impedance is the input impedance of the first stage only. The source drives the first stage alone.

The first stage, in turn, drives the second stage, and so on. Therefore the source only "sees" the first stage because it is the only stage to which it delivers current. In a similar fashion, the output impedance of the system is the Zout of the last stage. An example is shown in Figure 7.30.

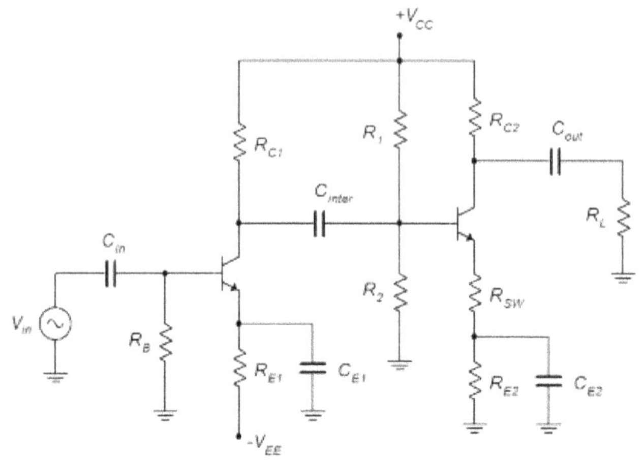

Figure 7.30 Two stage amplifier

In this circuit, stage one is a non-swamped common emitter amplifier utilizing two- supply emitter bias. Stage two is a swamped common emitter amplifier using voltage divider bias. As far as the DC analysis is concerned, these are two separate circuits. The inter-stage coupling capacitor, Cinter, prevents the DC potential at the collector of the first transistor from interfering with the bias established by R1 and R2 for transistor number two. For the AC computation, the first stage is analyzed in normal fashion except that its load resistance is comprised of R1 || R2 || Zin-base2(i.e., Zin of stage 2). The second stage is analyzed without changes and its gain is multiplied by the first stage's gain to arrive at the final gain for the pair. The input impedance of the system is RB || Zin-base1 (i.e., Zin of stage 1).

It should be obvious that by cascading several stages it is possible to achieve very high system gains, even if each stage is heavily swamped in order to reduce distortion. For example, three swamped common emitter stages with voltage gains of just 10 each would produce a system voltage gain of 1000.

Direct Coupling

With a little creativity, it is possible to create multi-stage designs that use fewer components but which achieve higher performance. One technique is to employ direct coupling of the stages. Direct coupling allows DC to flow from stage to stage. As such, it is possible to design an amplifier that has no lower frequency limit. An example is shown in Figure 7.31.

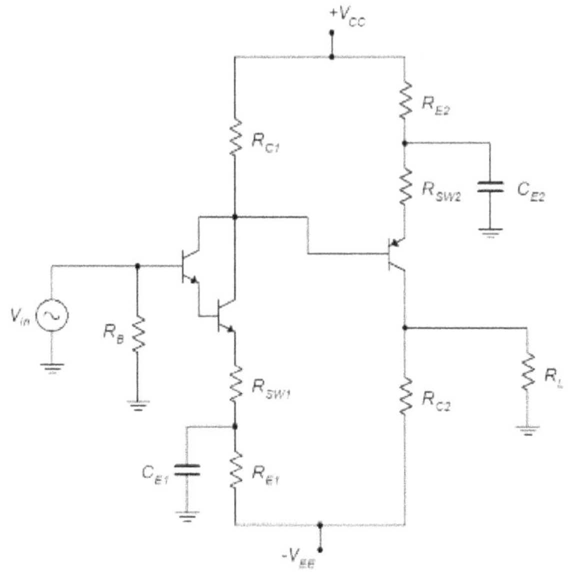

Figure 7.31 Direct coupled amplifier

This two-stage amplifier uses no coupling capacitors nor does it rely on voltage divider resistors for the second stage[29]. Here is how it works: The first stage is a fairly ordinary swamped common emitter amplifier using two-supply emitter bias. It also uses a Darlington pair to maximize the input impedance. Because the base current is so low, the DC drop on *RB* could be small enough to ignore so we may dispense with the input coupling capacitor. The DC potential at the collector of the Darlington is applied directly to the base of the second stage. This is used to set up the bias of the second stage via the stage two emitter resistors. This is precisely what we did with the circuit of Figure 7.7. The only difference is that here the base voltage is derived from the preceding stage instead of from a voltage divider. The computations for *IC*, *r'e* and the like would proceed unchanged. In any event, this eliminates two biasing resistors and another coupling capacitor.

Note the use of the PNP device for the second stage. By using a PNP, its collector voltage must be less than its emitter voltage. As we're also using a bipolar power supply, we can eliminate the need for the final output coupling capacitor. All we need to do is set up the resistor values such that the drop across RC2 is the same as VEE. This will place the stage two DC collector voltage at 0 volts. If there's no DC voltage then there's nothing to block, and therefore no need for the coupling capacitor.

CONCISE SUMMARY

- A simplified AC model of the bipolar junction transistor consists of a controlled current source in the collector and a dynamic resistance in the emitter called *r'e*. This resistance is a function of the DC bias current, *IC*; the higher the biasing current, the lower the resistance. Fluctuations in this resistance can lead to waveform distortion. Swamping, also known as emitter degeneration, is a technique used to reduce distortion and stabilize gain. The basic idea is to add a fixed resistor in series with the emitter so as to buffer or "swamp out" the changes of *r'e*.

- There are three basic AC amplifier configurations: common emitter, common collector and common base. The common emitter configuration produces a voltage amplifier with high gain and intermediate input impedance. It also inverts the signal. Because it exhibits both voltage gain and current gain, it has a potential for high power gain. The common collector configuration is known as a follower because its output follows the input. It produces a non-inverting voltage gain of one and exhibits high input impedance and low output impedance. Therefore, it is useful as either an input buffer or as a final drive stage to a low impedance load. The common base configuration exhibits high non-inverting voltage gain. It has a low input impedance and a high output impedance.
- The Darlington pair is a two-transistor configuration that may be treated as a single device. As such, it exhibits a doubling of both *VBE* and *r'e*, and a very large β.
- In order to achieve higher gains, multiple stages may be cascaded. Their gains multiply together to produce the combined system gain. The stages may be coupled through capacitors or via a capacitor-less direct coupling technique that can improve performance while reducing component count.

CHAPTER-8

CLASS A POWER AMPLIFIERS BASED ON BIPOLAR JUNCTION TRANSISTORS (BJTS)

8.1 INTRODUCTION: AN OVERVIEW

Class A power amplifiers based on Bipolar Junction Transistors (BJTs) represent a key category in electronic amplification, renowned for their simplicity and high-quality signal reproduction. In Class A amplifiers, the BJT conducts over the entire input cycle, ensuring minimal distortion and a faithful reproduction of the input signal. The fundamental operating principle involves biasing the transistor at a point that keeps it conducting even when there is no input signal. This constant conduction results in a continuous output signal, making Class A amplifiers suitable for applications where signal integrity is critical, such as in high-fidelity audio systems. The inherent linearity and low distortion characteristics of Class A amplifiers make them a preferred choice in audio engineering, emphasizing purity and accuracy in signal amplification. Understanding the principles behind BJT Class A power amplifiers is essential for designing audio circuits that prioritize uncompromised signal fidelity.

Now that we have examined BJT voltage amplifiers in terms of their gain, input impedance and output impedance, it is time that we extend the small signal analysis to larger signals. Of primary importance will be determination of the maximum output voltage swing, or compliance, along with maximum load power, device dissipation requirements and amplifier efficiency. To assist with this, we introduce the concept of the AC load line. In general, we will not concern ourselves with input impedance, voltage gain or even r'e. Indeed, we shall simply consider r'e as a source of distortion. Most power amplifiers are configured as voltage followers so we will focus largely on those.

8.2 AMPLIFIER CLASSES

There are several classes of amplifier operation. The class of an amplifier has nothing to do with the fidelity or quality of the amplifier. Rather, the class indicates the fundamental operational principle of the circuit. In general, as the class letter increases, the designs become more complicated but also more efficient. For audio and other linear applications, classes A, B and D are relatively common these days. Class C is largely relegated to high power telecommunications while classes G and H are essentially variations of class B.

The definition of class A is that signal current in the collector flows 360° out of the cycle. In other words, it flows for the entire cycle without interruption. All of the amplifiers that were presented in the prior chapter are class A designs. In class B, *iC* flows for just 180°, and for class D, *iC* is discontinuous; the transistor is used as a switch. Class B and D designs are examined in later chapters.

8.3 CLASS A OPERATION AND LOAD LINE

The signal current in the class a amplifier flows continuously throughout the entire cycle of the waveform. Ultimately, we would like to know just how large this signal can be before it is limited and grossly distorted. To do so, we need to examine the AC equivalent of the amplifier. A generic AC equivalent is shown in Figure 8.1. This includes both AC collector and emitter resistances so it can be used for either swamped or unswamped common emitter amplifiers or for emitter followers. If one of the resistances is not used (for example, rC in a follower), we can just substitute a value of zero for it.

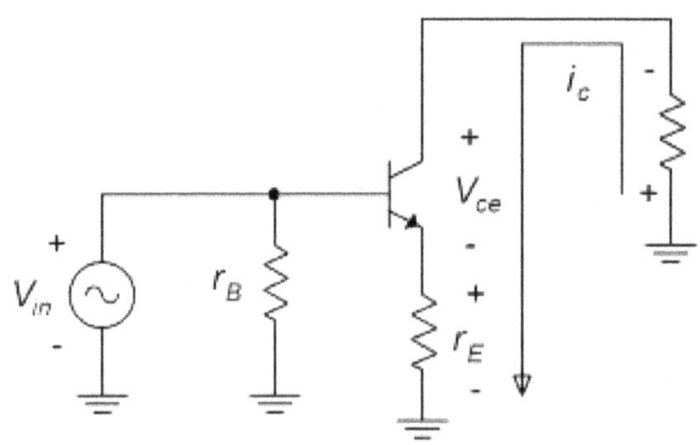

Figure 8.1 AC equivalent circuit.

The voltage polarities and current direction are shown for a positive input voltage. To determine the maximum load voltage swing (compliance), we will need to construct an AC load line as shown in Figure 8.2.

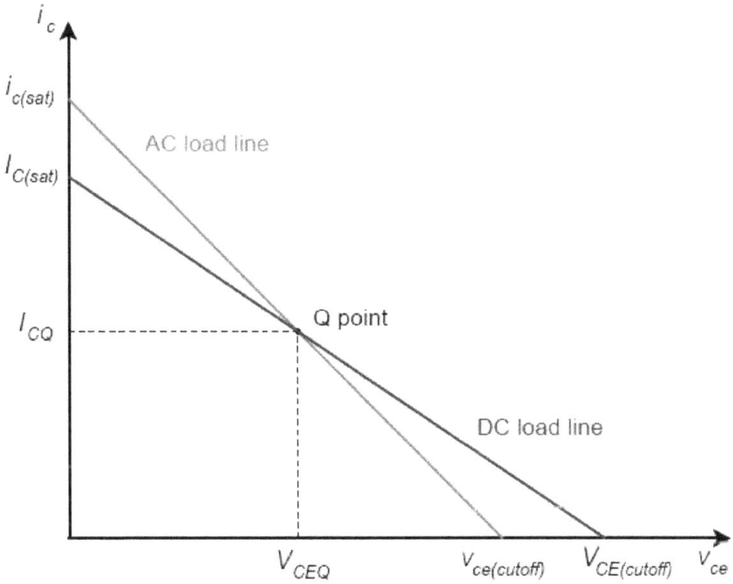

Figure 8.2 AC and DC load lines.

The AC load line is similar to the DC load line that was used for analyzing biasing circuits. As in the DC version, there will be a cutoff voltage, vCE(cutoff), and a saturation current, iC(sat). The AC and DC load lines normally are not the same, however, they must share one point in common, and that's the Q point. Usually, the slope of the AC load line is steeper than that of the DC load line. This is because the AC resistance tends to be less than the DC resistance due to loading and capacitor bypassing. Consequently, vCE(cutoff) tends to be smaller than VCE(cutoff) and iC(sat) tends to be larger than IC(sat).

To determine expressions for the AC load line endpoints, let's examine the AC equivalent circuit. Because both load lines share the Q point, we can consider the circuit of Figure 8.1 as having a no-signal current of ICQ and a no-signal transistor voltage of VCEQ. As the input signal grows, iC increases. The effect of this is to increase the voltage drops across rE and rC due to Ohm's law. This, in turn, forces vCE to decrease due to KVL. The collector current can only increase to the point where vCE drops to 0 V. This is a maximum increase of VCEQ/(rC+rE). Therefore

$$i_{C(sat)} = I_{CQ} + \frac{V_{CEQ}}{r_E + r_C} \tag{8.1}$$

In terms of cutoff voltage, the transistor starts with *VCEQ* and *ICQ*. The largest *vCE* increase that can occur is if the current falls to zero. Then, all of the potential originally developed across *rE* and *rC* by *ICQ* must be absorbed by the transistor.

Therefore

$$v_{CE(cutoff)} = V_{CEQ} + I_{CQ}(r_E + r_C) \quad (8.2)$$

There are three possible ways this can be configured: Q point closer to saturation, Q point closer to cutoff, or Q point centered on the AC load line. Let's first consider the Q point closer to saturation. This is shown in Figure 8.3.

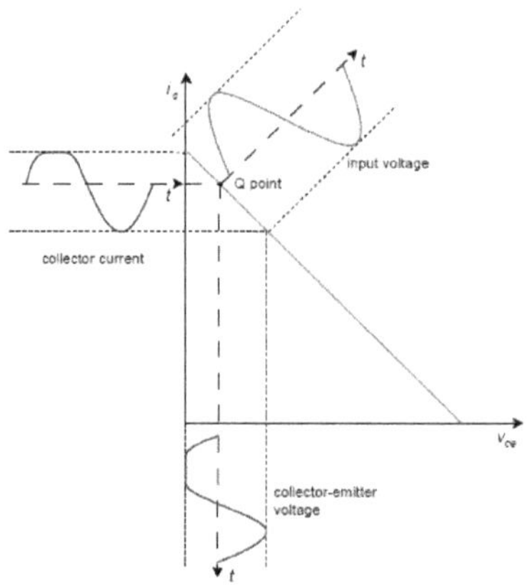

Figure 8.3 AC load line, Q point closer to saturation.

Here we have plotted the input voltage in red and drawn the corresponding collector current and collector-emitter voltage in blue. It is apparent that as the input signal increases, eventually, the output signal is limited at zero for vCE and at iC(sat) for iC. The two blue waveforms are severely clipped and distorted. The largest unclipped peak voltage swing is VCEQ and the largest peak current swing is iC(sat) - ICQ, or more conveniently, VCEQ/(rE+rC).

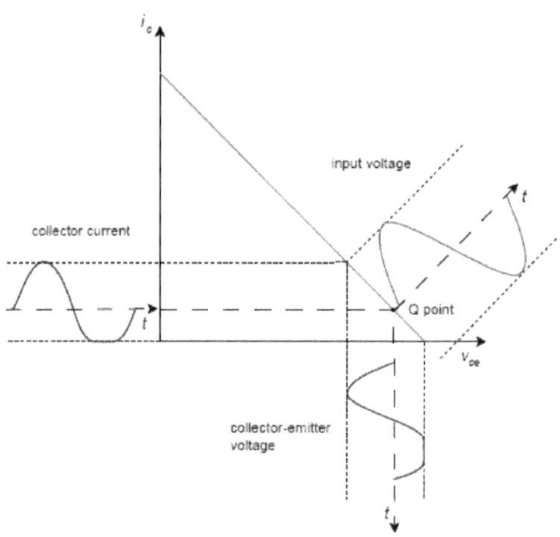

Figure 8.4 AC load line, Q point closer to cutoff.

If we shift the Q point toward cutoff, we solve the saturation clipping problem but now we have a new problem, as illustrated in Figure 8.4. It should come as no surprise that we now have cutoff clipping.

In this version the largest unclipped peak voltage swing is vCE(cutoff) - VCEQ (or alternately, ICQ(rE+rC)) and the largest peak current swing is ICQ. What's important here is that the waveform has been clipped. It doesn't really matter which side has been clipped, either way it's gross distortion. Eventually, every amplifier will have a limit but we will be able to produce the largest unclipped voltage swing if the Q point is centered on the AC load line. This is shown in Figure 8.5.

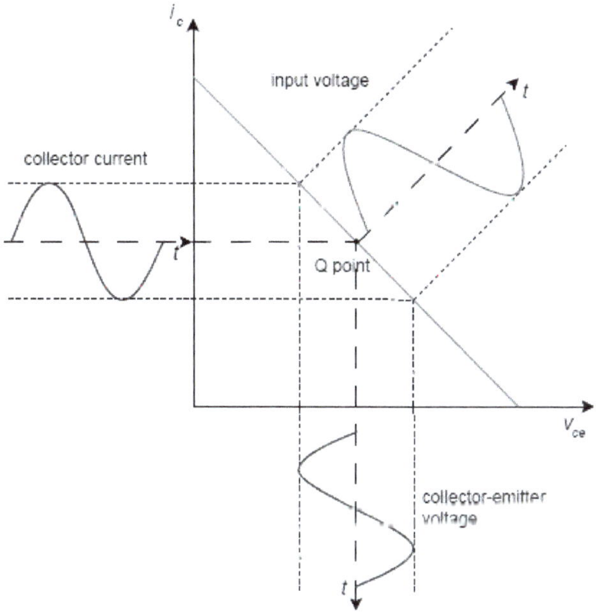

Figure 8.5 AC load line, centered Q point.

With a centered Q point, the largest unclipped peak voltage swing is VCEQ and the largest unclipped peak current swing is ICQ. By examining equations 8.1 and 8.2 it is apparent that in order to achieve a centered Q point on the AC load line, the following must be true:

$$\frac{V_{CEQ}}{I_{CQ}} = r_E + r_C \qquad (8.3)$$

Of course, while it is useful to determine the maximum voltage across the transistor, it is more important to determine the maximum voltage across the load. Looking back at the circuit of Figure 8.1, most times the maximum load voltage (i.e., the compliance) will equal the maximum transistor voltage. This will be the case in voltage followers and unswamped amplifiers. The only time there will be a noticeable reduction is with very heavily swamped amplifiers. In this case the compliance will be reduced by the voltage divider between the load and swamping resistors. For example, a swamped amplifier with a voltage gain of 4 would lose about 20% of the maximum swing. Swamping has to be very heavy resulting in very low gains before appreciable signal is lost.

Thus we arrive at the following general rule:

$$\textit{Peak compliance is the smaller of } V_{CEQ} \textit{ or } I_{CQ}(r_E+r_C) \qquad (8.4)$$

Knowing the compliance, the maximum load power may be determined using power law. Power is determined using RMS values, so the peak compliance will need to be divided by √2 (or multiplied by 0.707) before continuing.

$$P_{load(max)} = \frac{Compliance_{RMS}^{2}}{R_L} \qquad (8.5)$$

There is something important to note about this equation. It uses the load resistance value, not the total AC effective value (i.e., not rL which is RL in parallel with a biasing resistor). If rL was used, we'd be calculating the power in the load plus the power in the biasing resistor.

We would also like to determine the maximum power dissipated by the transistor. Because the transistor's current and voltage are fluctuating with the input signal, we need to determine the magnitude of the load voltage that produces maximum power in the transistor. Intuitively, we might guess that this occurs at maximum load power but it turns out that this guess is incorrect. Under no-signal conditions the transistor is operating statically at the Q point. Therefore, quiescent power dissipation is

$$P_{DQ} = V_{CEQ} I_{CQ} \qquad (8.6)$$

In contrast, at full load for a centered Q point, we have

$$v_{CE} = V_{CEQ}(1-\sin 2\pi ft)$$
$$i_C = I_{CQ}(1+\sin 2\pi ft)$$

$$P_D = v_{CE} i_C$$
$$P_D = V_{CEQ}(1-\sin 2\pi ft) \times I_{CQ}(1+\sin 2\pi ft)$$
$$P_D = V_{CEQ} I_{CQ}(1-\sin^2 2\pi ft)$$
$$P_D = V_{CEQ} I_{CQ}(.5+.5\cos 4\pi ft)$$
$$P_D = \frac{P_{DQ}}{2} + \frac{P_{DQ}}{2}\cos 4\pi ft \qquad (8.7)$$

The first term of Equation 8.7 is a fixed offset while the second term is a sinusoid at twice the signal frequency. Because the peak amplitude of this sinusoid is the same as the fixed offset, the average over time is simply the offset value. These waveforms are illustrated in Figure 8.6.

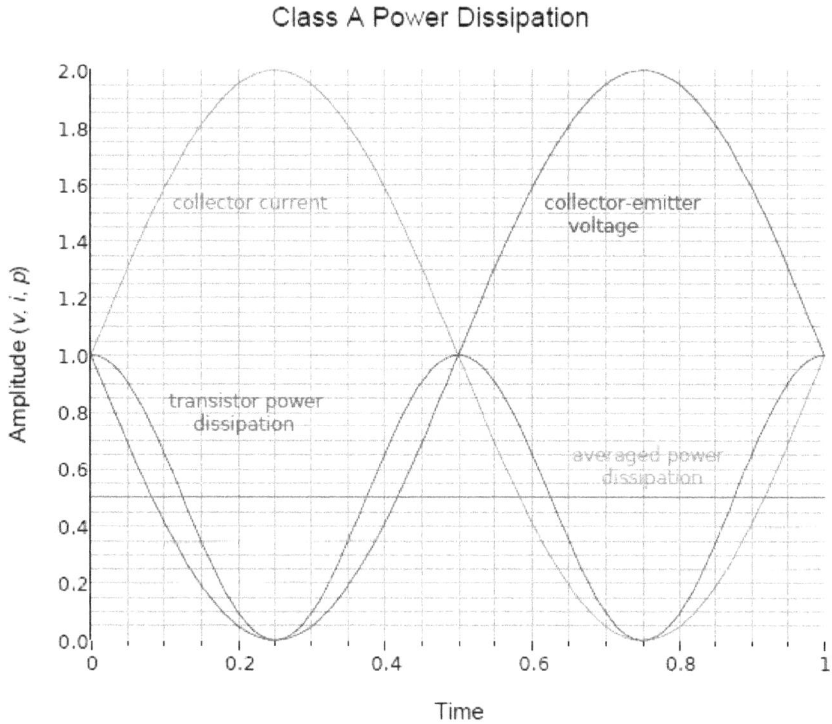

Figure 8.6 Transistor power dissipation at full load power.

The result is that the transistor only dissipates half the power at full load that it dissipates under idle conditions. This makes perfect sense if you stop to consider that the class A amplifier always draws the same power from the DC supplies, regardless of the size of the load signal. Without clipping, the average current will be *ICQ*. That current times the supply voltage yields the supplied power. What's happening is that as the signal increases in amplitude, more and more of the power dissipated by the transistor is shifted to the load. At maximum load swing, both the transistor and the load will be dissipating *PDQ*/2. As strange as it might seem, if you want to keep the output transistor of a class A amplifier cool, don't turn the volume down, turn it up.

The foregoing implies that class A designs are not power efficient. This is indeed the situation. As we have just seen, the best case maximum load power will be one half of *PDQ*, assuming a centered Q point (non-centered will be worse). To achieve this swing, the power supply will have to be at least twice as large as *VCEQ* because it has to cover the peak-to-peak swing, while *VCEQ* represents the peak swing for a centered Q point.[30] In any event, the best case efficiency turns out to be dismal, as follows.

$$\eta = \frac{P_{out}}{P_{in}} = \frac{P_{load}}{P_{DC}}$$
$$\eta = \frac{P_{DQ}/2}{2V_{CEQ}I_{CQ}}$$
$$\eta = \frac{P_{DQ}/2}{2P_{DQ}}$$
$$\eta = 25\%$$

This represents the maximum or best case efficiency for an RC coupled class A amplifier. It may be considerably less depending on precisely how it is biased. This, truly, is the Achilles heel of the class A topology: it is wasteful. It draws full power from the supply regardless if signal is present and, at best, will translate only one quarter of that power into useful load power. At the same time, the power dissipation of the transistor will need to be at least twice that of the delivered load power, and might need to be much greater. Why use it then? To its advantage, it is a relatively simple design so if large output powers are not needed, it can prove useful. This is most definitely the case for the early stages of a multi-stage amplifier where the amount of load power is very small (basically the power delivered to the following stage). In that instance, the increased complexity of more power efficient designs is not warranted or cost effective.

Example 8.1

For the amplifier shown in Figure 8.7, determine the compliance, maximum load power, worst case transistor dissipation and efficiency.

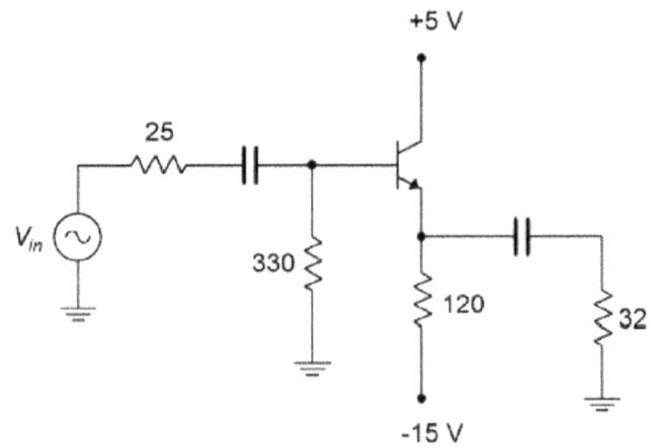

Figure 8.7 Schematic for Example 8.1.

$$I_{CQ} = \frac{|V_{EE}| - V_{BE}}{R_E}$$
$$I_{CQ} = \frac{15\,\text{V} - 0.7\,\text{V}}{120\,\Omega}$$
$$I_{CQ} = 119\,\text{mA}$$

By inspection, V_{CEQ} = 5.7 V. The AC cutoff voltage is

$$v_{CE(cutoff)} = V_{CEQ} + I_{CQ}(r_C + r_E)$$
$$v_{CE(cutoff)} = 5.7\,\text{V} + 119\,\text{mA}(0 + 120\,\Omega \parallel 32\,\Omega)$$
$$v_{CE(cutoff)} = 5.7\,\text{V} + 119\,\text{mA}(25.3\,\Omega)$$
$$v_{CE(cutoff)} = 5.7\,\text{V} + 3\,\text{V}$$
$$v_{CE(cutoff)} = 8.7\,\text{V}$$

The smaller of V_{CEQ} and ICQ(rC+rE) is the peak compliance, so *compliance* = 3 V peak

Given the compliance, we can use power law to find the load power

$$P_{load(max)} = \frac{Compliance_{RMS}^2}{R_L}$$
$$P_{load(max)} = \frac{(.707 \times 3\,\text{V})^2}{32\,\Omega}$$
$$P_{load(max)} = 141\,\text{mW}$$

This is not a lot of power for something like a loudspeaker but is a fair amount to drive something like a pair of headphones.

The transistor's worst case power dissipation is

$$P_{D(max)} = P_{DQ} = I_{CQ} V_{CEQ}$$
$$P_{D(max)} = 119\,\text{mA} \times 5.7\,\text{V}$$
$$P_{D(max)} = 678\,\text{mW}$$

The supplied circuit power is the average current draw times the total supplied voltage differential

$$P_{DC} = I_{CQ}(V_{CC} - V_{EE})$$
$$P_{DC} = 119\,\text{mA} \times 20\,\text{V}$$
$$P_{DC} = 2.38\,\text{W}$$

This is much worse than the theoretical best case. This is due, at least in part, to the fact that the Q point is not centered on the AC load line.

$$\eta = \frac{P_{load(max)}}{P_{DC}}$$
$$\eta = \frac{141\,\text{mW}}{2.38\,\text{W}}$$
$$\eta = 5.9\%$$

To complete the analysis, note that the transistor's breakdown rating (BV_{CEO}) should be at least as large as $v_{CE(cutoff)}$ (8.7 volts), and the maximum current rating should be at least as large as $i_{C(sat)}$ (119 mA+5.7 V/25.3 Ω = 344 mA).

Computer Simulation

A computer simulation of a class A emitter follower using a Darlington pair is examined next. Of primary interest here is the verification of the output compliance so a transient analysis will be used. The simulator schematic is shown in Figure 8.8.

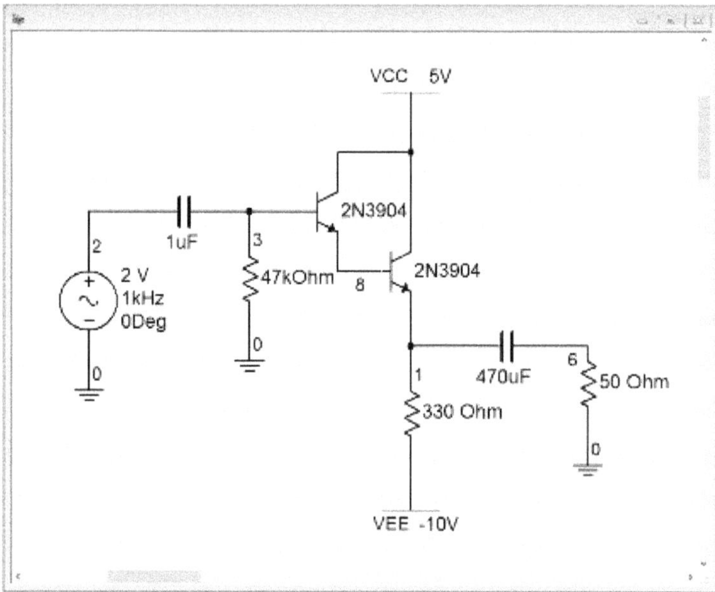

Figure 8.8 Class A follower in simulator.

We can make a few quick computations to determine the compliance. First, we find the collector Q point current.

$$I_{CQ} = \frac{|V_{EE}| - V_{BE}}{R_E}$$

$$I_{CQ} = \frac{10\,\text{V} - 1.4\,\text{V}}{330\,\Omega}$$

$$I_{CQ} = 26\text{mA}$$

By inspection, the emitter is two base-emitter junction potentials below ground, or−1.4 V. As the collectors are tied to VCC, this means that VCEQ = 6.4 V. The other half of the swing, from VCEQ to vCE(cutoff) is

$$v_{CE(cutoff)} - V_{CEQ} = I_{CQ}(r_C + r_E)$$

$$v_{CE(cutoff)} - V_{CEQ} = 26\,\text{mA}(0 + 330\,\Omega \,||\, 50\,\Omega)$$

$$v_{CE(cutoff)} - V_{CEQ} = 26\,\text{mA}(43.4\,\Omega)$$

$$v_{CE(cutoff)} - V_{CEQ} = 1.13\,\text{V}$$

The Q point is not centered and is closer to cutoff. This means that the amplifier will produce cutoff clipping around 1.1 volts and saturation clipping around 6 volts. In other words, there is more room for the current to swing up to saturation than to swing down to zero. As this is the current flowing through the load and we have a non-inverting follower, we expect to see the load voltage echo this. That is, the negative portion of the load voltage should clip before the positive portion.

The transient analysis results are shown in Figure 8.9. A two volt peak input signal is applied (blue trace). The negative portion of the load voltage clips at approximately 1.1 volts as expected (red trace). The input signal is not large enough to cause saturation clipping. This was done on purpose to verify the voltage gain of the follower. It should be very close to unity. In fact, the trace shows that the gain is around 0.95 or so.

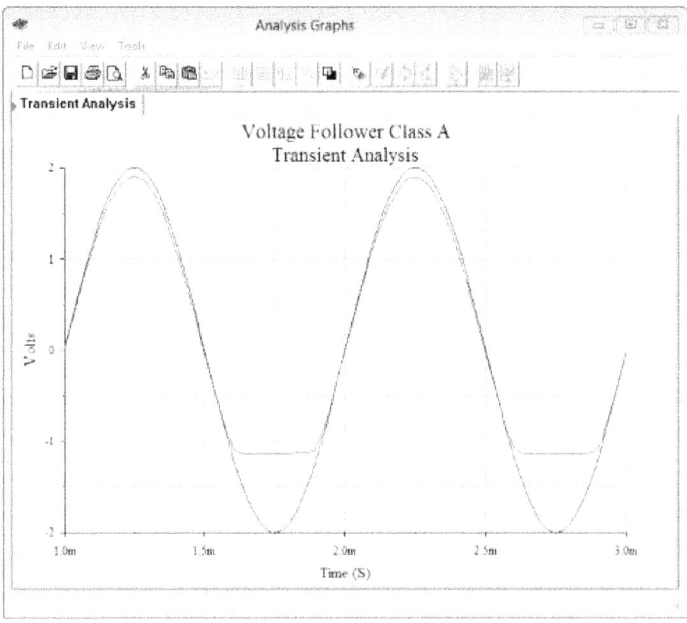

Figure 8.9 Class A follower transient analysis.

If this had been a voltage amplifier instead of a follower, these waveforms would appear flipped vertically. To verify this, the circuit is modified to produce a voltage amplifier with a gain of approximately one. This is achieved by moving the load to the collector and adding a 330 Ω biasing resistor. This will result in the same AC load impedance. To maintain a similar VCEQ, VCC is raised by 10 volts. Finally, the original 330 Ω emitter biasing resistor is split in two: 287 Ω and 43 Ω. This will yield the same ICQ and achieve a voltage gain of unity. As a result, we expect to see clipping at approximately 1.1 volts on the positive portion. The modified circuit is shown in Figure 8.10 and the resulting transient simulation in Figure 8.11.

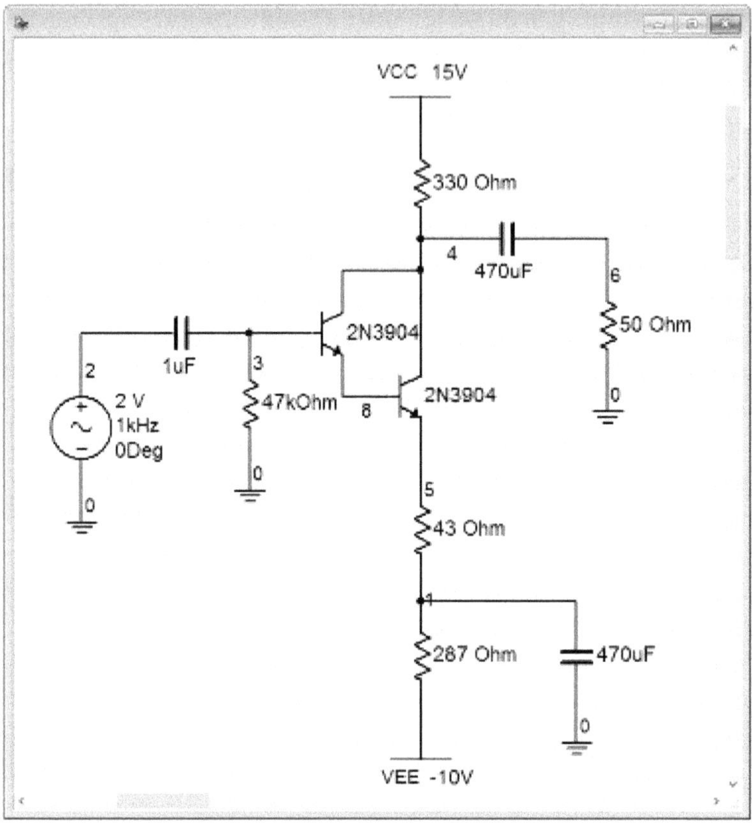

Figure 8.10 Class A amplifier in simulator.

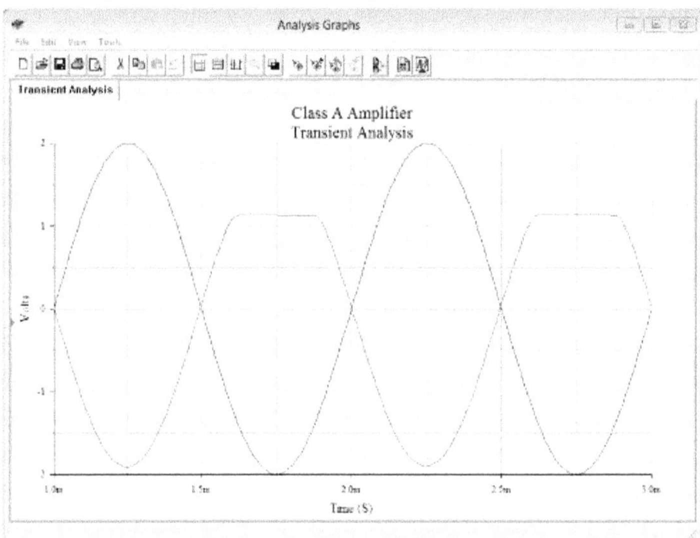

Figure 8.11 Class A amplifier transient analyses.

One final item of interest regarding the simulations: If the input level is increased in an attempt to see clipping on the other half of the waveform, something strange happens. At first it will appear as though it never clips. A careful examination reveals something different, though. Given the values in these circuits, they will exhibit a certain amount of clamping action (clamping was presented in Chapter 3). This will cause the waveform to shift. If you inspect the peak-to-peak value, it will be close to the value of vCE(cutoff). It will be a little less due to the fact that, particularly for a Darlington pair, VCE(sat) is not 0 V.

8.4 LOUDSPEAKERS

One of the more common loads for amplifiers is a loudspeaker. It makes sense then to look at how they are constructed and note anything interesting or peculiar as far as their electrical characteristics are concerned. The most common form of loudspeaker is the dynamic loudspeaker31. All dynamic loudspeakers share certain common elements regardless of size or acoustic output capability. A cutaway view of a low frequency driver is shown in Figure 8.12.

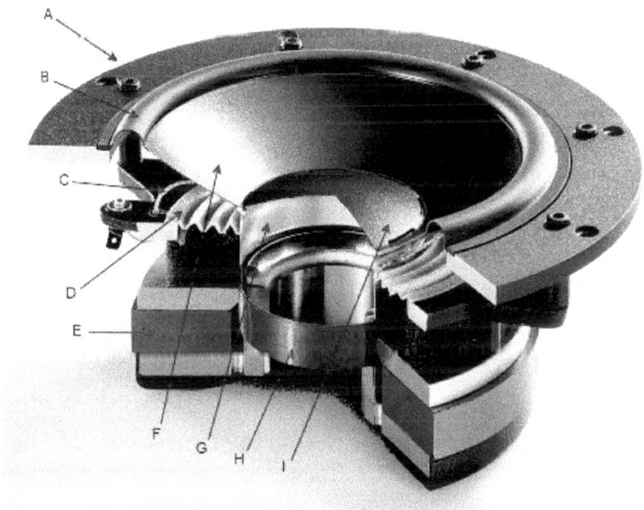

Figure 8.12 Dynamic loudspeaker. (A) Frame, (B) Suspension, (C) Lead wire, (D) Spider, (E) Magnet, (F) Diaphragm, (G) Voice coil former, (H) Voice coil, (I) Dust cap Image courtesy of Audio Technology.

The idea behind its operation is magnetic repulsion and attraction. The heart of the unit is the voice coil (H). This is a coil of magnet wire wound around a former (G) that typically is made of aluminum or some other high temperature material. The voice coil might be a single layer of edge-wound ribbon wire or perhaps several layers of ordinary round wire. Depending on the design, the voice coil might be anywhere from a fraction of an inch to several inches in diameter. The coil ends are connected to flexible lead wires (C) that terminate on the loudspeaker frame (A). Ultimately, that's what the amplifier will connect to.

The voice coil is fixed to a diaphragm (F) and is freely suspended by an outer edge suspension (B) and an inner element known as a *spider* (D). The voice coil sits in a strong magnetic field that is created by a powerful permanent magnet (E) that commonly uses ceramic, alnico or rare earth construction. When current from the amplifier flows through the coil, it will create its own magnetic field that will either aid or oppose the fixed field created by the permanent magnet, depending on the direction of the current. This results in a force that causes the coil to move within the fixed field. As the coil moves, the diaphragm moves with it, pushing on the surrounding air and creating sound. The larger the current, the stronger the newly created field and the greater the resulting aid or opposition, which results in greater movement of the diaphragm and a larger sound pressure. This fundamental design has changed little since its invention in the 1920s. Modern magnets, suspension and diaphragm materials have improved considerably in the intervening years but the operational principle is pretty much the same.

It is very difficult to create a driver that can cover the full audio spectrum of 20 Hz to 20 kHz while achieving sufficient listening volume at low distortion. Consequently, drivers are often designed to cover a limited portion of the audio spectrum. Low frequency drivers are commonly referred to as *woofers* while high frequency drivers are called *tweeters*. Drivers that cover the middle range of frequencies are given the highly inventive name *midranges* (although once upon a time they were called *squawkers*). A combination of these devices will be wired together with other components to create a complete home or auto loudspeaker system. Although very high quality systems can be produced, virtually all direct radiating dynamic loudspeaker systems suffer from low conversion efficiency. For a typical consumer system, only about 1% to 2% of the applied electrical power is turned into useful acoustic output power. The vast majority of the applied power simply makes the voice coil hot.

Loudspeaker Impedance

Loudspeakers are given a nominal impedance value. The most common impedance for home use is 8 Ω while 4 Ω is common in automotive systems. It is important to remember that this is a nominal value and the true value varies with frequency. While it is common to test power amplifiers with large power resistors, they are only a coarse approximation of a real loudspeaker.

The electrical and mechanical characteristics of a loudspeaker combine to create an equivalent circuit with resistive, inductive and capacitive elements. A typical electrical circuit model32 of a single loudspeaker driver is shown in Figure 8.13. RVC and LVC are the resistance and inductance of the voice coil, respectively. The other components are electrical equivalents of mechanical properties such as suspension losses.

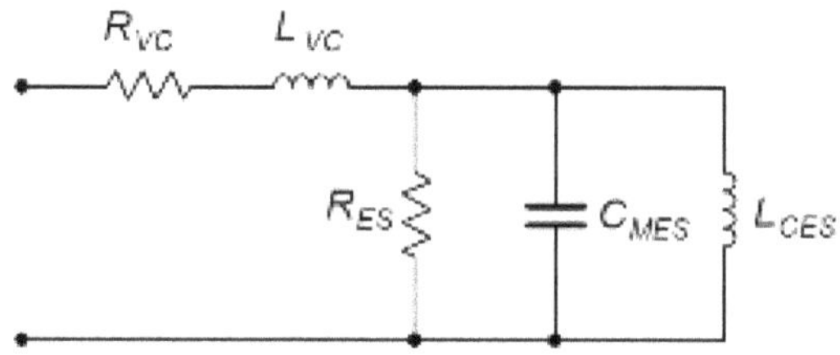

Figure 8.13 Dynamic loudspeaker electrical model.

Clearly, this is not a simple 8 Ω resistor. In fact, we see a very complex impedance: The three parallel elements will create a resonant peak and the series inductance will cause the impedance to rise with frequency. Typically, the resonant peak occurs at the lower end of the spectrum and the associated resonant frequency is denoted on a spec sheet as fS, the free-air resonance. For a nominal 8 Ω woofer, the peak impedance can be over 30 Ω. An example of a loudspeaker impedance plot is shown in Figure 8.14.

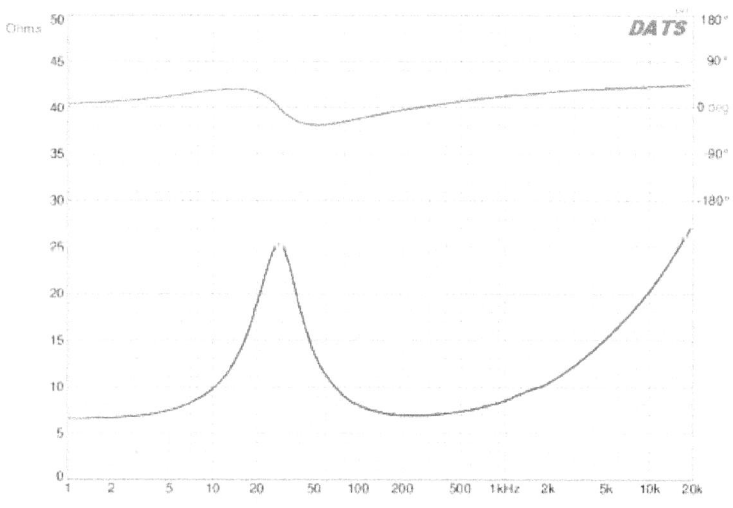

Figure 8.14 Dynamic loud speaker impedance plots.
Courtesy of Dayton Audio.

This loudspeaker is a nominal 8 Ω unit yet the impedance can be many times this value, and at some frequencies, less than 7 Ω. This plot also includes the phase angle of the loudspeaker and we can see that it can be upwards of 40° capacitive or inductive, depending on the frequency. What makes this more interesting is that a consumer loudspeaker system is a combination of multiple drivers plus other electrical components, and this can result in an even more complex impedance plot.

The obvious question is, "Does this have any effect on the analysis of the power amplifier?" The simple answer is, "Yes". Areas of the spectrum where the impedance magnitude drops below the nominal value will require more current for any given load voltage. Further, the phase shift caused by a partly reactive load will impact the power dissipation of the transistor.

Consider the power graph shown in Figure 8.6 for a purely resistive load. If we repeat the plot but add a noticeable phase shift to simulate a partly reactive load, something interesting happens, as shown in Figure 8.15.

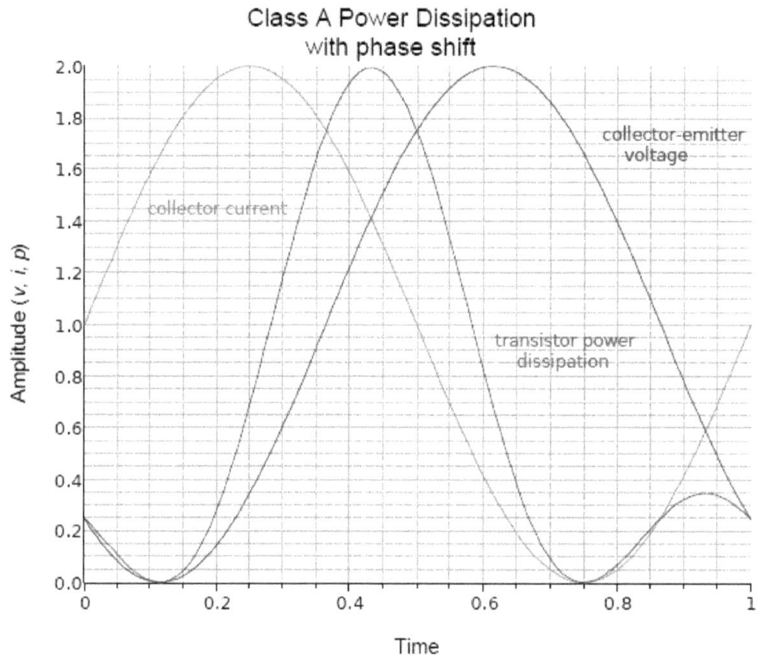

Figure 8.15 Power dissipation with reactive load.

The purple trace represents the power dissipation of the transistor. The peak current- voltage product is twice the value seen with a purely resistive load. This combination might lay outside the safe operating area of the transistor. Ultimately, reactive loads are somewhat more "challenging" than simple resistive loads. Therefore, the transistors may need to be rated higher than the values computed for an idealized resistive load. Another way of looking at the issue of phase shift induced by loudspeakers and other complex loads is to examine the AC load line. Our previous work with load lines always assumed that the load was purely resistive. What happens in the complex impedance case?

If we examine a generic complex load at a single frequency, our former straight line load line turns into an ellipse, as shown in Figure 8.16.

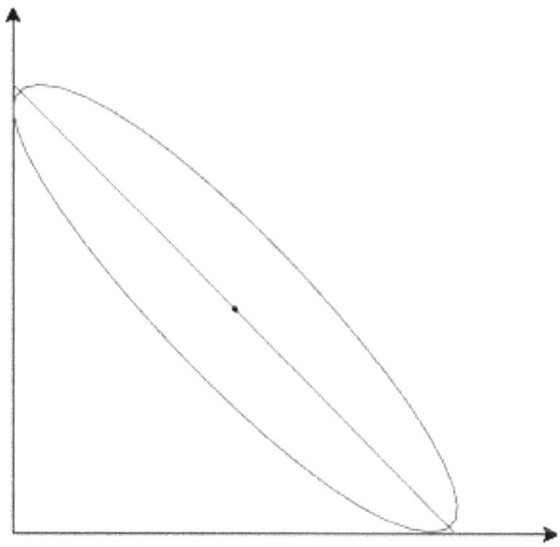

Figure 8.16 AC load line with complex load.

This plot assumes that the circuit has a centered Q point. The normal resistive load line is shown in green. We can visualize the signal starting at zero amplitude, meaning all we see is the Q point. As the signal gets larger and larger, it swings along the green line until, eventually, it maxes out at the two axes. In the complex impedance case, we also start at the Q point. As the signal increases, it traces out an ellipse around the Q point. Further increases would create a larger ellipse, and then a larger ellipse, and so on. Eventually, we would see the maximum swing just touching the axes. This is what is plotted above in red, the maximum case (i.e., at full compliance).

If the impedance angle changes, the aspect ratio of the ellipse changes in reaction to it. The larger the angle, the more open the ellipse becomes. The extremes are 0°, the purely resistive case that yields a collapsed ellipse or straight line; and 90°, the purely reactive case that yields a fully open ellipse, or circle. We have already seen that the phase angle of a loudspeaker changes with frequency, therefore, the load line also changes with frequency. As we sweep the input frequency from low to high, we can imagine the load line wavering back and forth between straight lines and various elliptical shapes. The important thing, though, is that some of these new operating regions (the areas where the red curve is above and to the right of the green line) may go outside the safe operating area of the transistor.

8.5 POWER TRANSISTOR DATA SHEET INTERPRETATION

The data sheet for a popular NPN power transistor, the 2N3055, is shown in Figure 8.17. This model is available from several different manufacturers. Due to the high power dissipation, the TO-92 plastic case that is used for small signal devices is not appropriate. Instead, this device uses the all-metal TO-3 case. Under the maximum ratings we find the device has a maximum power dissipation of 115 W at a case temperature of 25 °C, a maximum collector current of 15 A and a maximum collector-emitter voltage of 60 V. Obviously, the device cannot withstand maximum current and voltage simultaneously.

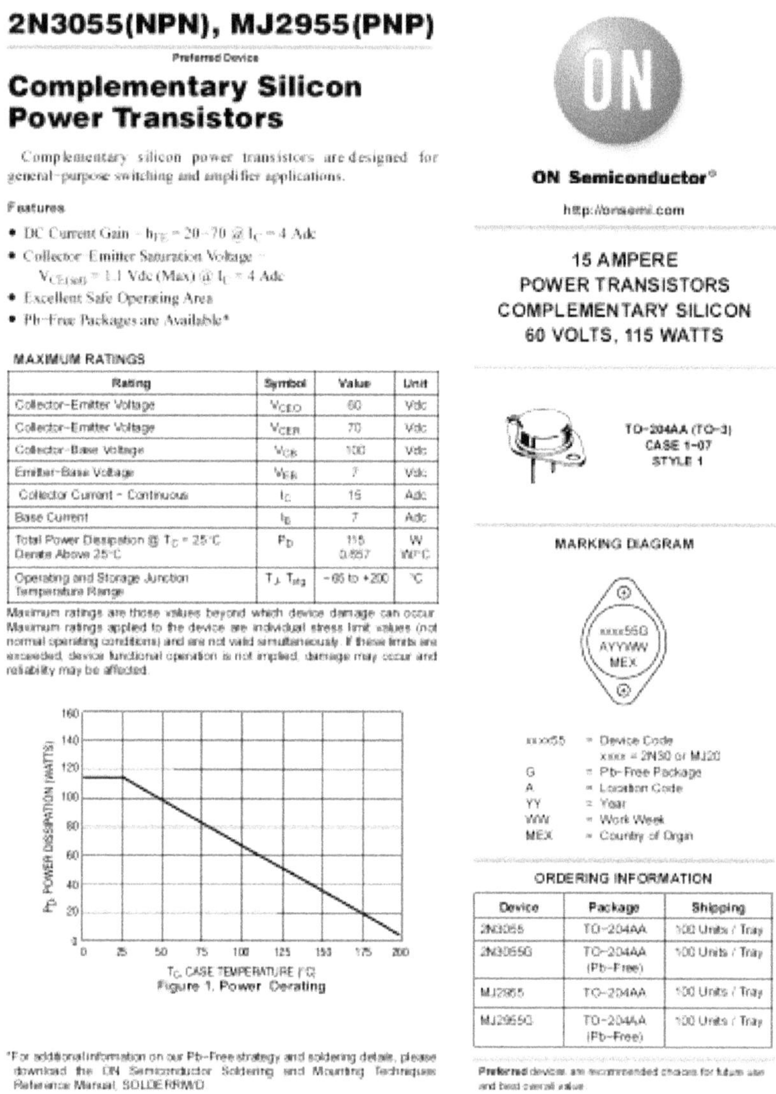

2N3055(NPN), MJ2955(PNP)

Preferred Device

Complementary Silicon Power Transistors

Complementary silicon power transistors are designed for general-purpose switching and amplifier applications.

Features

- DC Current Gain – h_{FE} = 20–70 @ I_C = 4 Adc
- Collector–Emitter Saturation Voltage – $V_{CE(sat)}$ = 1.1 Vdc (Max) @ I_C = 4 Adc
- Excellent Safe Operating Area
- Pb–Free Packages are Available*

MAXIMUM RATINGS

Rating	Symbol	Value	Unit
Collector–Emitter Voltage	V_{CEO}	60	Vdc
Collector–Emitter Voltage	V_{CER}	70	Vdc
Collector–Base Voltage	V_{CB}	100	Vdc
Emitter–Base Voltage	V_{EB}	7	Vdc
Collector Current – Continuous	I_C	15	Adc
Base Current	I_B	7	Adc
Total Power Dissipation @ T_C = 25°C Derate Above 25°C	P_D	115 0.657	W W/°C
Operating and Storage Junction Temperature Range	T_J, T_{stg}	–65 to +200	°C

Maximum ratings are those values beyond which device damage can occur. Maximum ratings applied to the device are individual stress limit values (not normal operating conditions) and are not valid simultaneously. If these limits are exceeded, device functional operation is not implied, damage may occur and reliability may be affected.

Figure 1. Power Derating

*For additional information on our Pb–Free strategy and soldering details, please download the ON Semiconductor Soldering and Mounting Techniques Reference Manual, SOLDERRM/D.

ON Semiconductor®

http://onsemi.com

15 AMPERE
POWER TRANSISTORS
COMPLEMENTARY SILICON
60 VOLTS, 115 WATTS

TO–204AA (TO–3)
CASE 1–07
STYLE 1

MARKING DIAGRAM

xxxx55 = Device Code
xxxx = 2N30 or MJ29
G = Pb–Free Package
A = Location Code
YY = Year
WW = Work Week
MEX = Country of Origin

ORDERING INFORMATION

Device	Package	Shipping
2N3055	TO–204AA	100 Units / Tray
2N3055G	TO–204AA (Pb–Free)	100 Units / Tray
MJ2955	TO–204AA	100 Units / Tray
MJ2955G	TO–204AA (Pb–Free)	100 Units / Tray

Preferred devices are recommended choices for future use and best overall value.

Figure 8.17a 2N3055 data sheet. Used with permission from SCILLC dba ON Semiconductor.

In the drawing of the TO-3 case, only two leads are shown. These are for the emitter and base. The entire body of the device is the collector. This is because the device will most likely be attached to a metal heat sink (see next section) to help dissipate the heat generated. The greater the contact area, the more effective the heat flow will be. The curves presented in Figure 8.17b indicate that β is considerably lower than what we saw for small signal devices. Further, IC(sat) tends to be larger for higher power transistors. For very high currents, β might fall to less than 20 while IC(sat) can be upwards of half of a volt.

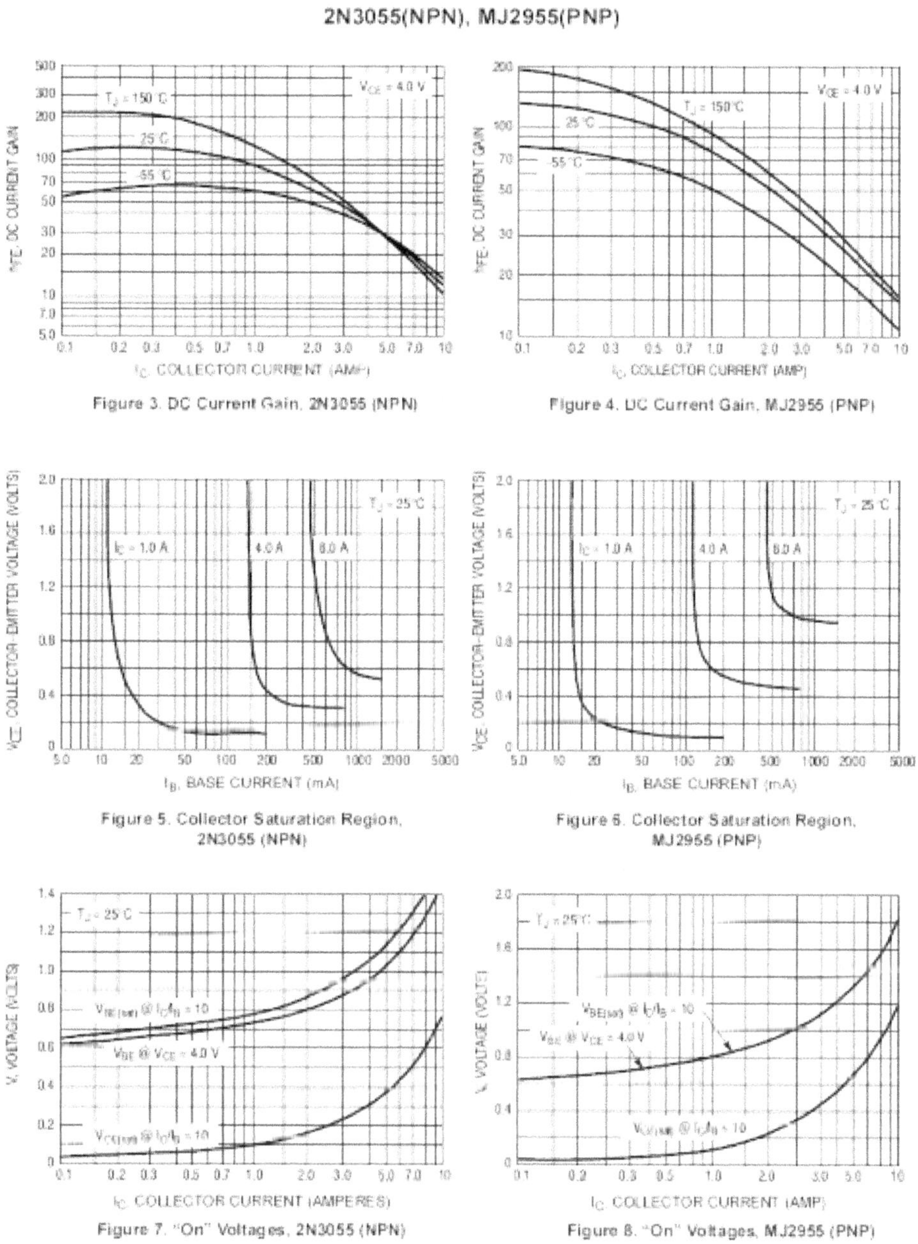

Figure 8.17b 2N3055 data sheet (cont).

THERMAL CHARACTERISTICS

Characteristic	Symbol	Max	Unit
Thermal Resistance, Junction-to-Case	$R_{\theta JC}$	1.52	°C/W

ELECTRICAL CHARACTERISTICS (T_C = 25°C unless otherwise noted)

Characteristic	Symbol	Min	Max	Unit
OFF CHARACTERISTICS*				
Collector-Emitter Sustaining Voltage (Note 1) (I_C = 200 mAdc, I_B = 0)	$V_{CEO(sus)}$	60	-	Vdc
Collector-Emitter Sustaining Voltage (Note 1) (I_C = 200 mAdc, R_{BE} = 100 Ω)	$V_{CER(sus)}$	70	-	Vdc
Collector Cutoff Current (V_{CE} = 30 Vdc, I_B = 0)	I_{CEO}	-	0.7	mAdc
Collector Cutoff Current (V_{CE} = 100 Vdc, $V_{BE(off)}$ = 1.5 Vdc) (V_{CE} = 100 Vdc, $V_{BE(off)}$ = 1.5 Vdc, T_C = 150°C)	I_{CEX}	 - -	 1.0 5.0	mAdc
Emitter Cutoff Current (V_{BE} = 7.0 Vdc, I_C = 0)	I_{EBO}	-	5.0	mAdc
ON CHARACTERISTICS* (Note 1)				
DC Current Gain (I_C = 4.0 Adc, V_{CE} = 4.0 Vdc) (I_C = 10 Adc, V_{CE} = 4.0 Vdc)	h_{FE}	 20 5.0	 70 -	-
Collector-Emitter Saturation Voltage (I_C = 4.0 Adc, I_B = 400 mAdc) (I_C = 10 Adc, I_B = 3.3 Adc)	$V_{CE(sat)}$	-	 1.1 3.0	Vdc
Base-Emitter On Voltage (I_C = 4.0 Adc, V_{CE} = 4.0 Vdc)	$V_{BE(on)}$	-	1.5	Vdc
SECOND BREAKDOWN				
Second Breakdown Collector Current with Base Forward Biased (V_{CE} = 40 Vdc, t = 1.0 s, Nonrepetitive)	$I_{s/b}$	2.87	-	Adc
Second Breakdown Collector Current with Base Forward Biased (V_{CE} = 40 Vdc, t = 1.0 s, Nonrepetitive)	$I_{s/b}$	2.87	-	Adc
DYNAMIC CHARACTERISTICS				
Current Gain - Bandwidth Product (I_C = 0.5 Adc, V_{CE} = 10 Vdc, f = 1.0 MHz)	f_T	2.5	-	MHz
*Small-Signal Current Gain (I_C = 1.0 Adc, V_{CE} = 4.0 Vdc, f = 1.0 kHz)	h_{fe}	15	120	-
*Small-Signal Current Gain Cutoff Frequency (V_{CE} = 4.0 Vdc, I_C = 1.0 Adc, f = 1.0 kHz)	f_{hfe}	10	-	kHz

*Indicates Within JEDEC Registration. (2N3055)
1. Pulse Test: Pulse Width ≤ 300 μs, Duty Cycle ≤ 2.0%.

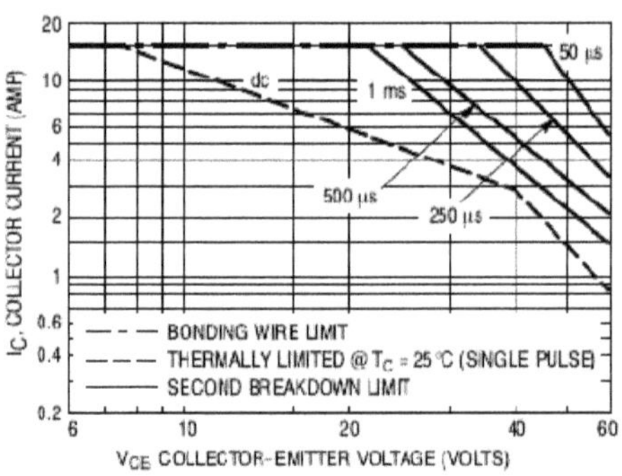

Figure 2. Active Region Safe Operating Area

There are two limitations on the power handling ability of a transistor: average junction temperature and second breakdown. Safe operating area curves indicate $I_C - V_{CE}$ limits of the transistor that must be observed for reliable operation; i.e., the transistor must not be subjected to greater dissipation than the curves indicate.

The data of Figure 2 is based on T_C = 25°C; $T_{J(pk)}$ is variable depending on power level. Second breakdown pulse limits are valid for duty cycles to 10% but must be derated for temperature according to Figure 1.

Figure 8.17c 2N3055 data sheet (cont).

One item of note in Figure 8.17c is the small graphic at the bottom of the sheet. This is a plot of safe operating area. Basically, the combination of VCE and IC must fall within the lower-left zone. What is of particular interest is that the safe zone extends out further if the current/voltage combination is the result of a short pulse rather than a continuous condition.

Power Derating

One final item of concern is the graph found in Figure 8.17a, and magnified in Figure 8.18. This is a power derating curve.

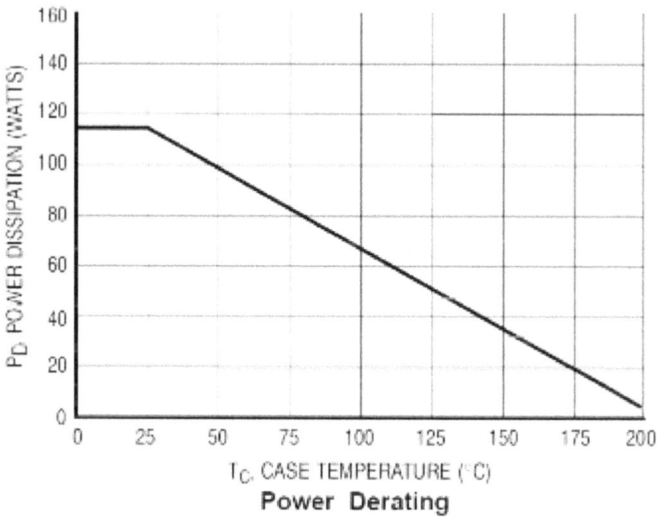

Figure 8.18: Power derating curve for 2N3055. Used with permission from SCILLC dba ON Semiconductor.

Although the device is rated for 115 watts, that is only true at case temperatures of 25°C or lower. At higher temperatures, the power dissipation capability decreases. For example, at 100°C this device can only dissipate about 65 watts. A precise value can be computed via the following formula:

$$P_D = P_{25} - D(T_{case} - 25°C) \quad (8.8)$$

Where

PD is the power dissipation at the new case temperature,

P25 is the power dissipation at 25°C,

D is the derating factor (units of W/C°),

Tcase is the new case temperature.

Example 8.2

Determine the power dissipation of a 2N3055 at 75°C.

Using the graph, this is a little over 80 watts.

For a more accurate result, we'll use Equation 8.8. From the data sheet the dissipation at 25°C, P25 , is 115 watts. The derating factor, D, is 0.657 W/C° (the derating factor is found directly above the graph of power derating in Figure 8.17a).

Example 8.2

Determine the power dissipation of a 2N3055 at 75°C.

Using the graph, this is a little over 80 watts.

For a more accurate result, we'll use Equation 8.8. From the data sheet the dissipation at 25°C, P25 , is 115 watts. The derating factor, D, is 0.657 W/C°(the derating factor is found directly above the graph of power derating in Figure 8.17a).

$$P_D = P_{25} - D(T_{ambient} - 25\,°\mathrm{C})$$
$$P_D = 115\,\mathrm{W} - 0.657\,\mathrm{W/C°}(75\,°C - 25\,°\mathrm{C})$$
$$P_D = 82.1\,\mathrm{W}$$

8.6 HEAT SINKS

The issue with power transistors is always heat. As noted in Example 8.2, as the transistor heats up due to internal power dissipation, its ability to dissipate heat is compromised. The trick, then, is to efficiently move the heat from the transistor to someplace else. This is normally achieved through the use of a heat sink.

A heat sink is a metal device that is attached to the power transistor. Typically, they are made of aluminum and feature an array of fins. By increasing the surface area, heat can be moved away from the transistor more efficiently than by the transistor alone.

Heat sinks are designed to mount specific device case styles. The most common case styles include the TO-3 "can" along with the various "power tab" styles such as the TO-220 and TO-202. Special mounting hardware and insulation spacers are also required in order to maintain electrical isolation between the transistor and the heat sink as we do not want the heat sink to be electrically live. This usually takes the form of a mica sheet, and plastic washers and bushings for the mounting machine screws (for small heat sinks, sometimes nylon machine screws are used).

There are a few general rules that should be followed when using heat sinks:

- Always use some form of heat sink grease or thermally conductive pad between the heat sink and the device. This will increase the thermal transfer between the two parts, however, excessive quantities of heat sink grease will decrease performance.
- Mount fins in the vertical plane for optimum natural convective cooling.

- Do not overcrowd or obstruct devices that use heat sinks.
- Do not block air flow around heat sinks – particularly directly above and below items that rely on natural convection.
- If thermal demands are particularly high, consider using forced convection (i.e., a small fan directed at the heat sink).
- Some typical heat sinks are shown below. Figure 8.19 shows a heat sink and thermal data plot for use with a single TO-3 case device.

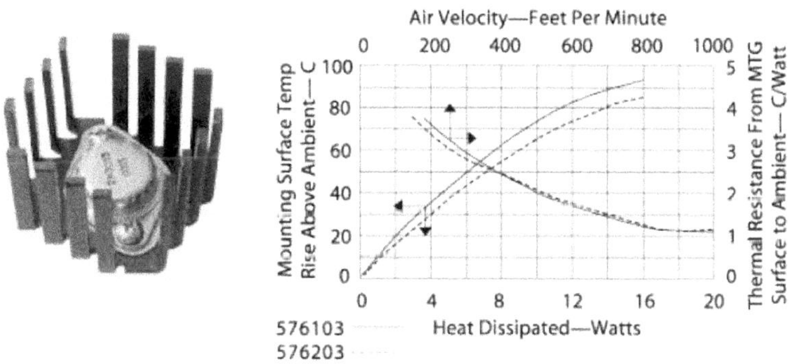

Figure 8.19 Heat sink for TO-3.Reprinted courtesy of AavidThermalloy, Inc.

Figure 8.20 shows a heat sink designed for a pair of transistors using TO-220 cases. In this photo, the white insulating pads can be seen between the transistors and heat sink.

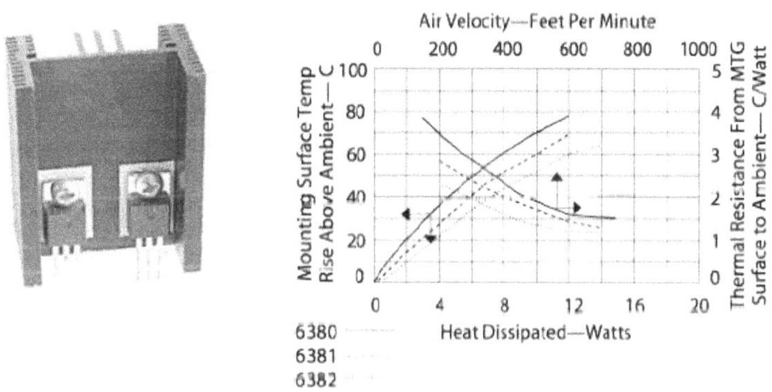

Figure 8.20 Heat sink for dual TO-220. Reprinted courtesy of AavidThermalloy, Inc.

Thermal Resistance

In order to specify a particular heat sink for a given application, a more technical explanation is in order. What we are going to do is create a *thermal circuit equivalent.*

In this model, the concept of *thermal resistance* is used. Thermal resistance denotes how easy it is to transfer heat energy from one mechanical part to another. The symbol for thermal resistance is ϑ, and the units are Celsius degrees per watt. In this model, temperature is analogous to voltage, and thermal power dissipation is analogous to current. A useful equation is,

$$P_D = \frac{\Delta T}{\theta_{total}} \tag{8.9}$$

Where *PD* is the power dissipated by the semiconductor device in watts, ΔT is the temperature differential, and ϑ*total* is the sum of the thermal resistances. Basically, this is a thermal version of Ohm's law.

In order to construct our model, let's take a closer look at the power-device/heat-sink combination. This is shown in Figure 8.21. The subscript *j* stands for junction, *c* is for case (of the transistor), *s* is for heat sink and *a* is for the ambient air. *Tj* is the semiconductor junction temperature and is created by the product of the transistor's current and voltage. This thermal source heats the device case to *Tc*. The thermal resistance between the two entities is ϑ*jc*. The case, in turn, heats the heat sink via the interconnection. This thermal resistance is ϑ*cs*, and the resulting temperature is *Ts*.

Finally, the heat sink passes the thermal energy to the surrounding air which is sitting at *Ta*. The thermal resistance of the heat sink to the air is ϑ*sa*. The equivalent thermal model is shown in Figure 8.22. Although this "thermal circuit" does not have perfect correspondence with normal circuit analysis, it does illustrate the main points.

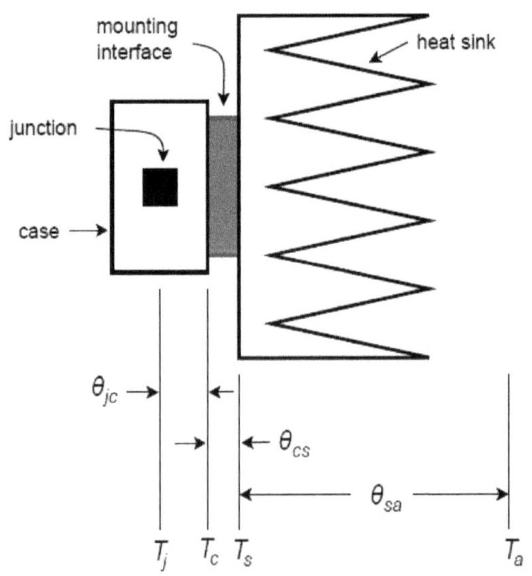

Figure 8.21 Device and heat sink.

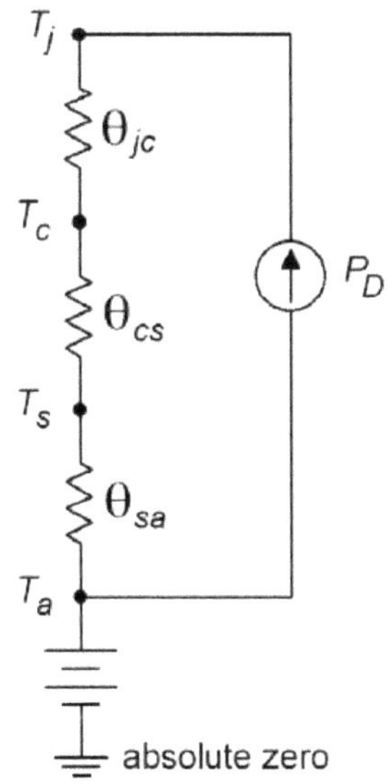

Figure 8.22 Equivalent thermal model of Figure 8.21.

In this model, ground represents a temperature of absolute zero. The circuit is sitting at an ambient temperature Ta, thus a voltage source of Ta is connected to ground and the heat sink. The three thermal resistances are in series and are driven by a current source that is set by the present power dissipation of the device. Note that if the power dissipation is high, the resulting "voltage drops" across the thermal resistances are high. Voltage is analogous to temperature in this model, so this indicates that a high temperature is created. Because there is a maximum limit to Tj, a higher power dissipation requires lower thermal resistances. As ϑjc is set by the device manufacturer, we have no control over that element. However, ϑcs is a function of the case style and the insulation material used, so we do have some control (but not a lot) over that. On the other hand, as the person who specifies the heat sink, we have a great deal of control over ϑsa. Values for ϑsa are given by heat sink manufacturers. A useful variation of Equation 8.9 is

$$P_D = \frac{T_j - T_a}{\theta_{jc} + \theta_{cs} + \theta_{sa}} \qquad (8.10)$$

Normally, power dissipation, junction and ambient temperatures, ϑjc and ϑcs are known. The idea is to determine an appropriate heat sink. Both Tj and ϑjc are given by the semiconductor device manufacturer. The ambient temperature, Ta, may be determined experimentally. Due to localized warming, it tends to be higher than the actual "room temperature". Standard graphs, such as those found in Figure 8.23, may be used to determine ϑcs. Note the generally lower values of ϑcs for the TO-3 case relative to the TO-220. This is one reason why TO-3 cases are used for higher power devices. This case also makes it easier for the manufacturer to reduce ϑjc.

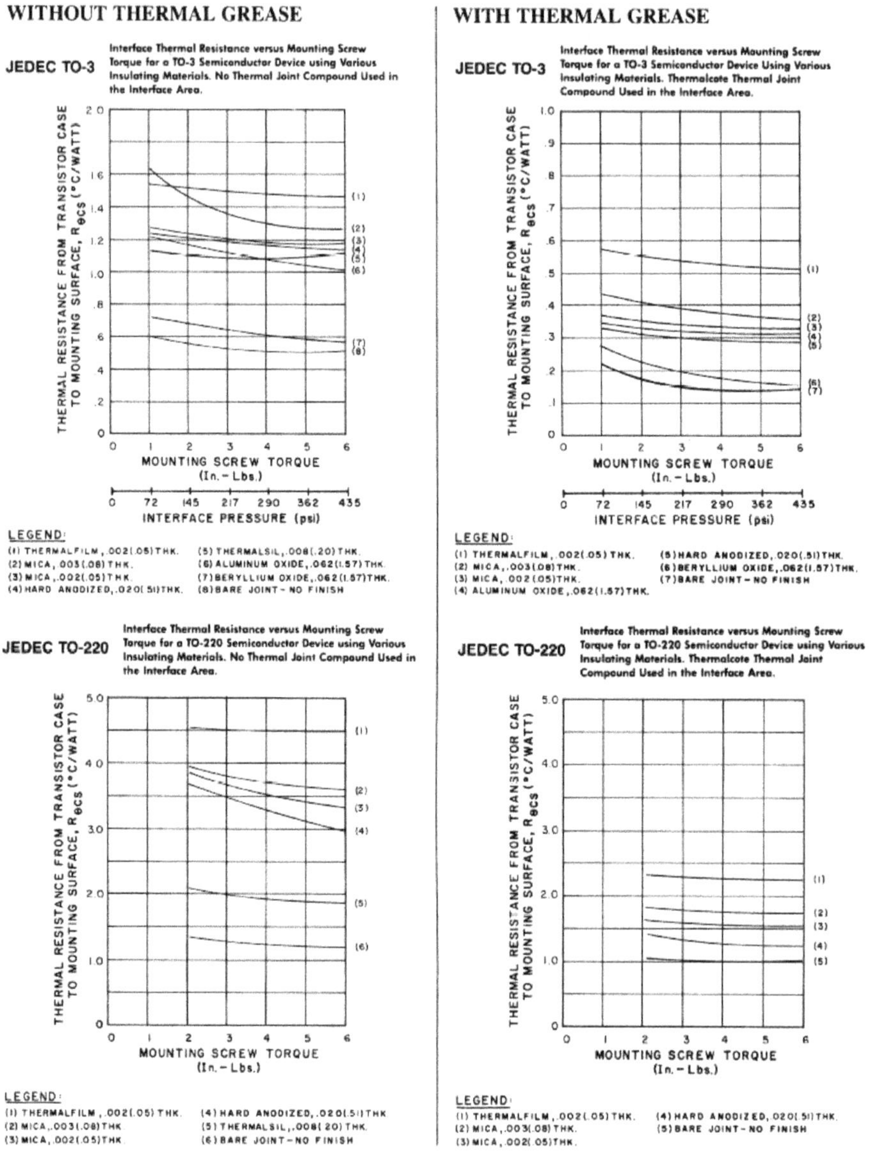

Figure 8.23 θ_{cs} for TO-3 and TO-220 Reprinted courtesy of Thermalloy, Inc.

Example 8.3

Determine the appropriate heat sink rating for a power device rated as follows: Tj(max) = 175°C, TO-3 case style, ϑjc = 1.5 C°/W. The device will be dissipating a maximum of 15 W in an ambient temperature of 40°C. Assume that the heat sink will be mounted with heat sink grease and a 0.002 mica insulator.

First, find ϑcs from the TO-3 "With Thermal Grease" graph in Figure 8.23. Curve 3 is used. The approximate value is 0.35 C°/W.

$$P_D = \frac{T_j - T_a}{\theta_{jc} + \theta_{cs} + \theta_{sa}}$$

$$\theta_{sa} = \frac{T_j - T_a}{P_D} - \theta_{jc} - \theta_{cs}$$

$$\theta_{sa} = \frac{175\,°C - 40\,°C}{15\,W} - 1.5\,C°/W - 0.35\,C°/W$$

$$\theta_{sa} = 7.15\,C°/W$$

This is the maximum acceptable value for the heat sink's thermal resistance. Note that the use of heat sink grease gives us an extra 0.8 C°/W or so. For this application, the heat sink pictured in Figure 8.19 will most likely be sufficient without added forced air cooling (the graph stops at less than4 C°/W with an air flow of under 200 feet/minute).

If we repeat this problem with a much higher *PD*, things work out a little differently. Let's use 40 W this time.

$$P_D = \frac{T_j - T_a}{\theta_{jc} + \theta_{cs} + \theta_{sa}}$$

$$\theta_{sa} = \frac{T_j - T_a}{P_D} - \theta_{jc} - \theta_{cs}$$

$$\theta_{sa} = \frac{175\,°C - 40\,°C}{40\,W} - 1.5\,C°/W - 0.35\,C°/W$$

$$\theta_{sa} = 1.53\,C°/W$$

If we hope to use that same heat sink, we will have to add forced air cooling of at least 700 feet/minute. The other option would be to find a more thermally efficient (and probably much larger) heat sink if we hope to use natural convection alone.

CONCISE SUMMARY

- Class A operation is defined as having collector current flow for 360° of the cycle. This means that a single output device can be used to amplify the entire input waveform. To determine the maximum signal swing, or compliance, an AC load line is used. This is similar to a DC load line and plots all possible transistor voltage and current coordinate pairs. The efficiency of the class A amplifier tends to be low. The maximum theoretical efficiency is only 25%. Further, the amplifier draws full current from the power supply regardless of whether or not a signal is present. As a consequence, the transistor runs hottest when there is no signal. With an applied signal, some of the power formerly dissipated within the transistor is shifted to the load.

- Loudspeakers offer a complex impedance as a load. As such, they are more challenging than simple resistive loads and may require the output transistor to be rated for a higher-than-normal power dissipation.
- Heat sinks are used to efficiently move heat from the transistor's internal structure to the surrounding air. The thermal effectiveness of a heat sink is measured by thermal resistance, θ. The lower the value of θ, the more effective the heat sink is at transferring heat from the transistor to the surrounding air. For high power applications where a good deal of heat is generated, heat sinks are augmented with forced air cooling.

CHAPTER-9

CLASS B POWER AMPLIFIERS BASED ON BIPOLAR JUNCTION TRANSISTORS (BJTS)

9.1 INTRODUCTION: AN OVERVIEW

Class B power amplifiers, utilizing Bipolar Junction Transistors (BJTs), represent a significant category in the realm of electronic amplification, particularly in applications where efficiency is paramount. In Class B amplifiers, each BJT conducts only for a half-cycle of the input signal, dividing the workload between two transistors. This configuration enhances efficiency as it reduces power dissipation during idle periods when no input signal is present. Class B amplifiers are commonly employed in audio systems, such as those used in speakers and public address systems, where power efficiency is crucial. While achieving higher efficiency, Class B amplifiers can introduce some distortion due to the transition between transistors. Striking a balance between efficiency and signal fidelity, Class B power amplifiers find extensive use in various electronic systems, highlighting their importance in modern audio and communication technologies. Understanding the operational characteristics and design principles of Class B power amplifiers is key for engineers working on applications that demand a balance between power efficiency and audio signal quality.

Class B amplifiers have long been a mainstay of linear amplifier design. Compared to class A operation, class B amplifiers offer much greater power efficiency. That is, a much larger percentage of the applied DC power can be turned into useful AC output to the load. This also means that the power dissipation requirements for the transistors are lowered. Further, unlike class A operation, class B designs do not continuously draw full power from their DC supplies. Instead, they draw current as it is needed, and therefore run relatively cool at idle and at low output power.

The downside for this improved efficiency is added complexity. For starters, two transistors are required for linear class B operation. Also, the biasing can be a little trickier which requires modifications to the relatively straightforward class A biasing circuits we have already examined. Also, class B amplifiers suffer from a unique form of distortion that class A amplifiers do not.

In this chapter we shall also examine the use auxiliary device configurations and sub-circuits to improve performance. These include the current mirror, Sziklai pair, VBE multiplier, and overload protection and prevention circuitry. As with the class A amplifier, the common collector or voltage follower configuration tends to be the most widely used configuration for class B circuits, hence, we shall focus on followers and not on voltage amplifiers.

9.2 THE CLASS B CONFIGURATION

Class B operation is defined as having AC collector current flow 180° out of the cycle. Consequently, in order to amplify the entire signal, two devices will be needed. Further, we will need to pay attention to how the two waveform halves are "stitched together" as this could be a problem area. The obvious question at this point is, why do we bother separating the positive and negative half-waves if it leads to circuit complexity and possible waveform issues? The answer is improved efficiency.

In the previous chapter we discovered that class A amplifiers are not efficient. In fact, at best they only transform 25% of the DC input power into useful load power. Why does this occur and how does the class B topology address this situation?

With a class A design, we determined that the optimal location for the Q point was midway along the AC load line. In this way, the signal could swing up toward saturation and down toward cutoff equally. Unfortunately, this also means that if no signal is present, then the transistor is still pulling half of the maximum current. It would make more sense if the transistor idled at zero current (or some small nominal value) than idling at half of the maximum33.

The basic idea of class B is to push the Q point down so that it is sitting right at cutoff on the AC load line. This means that *ICQ* is 0 A and virtually no power is drawn from the supply at idle. Locating the Q point at cutoff also means that the transistor will immediately clip the negative portion of the wave. Consequently, we will need a mirror image circuit to produce that portion (and which will clip the positive portion).

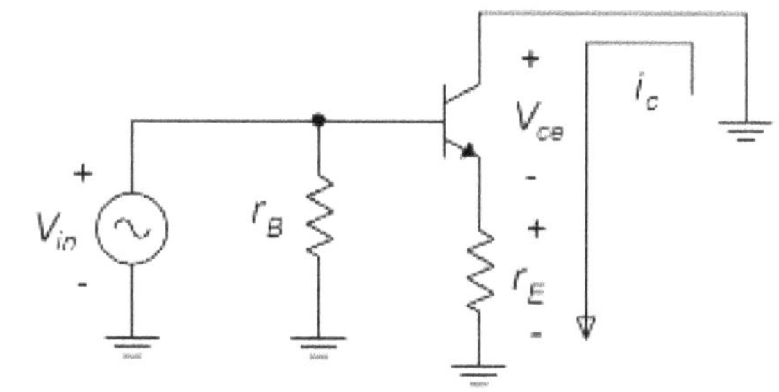

Figure 9.1 Voltage follower simplified AC circuit.

To gain a better understanding of how class B operation works, consider the simplified AC circuit of a voltage follower shown in Figure 9.1. If we situate the Q point directly at vCE(cutoff) then the associated ICQ is 0 A. As the input signal swings positive, the collector current increases. As it does so, the voltage across the load (rE) begins to increase and the voltage across the transistor's collector-emitter begins to decrease (due to KVL). When the input signal swings negative, the transistor is turned off. As a result, no collector current is created, no voltage is developed across the load and vCE stays at cutoff. It is as if the input waveform has been half-wave rectified. This action is shown in Figure 9.2. As the input signal swings positive, the operating point slides up the load line, moving toward saturation, and this increased current creates a load voltage that follows the input signal. In contrast, when the input tries to swing negative, there is no place else to go on the load line and the negative portion of the wave is simply clipped.

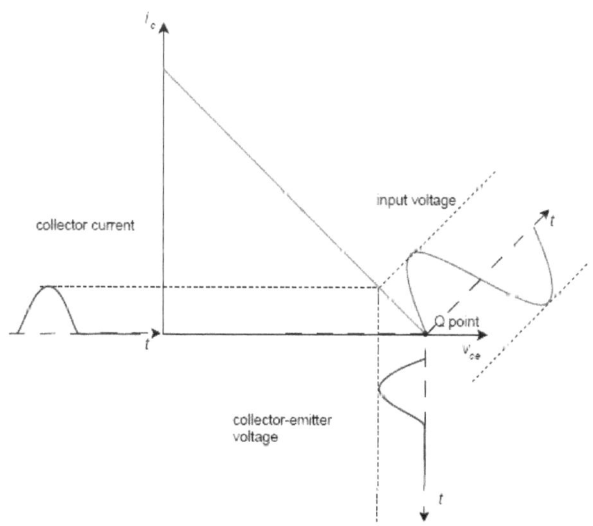

Figure 9.2 AC load line for class B operation.

If we made a PNP version of the circuit depicted in Figure 9.1, the exact opposite would happen: the amplifier would reproduce the negative portion of the wave and clip the positive portion. The next question is, how do we bias the transistor at cutoff and splice together the NPN and PNP versions into a workable whole?

Let's begin by making two bare-bones emitter followers, one NPN and the other PNP. We'll connect their emitters together and tie that to the load. We'll connect the NPN's collector directly to a DC supply and the PNP's collector to ground. Remember, collector resistors will not be needed because these are followers. We will include no biasing components on the base because we want to set ICQ to 0 A. We will also have to add input and output capacitors to prevent the source and load from inadvertently shorting out or shunting portions of the DC circuit. The result is seen in Figure 9.3.

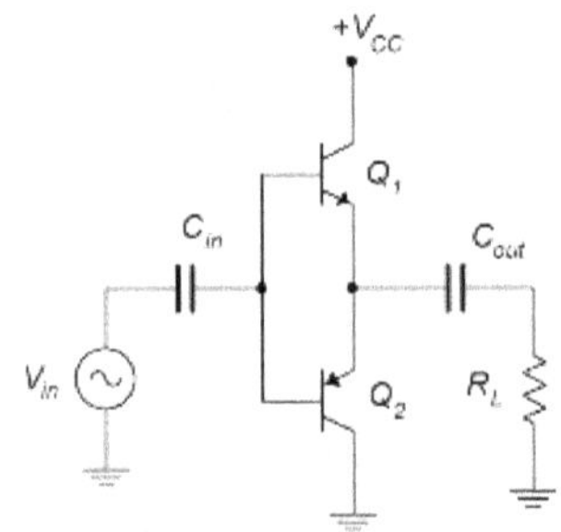

Figure 9.3 Prototype Class B circuit.

With no signal supplied, both transistors must be off. This is because their bases are tied together, and without some other applied potential, both base-emitter voltages must be zero. Assuming Q1 and Q2 are matched, the power supply voltage should split evenly between them, leaving half of VCC at the emitters. This potential also appears across Cout, preventing the DC voltage from reaching RL.

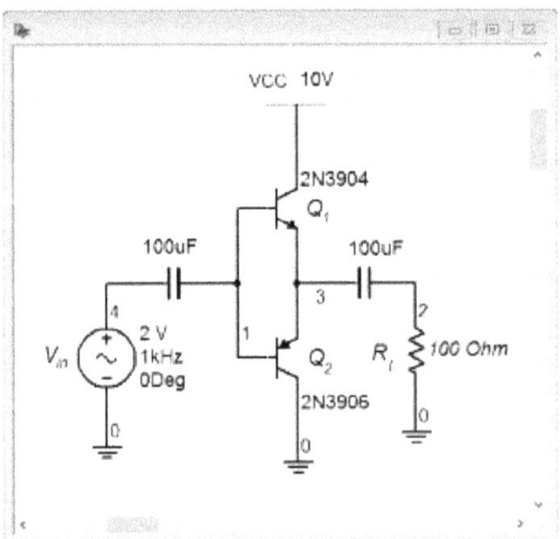

Figure 9.4 Prototype Class B circuit in simulator.

When the input signal goes positive, it raises VB1 and VB2 above 0.5VCC. This keeps Q2 off but turns on Q1. Current is now free to flow down through Q1 and into the load. When the input signal swings negative, the inverse happens: Q1 is turned off and Q2 is turned on. This allows current to flow up from the load and down through Q2 (if this is confusing, remember that a DC voltage had already been established across Cout that is equal to 0.5VCC and this is what allows the current to flow from ground up through RL and then down through Q2 as Q2 begins to conduct). You can think of Q1 pushing current into the load (sourcing) and Q2 pulling current from the load (sinking). Consequently, class B amplifiers are sometimes called a push-pull amplifiers.

To see how well this prototype works, we'll enter a version into a simulator, as shown in Figure 9.4. A transient analysis is performed with the results shown in Figure 9.5.

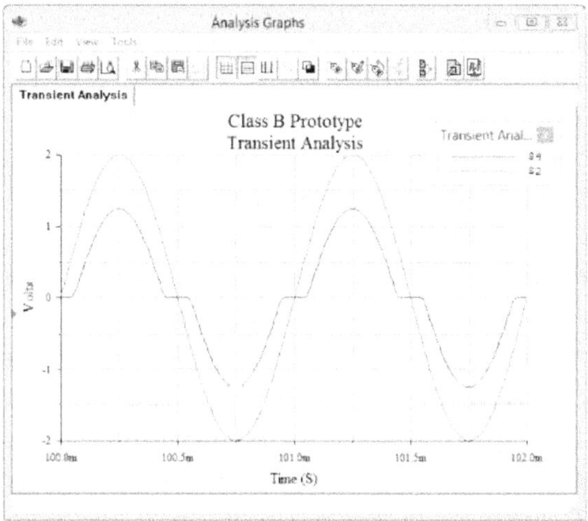

Figure 9.5 Transient analysis of prototype Class B circuit.

Clearly, there are issues with the output waveform (node 2, in blue). First off, the signal amplitude is noticeably smaller than the 2 volt peak input signal. The second issue is the bizarre "flat spotting" of the output waveform near the zero-crossing points. It turns out that these two problems are manifestations of the same root cause. If we look carefully at the peaks, we can get a clue as to what is going on. The peak output voltage is about 0.75 volts below the input, or just about one PN junction forward potential. The problem is that the input signal will not truly turn on the NPN transistor until the signal exceeds approximately 0.7 volts or drops below-0.7 volts for the PNP side. That region between -0.7 volts and +0.7 volts is a dead zone that the amplifier will not respond to. Essentially, the amplifier "rips out" anything between ±0.7 volts. This is a gross form of distortion and goes by many names including notch distortion and cross-over distortion. The particularly nasty part about this form of distortion is that it hits small signals worse than large signals. Most other forms of nonlinearities tend to get worse as the signal level increases.

The basic solution to this problem is to provide a small idle current so that the transistors are almost on. In this way, only a very small input signal will be needed to turn on the devices. As this would slightly increase the conduction angle, this form of operation is referred to as class AB operation. One potential solution is to add a voltage divider as depicted in Figure 9.6.

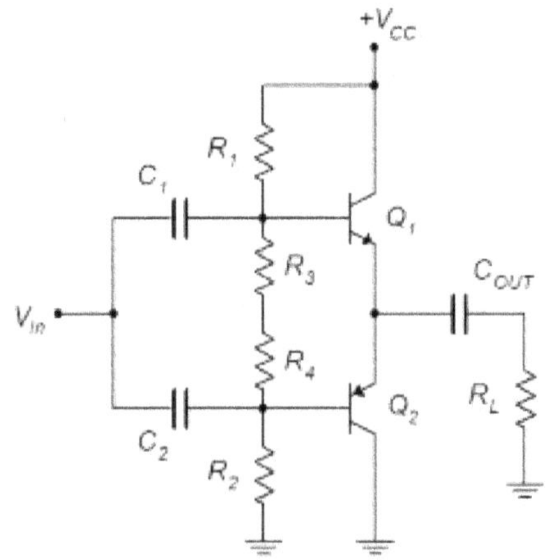

Figure 9.6 Prototype Class AB circuit.

This circuit uses a symmetrical layout: everything above a horizontal line drawn through the middle is echoed by a component below the line. In other words, R1 = R2, R3 = R4, C1 = C2, and Q1 and Q2 are a complimentary pair. The divider is configured so that the voltage drops across R3 and R4 are about 0.7 volts each.

Properly designed, the circuit of Figure 9.6 will reduce notch distortion. Unfortunately, it has other problems. The first issue involves the three capacitors. COUT in particular might be quite large. These can be removed if we went to a symmetrical bipolar power supply. Instead of running the PNP's collector to ground, we'll tie it to a negative DC supply. To keep the same total voltage, we'll set Q1's collector to half of the original VCC and Q2's collector to half of the original VCC but negative. On the input side, we could run the input signal to the junction of R3 and R4.

The second issue plaguing the circuit of Figure 9.6 is the stability of the bias. The voltage divider resistors have to be very accurate in order to set the transistors right where we want them and stability is a concern. The problem is that we are trying to use a device with a linear current-voltage characteristic (a resistor) to match the exponential current-voltage characteristic of a PN junction. This problem is exacerbated by the fact that these devices will drift with temperature, and drift in different ways. The solution to this problem is to use a device with better matching characteristics. What better device to match a PN junction than another PN junction?

The Current Mirror

Consider the circuit shown in Figure 9.7. This is called a current mirror. Here is how it works: First, look at the divider between R and D. The voltage across R must equal the supply voltage minus the diode drop, or approximately V - 0.7. This sets up a current, IR. If the base current is small enough to ignore, this same current flows down through the diode as ID. This diode current sets up a specific voltage across the diode (somewhere in the vicinity of 0.7 volts although the exact voltage is not important). Because the diode is in parallel with the base-emitter junction, then VBE = VD. If the trans conductance curve (I-V curve) of the transistor is identical to that of the diode, then the emitter current must be the same as the diode current. Any change in the diode current would cause a slight change in diode voltage, and since diode voltage and base-emitter voltage are the same, then the emitter current must change in response. In other words, the emitter current mirrors the diode current. We can program the diode current (and hence, the emitter current) by setting an appropriate value for R. Whatever the current through R is, that's also the collector current.

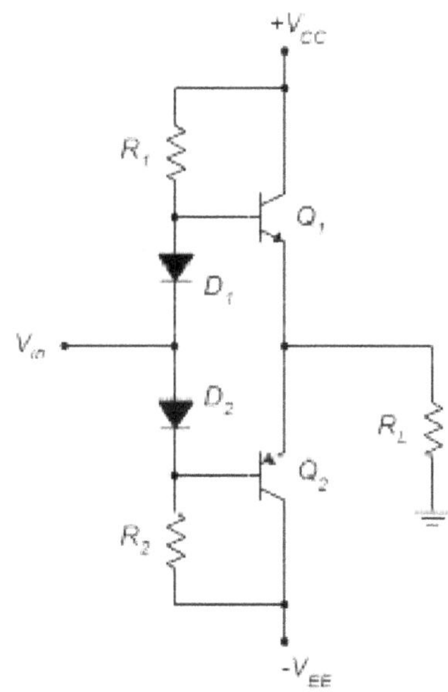

Figure 9.7 A simple current mirror.

The idea of the current mirror is used with great effect in integrated circuits where it is easy to match device characteristics. When it comes to discrete components, it is not nearly so easy to match a diode to a transistor. Fortunately, we don't have to have a perfect match. Simply using any signal diode will provide a much better match for VBE than using a resistor. Although the diode current won't match the collector current precisely, the bias will be more stable and the components will track much better than when using a resistor.

Combining these ideas leads us to the circuit of Figure 9.8. This is our first practical class B amplifier with the potential for decent values of distortion and stability.

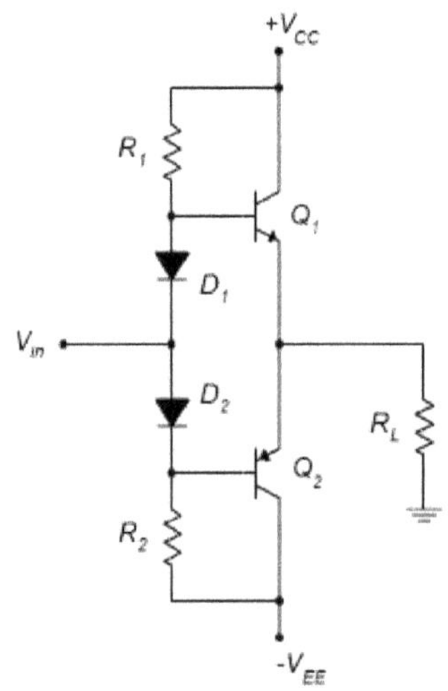

Figure 9.8 Class B amplifier with diode bias and bipolar supplies.

One thing that sometimes bothers people when they first see a diode biased amplifier is how the AC signal will pass through the diodes to the bases. At first glance, it appears that the positive portion of the input signal would be "going the wrong way" against diode D1. What we need to remember is that D1 is already forward-biased due to the DC supply and surrounding resistors. The signal won't see an open, it will see the dynamic resistance of the diode34.

Class B Circuit Maximums

Now that we have a workable circuit, we need to derive formulas for the endpoints of the AC load line, meaning vCE(cutoff) and iC(sat), and determine the compliance. The first item of note is cutoff. Because the two transistors will split the available supply, VCEQ will always equal half of the total supply. Further, we have biased these devices at cutoff, and therefore

$$V_{CE(cutoff)} = V_{CEQ} = 0.5 \cdot Total\ DC\ Supply \qquad (9.1)$$

In the case of a bipolar supply, that's the same as one of the two sides. Because the class B uses two transistors, peak compliance will be the same as cutoff.

$$Compliance_{peak} = V_{CEQ} = 0.5 \cdot Total\ DC\ Supply \qquad (9.2)$$

The maximum voltage rating of the transistors will occur when they are off. In that instance, if the opposite transistor is fully conducting it will have a negligible voltage across its collector-emitter. Consequently, the off-state transistor can see the entire power supply.

$$BV_{CEO} = Total\ DC\ Supply \qquad (9.3)$$

The saturation current is dictated by the compliance and the load. The only thing that limits AC current is the load. Therefore

$$i_{C(sat)} = \frac{Compliance_{peak}}{r_L} \qquad (9.4)$$

Based on that, we can say

$$P_{load(max)} = \frac{Compliance_{RMS}^2}{r_L} \qquad (9.5)$$

Before we go any further, there is an important item to note about the circuit of Figure 9.8 (and the variants we shall discuss). If you take another look at the circuit you will see that there is nothing in the collector-emitter line to limit DC current. Infact, if we were to plot the DC load alongside the AC load, we'd get something like Figure 9.9.

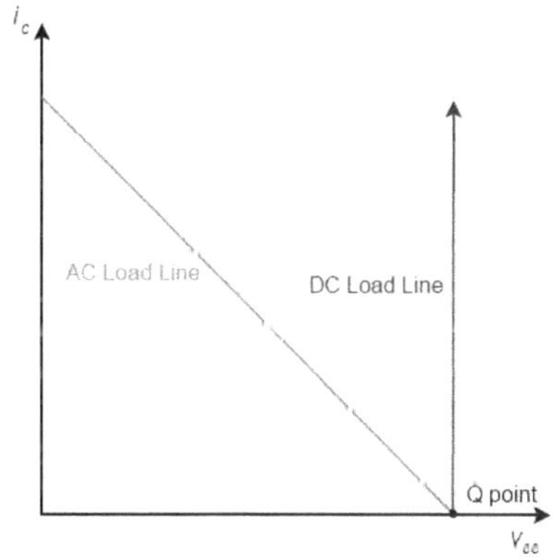

Figure 9.9 Comparison of AC and DC load lines for class B operation.

The DC load line goes straight up. There is no saturation current value short of infinity. What this really means is that if we're not careful with the bias, it is possible to destroy the transistors. The same is true if we accidentally short the load. A quick examination of Equations 9.4 and 9.5 shows that a shorted load condition would lead to huge currents and powers, and destroy the transistors in the process. With an audio amplifier, this could happen if one of the strands of loudspeaker wire unraveled and touched the adjacent lead. Obviously, not a happy situation unless you prefer the smell of burnt silicon over the sound of music. We will examine means of protecting the transistors from accidental overloads later in the chapter.

Class B Power Dissipation

Transistor power dissipation for the class B configuration is a bit trickier than it was for the class A. The idle (Q point) power dissipation is very low. It is found by multiplying ICQ by VCEQ. ICQ is generally set at a few percent of iC(sat) so it's obviously way below the maximum load power. Unlike the class A design, PDQ does not represent the worst case for class B.

To determine the worst case transistor power dissipation, we begin by describing the current and voltage waveforms during the conduction phase of the NPN. The collector current will appear as the positive half of a sine wave. The maximum case has the current starting at zero and peaking at *iC(sat)* (or alternately, *VCEQ/ rL*). For *vCE*, it starts at *VCEQ* and then swings down to zero as a negative half sine.

$$i_C = k\frac{V_{CEQ}}{r_L}\sin 2\pi ft \quad (9.6)$$

$$v_{CE} = V_{CEQ}(1-k\sin 2\pi ft) \quad (9.7)$$

Where $0 \le k \le 1$

Nothing says that the maximum load power case must cause the maximum transistor dissipation. In fact, we saw this wasn't the case for class A. Consequently, we will introduce a coefficient, *k*, that represents the percentage of the maximum current. We now arrive at our general equations for transistor current and voltage for the first half cycle.

For convenience, we'll set 2πf to 1. To get the power dissipation, we find the product of the transistor's current and voltage.

$$P_D = i_C v_{CE}$$

$$P_D = k\frac{V_{CEQ}}{r_L}\sin t \times V_{CEQ}(1-k\sin t)$$

$$P_D = \frac{V_{CEQ}^2}{r_L}k\sin t - \frac{V_{CEQ}^2}{r_L}k^2\sin^2 t$$

$$P_D = \frac{V_{CEQ}^2}{r_L}(k\sin t - k^2\sin^2 t)$$

Remove ugly $\sin_2$ term...

$$P_D = \frac{V_{CEQ}^{2}}{r_L}\left(k\sin t + \frac{k^2}{2}\cos 2t - \frac{k^2}{2}\right)$$

..and integrate to get:

$$P_D = \frac{V_{CEQ}^{2}}{r_L}\left(-k\cos t - \frac{k^2}{4}\sin 2t - \frac{k^2 t}{2}\right)\Bigg|_0^{\pi}$$

Note that $\frac{V_{CEQ}^{2}}{r_L} = 2P_{load(max)}$ This is a constant, so replace it to simplify and then evaluate the expression. Finally, divide by 2π to find the average over one full cycle.

$$P_D = \frac{2P_{load(max)}\left(2k - \frac{k^2\pi}{2} - \frac{k^2 t}{2}\right)}{2\pi}$$

$$P_D = 2P_{load(max)}\left(\frac{k}{\pi} - \frac{k^2}{4}\right) \qquad (9.8)$$

Equation 9.8 is the general case. For the worst case, we need the min/max k value. We will take the derivative of Equation 9.8 and then set it to zero to find the worst case value of k.

$$P_D = 2P_{load(max)}\left(\frac{k}{\pi} - \frac{k^2}{4}\right)$$

$$\frac{dP_D}{dk} = 2P_{load(max)}\left(\frac{1}{\pi} - \frac{k}{2}\right)$$

Worst case occurs at k = 2/π. This means that only 2/π, or 63.7%, of the load line is used at maximal heating of the transistor. 63.7% of the load line corresponds to a load power of about 40% of Pload(max) (i.e., 0.637^2). We now substitute this value back into Equation 9.8 to find the worst case PD.

$$P_D = 2P_{load(max)}\left(\frac{2/\pi}{\pi} - \frac{(2/\pi)^2}{4}\right)$$

$$P_D = \frac{2}{\pi^2}P_{load(max)} \approx \frac{P_{load(max)}}{5} \qquad (9.9)$$

The end result is that when the load is receiving about 40% of its maximum power, the transistors will be at their hottest and will be dissipating approximately 20% of the maximum load power (or about half of the power delivered to the load at that point). Thus, if a class B amplifier is rated to produce a maximum load power of 100 watts, the transistors will get their hottest when the load is receiving 40 watts, and each transistor will be dissipating 20 watts. The transistors will be dissipating *less* power when the load is at maximum. This is easily verified by substituting $k = 1$ into Equation 9.8. The result is a power dissipation of 13.7% of maximum load power, or 13.7 watts for the preceding example.

To help gain a deeper understanding of precisely what's happening here, the transistor waveforms are plotted in Figures 9.10 and 9.11. The maximum load power case is presented in Figure 9.10. Here, we see the full iC and vCE swings. Note that when iC is maximum, vCE is 0, hence the power is 0. In contrast, Figure 9.11 shows the worst case. Collector current peaks at just under 64% of maximum but vCE drops to only about 36% rather than 0% of its maximum. This results in a power dissipation curve with considerably greater area underneath it, indicating a higher average power.

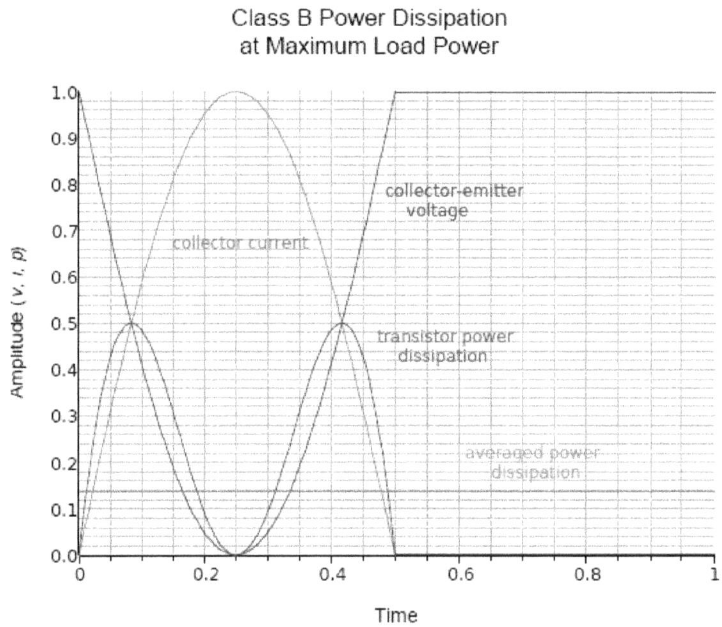

Figure 9.10 Class B transistor power dissipation at $P_{Load(max)}$.

Finally, it is worth remembering that the reactive load issues discussed for class A amplifiers still apply to class B amplifiers. Loads with a complex impedance may be harder to drive than the ideal purely resistive loads examined here. Therefore, we may have to over-rate the transistors above Pload(max) /5. As a side note, the AC load line for a class B amplifier with a reactive load will appear as an ellipse that has been cut in half (refer back to Figure 8.16 and imagine a horizontal cut line running through the Q point).

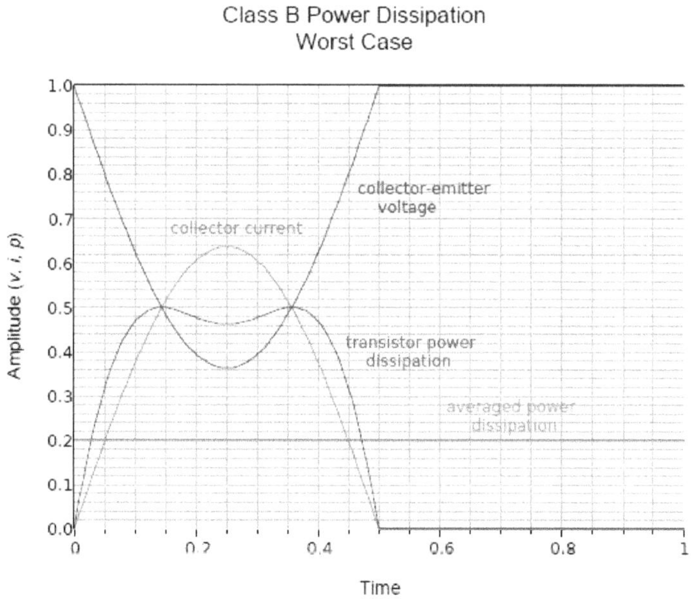

Figure 9.11 Class B transistor power dissipation, worst case.

Class B Efficiency

Efficiency is defined as useful output or load power versus supplied DC power.

$$\eta = \frac{P_{out}}{P_{in}} = \frac{P_{load(max)}}{P_{DC}}$$

This is dynamic for class B amplifiers. At *Pload(max)* the DC supply is delivering the full voltage of 2 *VCEQ* and the corresponding peak value of the current is *iC(sat)*.

$$i_{C(sat)} = \frac{V_{CEQ}}{r_L}$$

The average of this over a half-cycle is

$$i_{C(avg)} = \frac{1}{\pi} \times \frac{V_{CEQ}}{r_L}$$

Therefore, the supplied power must be

$$P_{DC} = 2V_{CEQ} i_{C(sat)}$$

$$P_{DC} = 2V_{CEQ} \times \frac{1}{\pi} \frac{V_{CEQ}}{r_L}$$

$$P_{DC} = \frac{2}{\pi} \times \frac{V_{CEQ}^2}{r_L}$$

As noted previously $\frac{V_{CEQ}^2}{r_L} = 2P_{load(max)}$ = 2 Pload (max) therefore

$$P_{DC} = \frac{4}{\pi} \times P_{load(max)}$$

Finally, substitute this expression back into the original definition for efficiency.

$$\eta_{max} = \frac{P_{load(max)}}{P_{DC}}$$

$$\eta_{max} = \frac{P_{load(max)}}{\frac{4}{\pi} P_{load(max)}}$$

$$\eta_{max} = \frac{\pi}{4}$$

$$\eta_{max} \approx 78.5\%$$

We find that the maximum theoretical efficiency of a class B amplifier is over three times that of a class A amplifier.

Time for an example.

Example 9.1

The amplifier shown in Figure 9.12 is driving a nominal 8 Ω loudspeaker. Determine the compliance, maximum load power and worst case transistor dissipation. Also estimate Zin assuming β = 50 and determine the transistor ratings for maximum current and BVCEO.

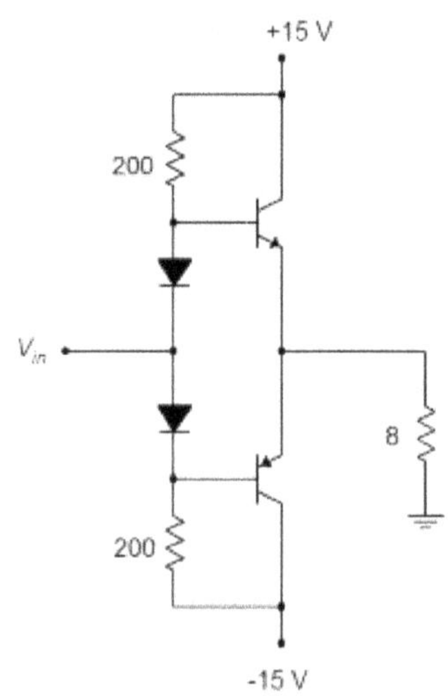

Figure 9.12 Schematic for Example 9.1.

By inspection, V_{CEQ} = 15 V. This is the peak compliance.

compliance = 15 V peak = 10.6 V RMS

Given the compliance, we can use power law to find the load power

$$P_{load(max)} = \frac{Compliance_{RMS}^{\ 2}}{R_L}$$

$$P_{load(max)} = \frac{(0.707 \times 15\,\text{V})^2}{8\,\Omega}$$

$$P_{load(max)} = 14\,\text{W}$$

That's not huge but it might be enough to irritate the neighbors. The transistors' worst case power dissipation is

$$P_D = \frac{P_{load(max)}}{5}$$

$$P_D = \frac{14\,\text{W}}{5}$$

$$P_D = 2.8\,\text{W}$$

The breakdown voltage is the entire supply so BV_{CEO} > 30 V. The maximum current through the transistors is the same as the maximum load current or $i_{C(sat)}$.

$$i_{C(sat)} = \frac{V_{CEQ}}{r_L}$$

$$i_{C(sat)} = \frac{15\,\text{V}}{8\,\Omega}$$

$$i_{C(sat)} = 1.88\,\text{A}$$

The input impedance is found in the usual manner but with a minor twist. $Z_{in(base)}$ is approximately equal to $\beta\, r_L$, or about 400 Ω. Only one transistor is on at any given time, though, so this is in parallel with the two 200 Ω biasing resistors but not the other $Z_{in(base)}$ (the "off" transistor has a very high input impedance because it is not conducting). This leaves a $Z_{in(base)}$ of about 80 Ω, or ten times the load impedance.

Computer Simulation

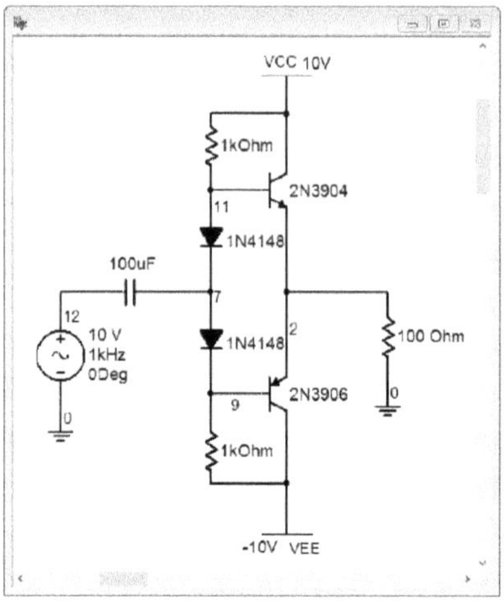

Figure 9.13 Class B amplifier in simulator.

To verify the basic operation of a class B amplifier, the circuit of Figure 9.13 is entered into a simulator. The amplifier should clip just below the ±10 volt power rails so a 10 volt peak source is used to verify this. Also, *Av* should be about 1.

The results of a transient analysis are shown in Figure 9.14. First, it is apparent that the voltage gain is approximately unity as the input and output waves are nearly coincident (with the exception of the clipped portion). Clipping occurs at about 8.5 volts. This is largely due to limiting from the biasing diodes. Once the input gets close in value to the power supply, the diodes become reverse-biased and the signal does not make it to the base. Thus, the output clips prematurely.

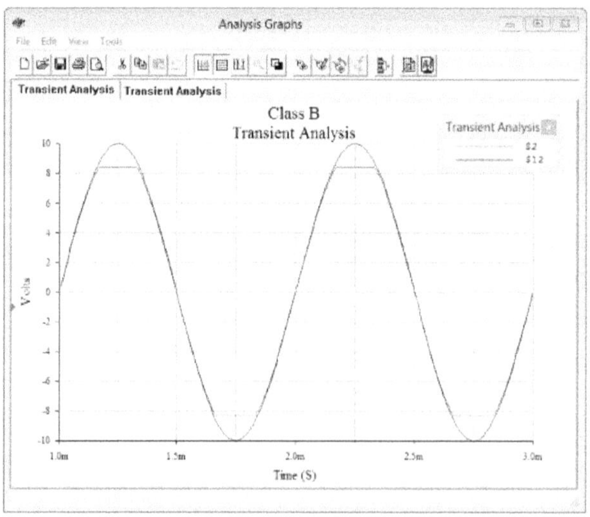

Figure 9.14 Class B amplifier transient analysis.

To verify earlier commentary regarding the biasing diodes, the circuit is modified so that the diodes are shorted out and the transient analysis is run again. The results are shown in Figure 9.15.

Two things should be apparent in the new simulation. First, the output waveform is suffering from obvious notch distortion. Therefore, we can be pleased that the biasing diodes have done their job to reduce this effect. The second item involves the compliance. The new output is not clipped. Granted, the lack of biasing diodes has reduced the signal by about 0.7 volts, but a careful examination of the output waveform shows that it has reached over 9 volts peak without clipping. In fact, if we increase the input to 12 volts peak, as in Figure 9.16, we can see that clipping occurs just under the power rails. Thus, the premature clipping is due to the diodes and their interaction with the power supplies and surrounding resistors. In short, we lose a little bit of compliance due to the diodes but that's much better than getting notch distortion.

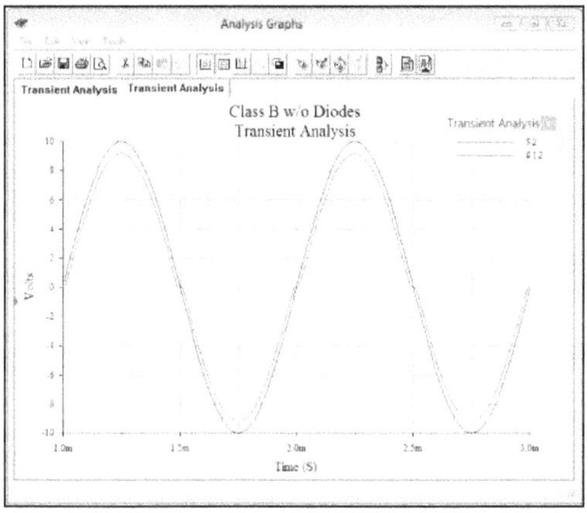

Figure 9.15 Class B amplifier transient analysis, without biasing diodes.

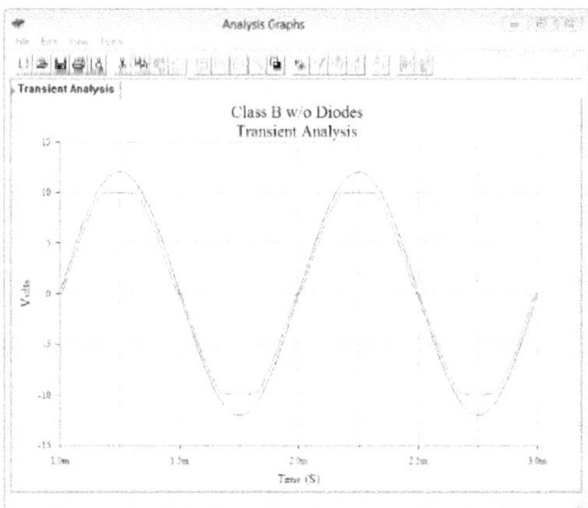

Figure 9.16 Class B amplifier transient analysis, without biasing diodes and showing clipping.

Direct Coupled Driver

The circuits we have examined offer both current gain and power gain but not voltage gain. In order to increase the signal voltage, some preceding voltage gain stages will most likely be needed. These stages can be connected to the previous class B follower circuits with coupling capacitors but this is not the most effective method. A more common technique is the use of a direct coupled driver.

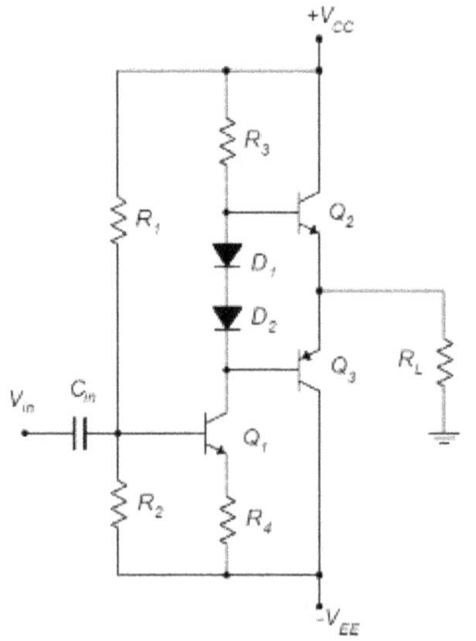

Figure 9.17 Class B amplifier with direct coupled driver.

A class B follower with a direct coupled driver stage is shown in Figure 9.17. What we've done here is combined an ordinary class A common emitter amplifier (in this case, using voltage divider bias) with a class B follower. The follower is positioned where the common emitter stage's collector resistor would be normally. This eliminates three components: the collector resistor, the inter stage coupling capacitor, and the lower base biasing resistor of the class B output stage. The removal of the resistors raises the effective load resistance for the first stage, thus producing a higher voltage gain from the driver stage.

Biasing the direct coupled driver is not difficult if we remember one thing: the DC voltage across R3 must be equal to approximately VCC – 0.7 V. If this is not the case, the class B stage will not be symmetrical, or in other words, VCEQ2 ≠ VCEQ3, and the final output will not be sitting at 0 VDC as it needs to. Knowing the value of R3 and its voltage, we can determine its current. This current flows down into Q1 as ICQ1.Knowing ICQ1, we can determine the voltage across R4 and eventually determine the appropriate divider ratio for R1 and R2 to achieve this value.

Example 9.2

Using the two-stage amplifier of Figure 9.17, first determine values for R1 and R2 to obtain proper system bias. Determine the output compliance, maximum load power and worst case transistor dissipation. Also estimate Av. Assume β = 50 for the output devices and 100 for the first stage. VCC = 20 V, VEE = -20 V, RL = 16 Ω, R3 = 560 Ω, R4 = 75 Ω.

For the output section (assuming it will be biased properly), by inspection, VCEQ = 20 V. This is the peak compliance.

compliance = 20 V peak = 14.1 V RMS

Given the compliance, we can use power law to find the load power

$$P_{load(max)} = \frac{Compliance_{RMS}^2}{R_L}$$

$$P_{load(max)} = \frac{(14.1\text{ V})^2}{16\,\Omega}$$

$$P_{load(max)} = 12.5\text{ W}$$

The transistors' worst case power dissipation is

$$P_D = \frac{P_{load(max)}}{5}$$

$$P_D = \frac{12.5\text{ W}}{5}$$

$$P_D = 2.5\text{ W}$$

To determine the biasing resistors, we start with *R3*. To achieve bias symmetry, all of *VCC* drops across *R3* with the exception of 0.7 volts for *D1*. This is the same as *ICQ1*.

$$I_{CQ_1} = \frac{V_{CC} - 0.7\text{ V}}{R_3}$$

$$I_{CQ_1} = \frac{19.3\text{ V}}{560\,\Omega}$$

$$I_{CQ_1} = 34.5\text{ mA}$$

The drop across R4 is found via Ohm's law

$$V_{R_4} = I_{CQ_1} R_4$$
$$V_{R_4} = 34.5\,\text{mA} \times 75\,\Omega$$
$$V_{R_4} = 2.6\,\text{V}$$

This implies that the voltage across R2 must be 0.7 volts more, or 3.3 volts. If we ignore the base current of Q1, then the ratio of R1 to R2 must be the same as the ratio of their voltages, 36.7 to 3.3, or 11.1 to 1. In other words, R1 must be 11.1 times larger than R2. For good bias stability we don't wish to set R2 too much larger than R4. If we set it to 200 Ω, for example, then R1 would need to be about 2.2 k Ω. Due to component tolerances, one of these resistors would need to be a potentiometer (connected as a rheostat) in order to "tweak" the final output to 0 VDC. A resistor/pot combo might be even better as it won't be so "touchy". For example, the 2.2 k Ω could be replaced with a series combination of a 1.8 k Ω and a 1 k Ω pot. This would be much easier to adjust than if the resistor was replaced with a standard 5 k Ω pot.

Now for the system voltage gain. The gain of the follower is approximately one so we need only concern ourselves with the first stage common emitter amplifier. This amplifier is swamped by R4 and given that the collector current is over 34 mA, r'e will be less than an ohm and can be ignored. All we need to do is find the effective load at the collector of Q1. This is R3 in parallel with a single Zin(base) (remember that only one transistor is on at any given time and the off transistor will appear as a high impedance).

$$Z_{in(base)} = \beta r_E$$
$$Z_{in(base)} = 100 \times 75\,\Omega$$
$$Z_{in(base)} = 7500\,\Omega$$

$$A_v = -\frac{r_L}{r_E}$$
$$A_v = -\frac{7500\,\Omega \,||\, 560\,\Omega}{75\,\Omega}$$
$$A_v = -6.95$$

Before leaving this section, there are a few items to note. First, the load power calculations have assumed that the entire power supply can be used by the output devices. As we saw with the diode bias version, this is not always the case. There is another situation that can limit the output swing. The class B stage is a follower and thus has a voltage gain of one. It the driver stage can't produce the full swing then the output stage can't either. Consequently, a class A analysis (i.e., AC load line) needs to be performed on the driver in order to determine just how large the signal can be before clipping. Indeed, it's quite likely that the driver stage will clip before the output stage.

Also, the direct coupled driver does not have to be an NPN as depicted in Figure 9.17. A PNP can be used instead, it simply needs to be shifted to the upper section rather than the lower section, as shown in Figure 9.18.

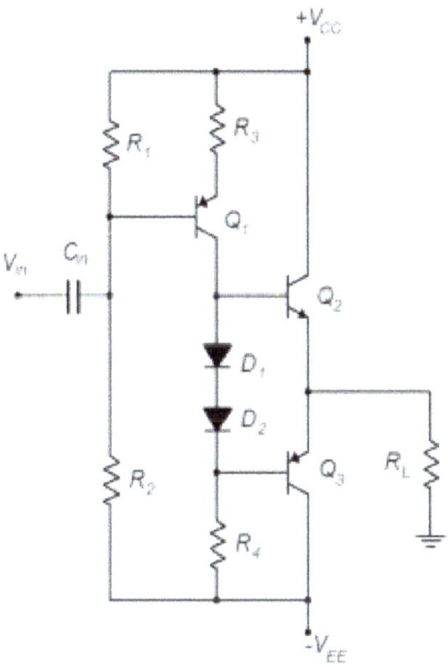

Figure 9.18 Class B amplifier with direct coupled driver, PNP version.

9.3 EXTENSIONS AND REFINEMENTS

The foregoing discussion has covered the basics of class B amplifier design and operation but there are a variety of things we might add.

Current Limiting

One of the more useful additions to an amplifier is some form of protection circuit. As noted earlier, there is nothing in the basic class B amplifier to limit current, so if the load is accidentally shorted, the transistor current will spike to very high levels and possibly damage the output transistors. How might we prevent this from happening?

Perhaps the most basic protection technique is to place a fuse in-line with the load. This is a somewhat tricky proposition because voice and music waveforms are very dynamic. Fuses are not fast response devices and selection of the proper current rating is a trade-off. If we want to catch large, fast transients, the fuse will have to be rated on the low side. Unfortunately, the fuse might then blow on moderately loud sustained low frequency content. Relays suffer from similar problems. Also, there is the issue that if the fuses are replaceable by the consumer, they may use the wrong value or some other item35that will allow too much current, resulting in blown output transistors. This can be avoided by having a second set of fuses mounted on the PC board, connected to the power rails, but this is also an imperfect solution.

One means of dealing with this situation is to employ an active current limiter, such as the circuit shown in Figure 9.19. In this circuit, Q1 is the main NPN output transistor, the PNP side not shown. At the top is the power supply connection and the load is connected off to the right. A small resistor, RE, is inserted into the emitter current path. The resistance is small enough that the voltage across it is normally less than 0.5 volts or so. Across the resistor is another transistor, Q2. Under normal operation Q2 is not on and does not affect the circuit. If the load current gets large enough (beyond the safe limit), the voltage drop across RE will reach 0.7 V. At this point Q2 turns on and begins to conduct current away from the base of Q1, limiting the amount of load current to approximately 0.7 V/RE. Unlike a fuse, once the fault is removed and the load current falls to safe levels, Q2 disengages and normal amplifier function resumes.

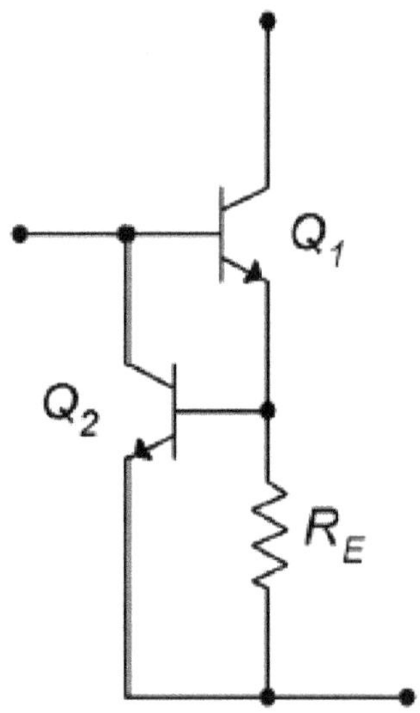

Figure 9.19 Active current limiter.

Two of these circuits are required for the amplifier; one for the positive half and one for the negative half. Figure 9.20 shows a class B amplifier with the added current protection circuits (within the dashed red box).

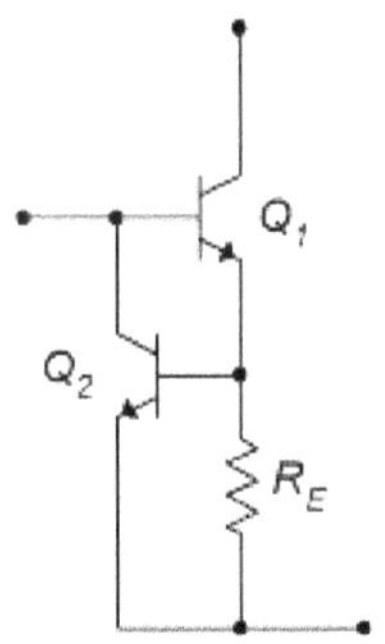

Figure 9.20 Class B amplifier with currentlimiters.

High Current Gain Configurations

For some applications the β of typical power transistors may prove to be insufficient. In such cases we can use high current gain configurations. There are several ways to configure the output devices. A Darlington scheme is shown in Figure 9.21. Q3 and Q5 are the main output devices. Q2 and Q4 can be thought of as drive transistors.

Although their BVCEO rating will need to be as high as that of Q3/Q5, they will be handling less current and therefore will dissipate less power. Sometimes Q2/Q4 are configured as seen here and sometimes emitter resistors may be added so that the output appears to be a cascade of emitter followers. Either way, there are now four base-emitter junctions to be compensated for, thus the addition of D3 and D4.

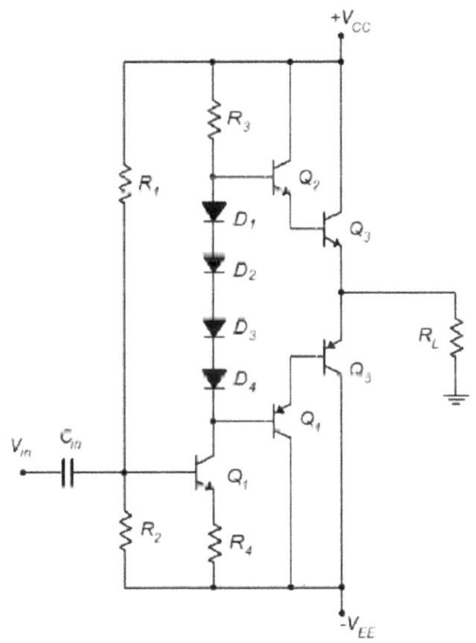

Figure 9.21 Darlington output devices for high current gain.

Some years ago, high power PNP transistors exhibiting decent audio quality were not available. Instead of using a PNP Darlington configuration, a Sziklai pair was used. This dual transistor configuration is named after Hungarian, and later, American, engineer George Sziklai. The Sziklai pair is also known as a composite pair. A composite PNP is shown in Figure 9.22. The operation is similar to that of a Darlington pair. The input transistor, Q1, drives its collector current into the base of Q2, the output transistor. Q2's base current is multiplied by its β, thus its collector current is the product of IB1, β1 and β2, just like a Darlington. The differences are that the main power device for the composite PNP is an NPN, and that there is only a single VBE to compensate for. An example amplifier based on the earlier direct coupled driver circuit is shown in Figure 9.23. This configuration is known as aquasi-complementary output36. Note the use of three compensating diodes; two for the Darlington NPN and one for the composite PNP/Sziklai pair. One advantage here is that the power devices, Q3 and Q5, can be identical models.

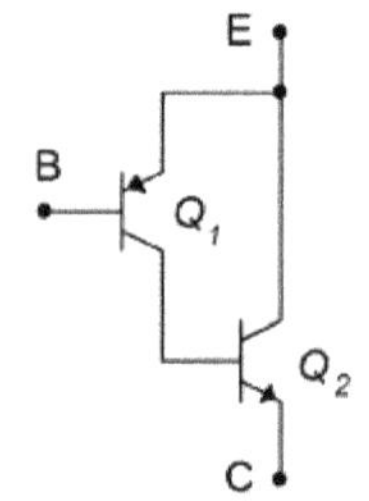

Figure 9.22 Sziklai pair, AKA composite PNP.

The current limiting circuit of Figure 9.19 can be added to both the Darlington and quasi-complementary output stages. Current limiting can also be added to the circuits presented in the next section, but have been left off for clarity.

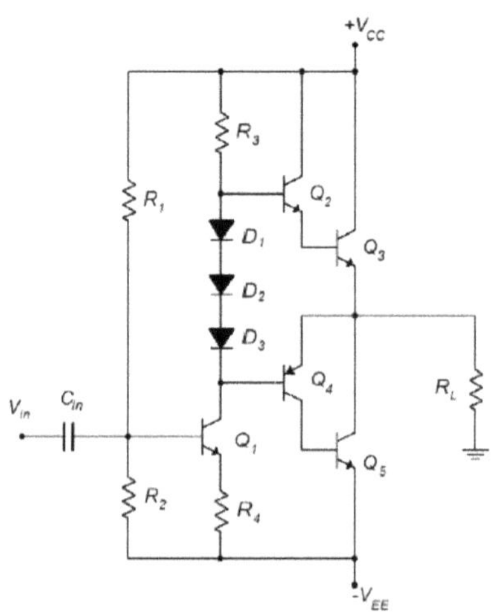

Figure 9.23 Quasi-complementary output.

Current Sharing

For higher output powers, and especially for amplifiers driving very low load impedances such 1 or 2 ohms, it may not be possible to use a single NPN and PNP for the output stage. Instead, the duties of each device will be shared among two or three transistors operating in parallel. Generally, current-sharing schemes will also require a Darlington-based approach due to the very high total load current.

Initially, it may be tempting to just place two or three transistors in parallel, each base terminal being fed by a common drive transistor. There is a fatal flaw with this approach. The issue is referred to as either thermal runaway or current hogging. The basic problem is that the temperature coefficient of trans conductance for a bipolar transistor is positive. In other words, r'e gets smaller as temperature rises. This means that, as the device heats up, it tends to more easily conduct current. Of course, if the device draws more current, it will dissipate more power, which means that it will get hotter, which means that it will conduct more current, which means that it will dissipate more power, and so on. This process spirals out of control and is particularly evil when we have multiple devices in parallel. Devices are never perfectly matched so the load current will never be perfectly shared among the devices. This means that one device will tend to get hotter than the other(s) which leads to it getting even hotter and grabbing a larger and larger share of the total current. Eventually, this device "hogs" all of the available current and winds up destroying itself. What we have is a positive thermal feedback loop. Transistor hara- kiri is a decidedly sub-optimal performance characteristic. To circumvent this problem we can add small resistors to the emitters of the devices. As all of the paralleled devices are being driven from a common drive transistor, they all see the same base voltage. If one output transistor starts to grab a larger share of the load current, the voltage drop across the emitter resistor will increase, thus forcing a reduction in that transistor's base-emitter voltage. This reduction compensates for the initial tendency of the current to increase. This technique is referred to as local negative feedback. We have seen this concept before, for example, a swamping resistor falls into this category as does the collector feedback bias scheme.

An example of a current-sharing output section is shown in Figure 9.24. This version uses a Darlington scheme. The newly added transistors are shown in red while the thermal/current feedback resistors are shown in blue.

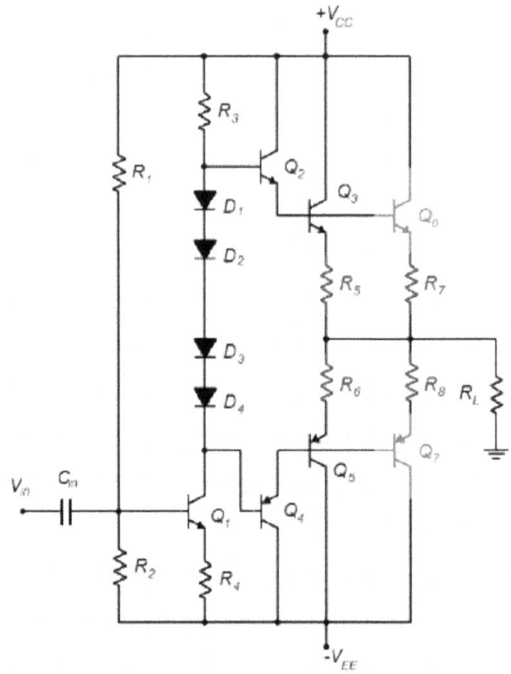

Figure 9.24 Current-sharing output section.

VBE Multiplier and Miller Capacitor

With the ever increasing complexity of the output section, the simple diode biasing scheme lacks the ability to set the optimal idle bias to achieve minimum crossover distortion. A more flexible approach is to use a VBE multiplier, as illustrated in Figure 9.25.

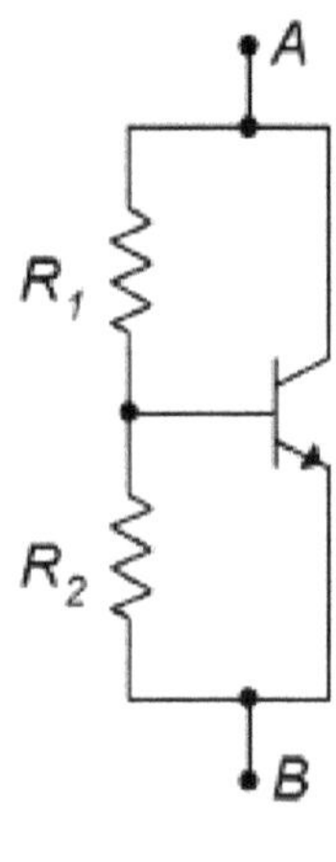

Figure 9.25 VBE multiplier

If we ignore base current, the currents through the two resistors are identical. Therefore, their voltages have the same ratio as their resistances. R2 is in parallel with VBE so its voltage must be equal to VBE. Consequently, the voltage across R2 must be a multiple of VBE. For example, if we want to generate the equivalent of four base-emitter drops from point A to point B, we make R1 three times as large as R2.

What makes this particularly useful as that we can set any ratio we want, and by replacing either resistor with a potentiometer, we can make that voltage adjustable. And example is shown in Figure 9.26, using the current-sharing amplifier of Figure

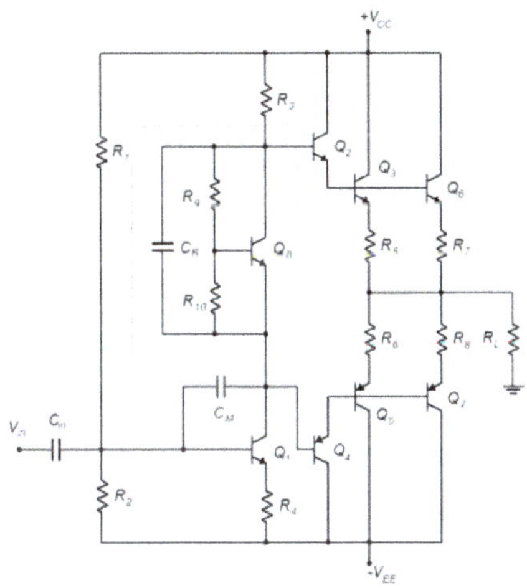

Figure 9.26 Amplifier with VBE multiplier and Miller compensation capacitor.

The circuit of Figure 9.26 also includes another capacitor, *CM*. This is a Miller compensation capacitor. It uses the Miller Effect (see Chapter 6, Miller's Theorem) to create a much larger equivalent input capacitance. This capacitance appears in parallel with the input of *Q1* and creates a lag network. The inclusion of *CM* allows the designer to tailor the high frequency response of the amplifier.

Bridging

Some amplifiers employ a bridged output scheme. This is particularly true where power supply voltages are limited, for example, in automobiles (nominally 12 VDC but closer to 13.8 VDC from the alternator). Without resorting to an expensive DC- to-DC converter, an amplifier designed for automotive use is in a bit of a bind. If we assume a +12 volt DC source, then a basic class B amplifier would use a capacitor coupled configuration like Figure 9.6. This would yield a VCEQ of 6 volts and a compliance of about 4.2 volts RMS. If we were to use a standard home loudspeaker of 8 Ω, the maximum load power would be a mere 2.25 watts (this is one reason why car audio systems typically use 4 Ω loudspeakers, because it doubles the output power, in this case to 4.5 watts). A bridged output can double this again.

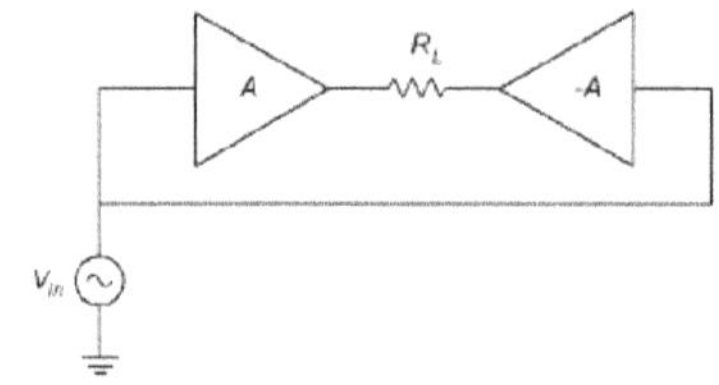

Figure 9.27 Block diagram of bridged amplifier.

A block diagram of a bridged drive is shown in Figure 9.27. The left half is what we would see normally. On the right half we have a second identical amplifier but we drive it with an inverted copy of the original signal. What ends up happening is that, as the left output goes positive, the right output goes negative by the same amount. The end result is that the load sees twice the voltage it would have from the left amplifier alone. Power varies as the square of voltage so doubling the voltage quadruples the power. The 2.25 watt output of that car amplifier jumps up to 9 watts37. If we want to go higher than this our only options are to further decrease the loudspeaker impedance (there are practical limits that will not let us go much further) or increase the DC power supply. In the home, increasing the supply is relatively easy as we have an AC power source. In an automotive system this is a much more expensive proposition because only DC is generally available.

An illustration of a bridged system at the transistor level is shown in Figure 9.28. For simplicity, this schematic uses basic class B stages. Input and right-side inversion circuitry are not shown.

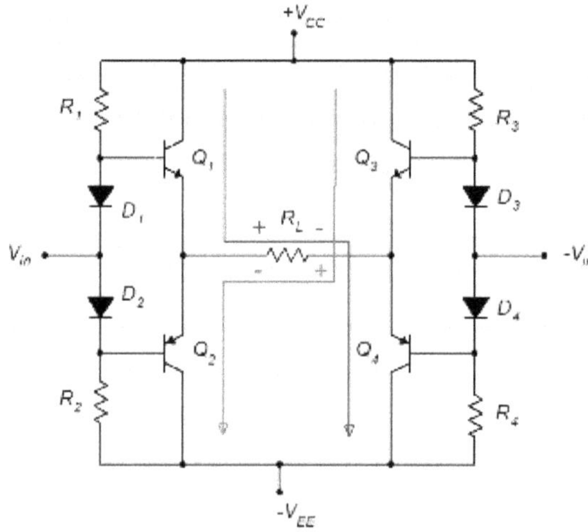

Figure 9.28 Bridged amplifier, transistor level.

The four transistors create a classic H bridge with the load in the center. For a positive Vin, Q1 turns on. At the same time, because it is being fed by an inverted version of Vin, Q4 turns on. Thus, current flows down from VCC, through Q1, across RL from left-to-right, down Q4, and finally to VEE (blue trace). At full swing, the voltage across the load will be VCC - VEE from left to right. For a negative Vin the opposite happens: Q3 and Q2 turn on, allowing current to flow through RL from right left. The maximum voltage will be the same but with inverted polarity, hence an effective doubling of load voltage and a quadrupling of load power.

This increased output does not come without downsides. The most obvious problem is a doubling of the output circuitry. The second issue may not be immediately apparent: the load is not grounded. In simple automobile audio systems, the chassis of the vehicle is used as the system common or "ground". Therefore, it is possible to just run a single wire to a loudspeaker from an amplifier's output. The other loudspeaker terminal is then connected to the chassis at a convenient location. With a bridged output, two wires must be run to the loudspeaker. If a single-lead loudspeaker system is upgraded with a bridged output amplifier, new loudspeaker wires must be installed. Otherwise, the chassis return lead winds up shorting out one side of the bridge38.

Power Supply Bypassing

A final comment regarding high power amplifiers, particularly audio amplifiers, is in order. All through this chapter we have assumed that the DC power supplies will act ideally, mainly that they will present themselves as good AC grounds. This is not always easy to achieve given the practical limitations of PC board layouts, wiring constraints and so forth. Consequently, power supply bypass capacitors are often used to ensure a good AC ground. While this was mentioned in Chapter 7 regarding small signal amplifiers, it is perhaps more important for power amplifiers. The power supply bypass capacitors used for power amplifiers tend to be larger and the quality more stringent. A simple 1 μF bypass capacitor would hardly be the norm for a high power audio amplifier. The practical problem is that large, high quality capacitors are not inexpensive. For example, a 10 μF polypropylene capacitor will be at least an order of magnitude more expensive than a similar size aluminum electrolytic (dollars versus cents). Unfortunately, the electrolytic will have higher leakage and ESR, and will not behave nearly as well at high frequencies (indeed, the impedance will actually start to increase due to inductive effects once the frequency gets high enough). To both improve performance and save money, bypass capacitors are sometimes doubled or tripled. For example, a large aluminum electrolytic might be placed in parallel with a much smaller polyester or polypropylene capacitor. The aluminum electrolytic will give the small XC needed at lower frequencies, and when it starts to behave less ideally at higher frequencies, the higher quality poly capacitor effectively shunts it and extends the operating range. The result is nearly as effective as the single large, high quality capacitor but much less expensive.

CONCISE SUMMARY

- Class B operation is defined as having transistor collector current active for 180° out of the waveform cycle. In order to amplify the entire 360° in linear fashion, two devices are required. Each transistor of a complementary pair is biased at cutoff to achieve a 180° conduction angle. This means that no-signal collector current is zero, leading to very low power consumption at idle and, unlike class A amplifiers, dynamic power consumption. Unfortunately, pure class B operation also results in notch or crossover distortion as the switch over from one transistor to the other is not seamless. This can be mitigated by biasing the devices slightly "on", which is known as class AB operation. While simple resistor voltage dividers may be used for this purpose, a generally superior method uses diodes as they mimic the base-emitter current-voltage characteristic.
- The compliance of a class B amplifier is based on its power supplies. Ideally, the peak-to-peak compliance of the amplifier will equal the total power supply differential. Worst case power dissipation and efficiency are far superior when compared to class A topology: Device power dissipation is only one-fifth of maximum load power and the theoretical efficiency at maximum load power is78.5%. Maximum heating of the transistors occurs at approximately 40% of maximum load power.
- A direct coupled drive stage is often used as it reduces parts count and improves performance. Amplifiers can also be extended through the use of Darlington pairs and current sharing schemes. Other refinements include the use of active current limiting for device protection, and VBE multipliers in place of simple biasing diodes for greater flexibility.
- Bridging is a technique used to increase output power. It relies on driving a floating load from both sides. Two amplifiers are needed for this configuration but it can offer a quadrupling of load power for the same power supplies.

CHAPTER-10

OPERATIONS OF JUNCTION FIELD EFFECT TRANSISTORS (JFETS)

10.1 INTRODUCTION: AN OVERVIEW

Junction Field Effect Transistors (JFETs) are semiconductor devices that play a significant role in electronic circuitry, particularly in amplification and signal processing applications. Operating on the principle of controlling current flow through a semiconductor channel via an electric field, JFETs come in two main types: N-channel and P-channel. Unlike bipolar transistors, JFETs are voltage-controlled devices, making them well-suited for low-power and high-input impedance applications. JFETs find widespread use in amplifiers, oscillators, and analog switches due to their simplicity, reliability, and ease of integration into various electronic circuits. Understanding the operational characteristics and applications of JFETs is essential for engineers designing circuits that require high input impedance, low power consumption, and reliable signal amplification. The field effect transistor, or FET, is a semiconductor device that serves as an alternative to the bipolar junction transistor. FETs are available in two broad types: the junction FET, or JFET, and the metal oxide semiconductor FET, or MOSFET.

It is best not to think of FETs as either better or worse than the BJT. They have different characteristics and lend themselves to applications where BJT performance might be wanting. The inverse is also true and for some applications the judicious use of a combination of BJTs and FETs can produce superior performance when compared to either device used alone. Like the BJT's NPN and PNP variants, FETs comes in two "flavors": the N- channel type and the P-channel type. We shall cover JFETs first and then discuss MOSFETs in subsequent chapters. In this chapter we will cover the internal structure of the JFET, its theory of operation and biasing techniques. In the next chapter we shall discuss small signal JFET amplifiers; both voltage amplifiers and voltage followers.

The JFET is fundamentally different from a bipolar junction transistor. While the JFET, like the BJT, relies on the PN junction for operation, the JFET is modeled as a voltage-controlled current source while the BJT is modeled as a current-controlled current source. Further, the BJT relies on a forward-biased base-emitter junction for proper operation while the JFET achieves current control via a reverse-biased junction. Consequently, JFET biasing circuits tend to be incompatible with BJT biasing schemes and one device cannot be swapped out for the other.

10.2 JFET INTERNALS

A simplified internal model of a JFET is shown in Figure 10.1. The main portion of the device is called the *channel*. The diagram illustrates an N-channel device. The channel is built upon a *substrate* (i.e., base layer) of oppositely doped material.

Attached to the opposing ends of the channel are two terminals; the *source* and the *drain*. Embedded within the channel is a region using the opposite material type. A lead is attached to this as well and is called the *gate*. Although there is not perfect correspondence between them, the drain, source and gate are roughly analogous to the BJT's collector, emitter and base, respectively.

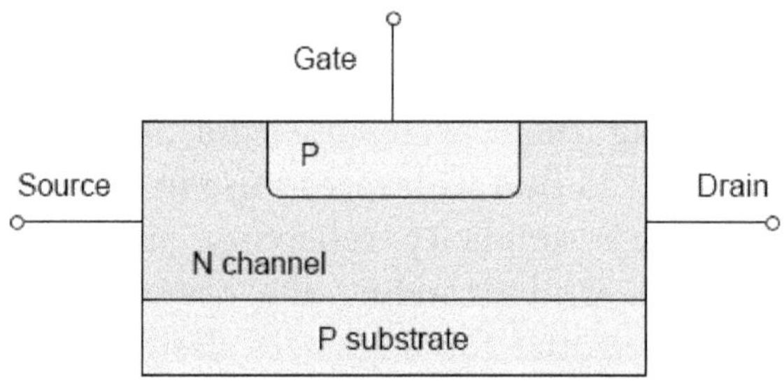

Figure 10.1 Simplified internal structure of P an N-channel JFET.

This diagram is drawn symmetrically. Some devices are designed in this fashion and their drain and source terminals can be swapped with no change in operation. This is not true for all devices, though. For small values of drain-source voltage, the channel exhibits a certain amount of resistance that is dependent on the doping level and physical layout of the device. Further, under normal operation, ID will equal IS.

To understand how the device behaves, refer to Figure 10.2. Here we shall consider electron flow (shown as a dashed line). First, a positive voltage, VDD, is attached to the drain terminal along with a current limiting resistor, RD. A negative supply, VGG, is applied to the gate terminal via resistor RG. Let's start with the gate supply set to zero. If we start VDD at zero, here is what happens as we increase its value. Initially, an increase in drain-source voltage will elicit a proportional increase in the current flowing through the channel. In other words, the channel acts like a resistor. As the voltage across the drain-source increases further, at some point the current will saturate, and no further increases in current will occur in spite of further increases in VDD and VDS. At this point the device is behaving as a constant current source. The drain-source voltage where this transition occurs is called the pinch-off voltage, Vp. If the drain-source voltage increases too much, breakdown will occur and current will begin to increase rapidly.

What's particularly interesting is what happens when the gate supply is increased in the negative direction. This reverse-biases the gate-source PN junction and results in a larger depletion region being formed. The depletion region widens into the channel, thus restricting current flow sooner and at a lower level. The more negative we make VGG, the lower ID becomes. Eventually, when VGG goes negative enough, the drain current will turn off. This voltage is called VGS(off) and it has the same magnitude as VP (i.e., VP = |VGS(off)|). The action can be thought of as operating like a water valve: turning the gate source voltage more negative is like turning off the spigot and decreasing the flow.

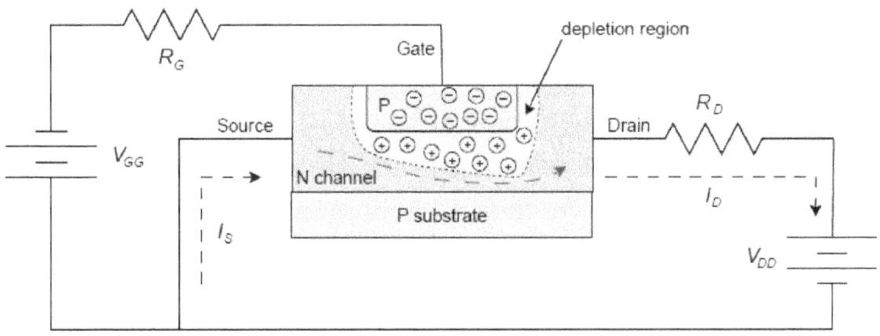

Figure 10.2 Electron flow in an N-channel JFET.

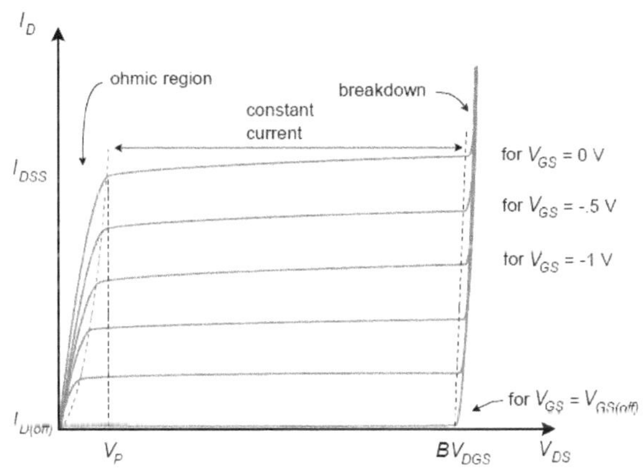

Figure 10.3 JFET drain curves.

The operation of the JFET can visualized nicely by plotting a set of drain curves, as shown in Figure 10.3.

The drain curve family plots drain current, *ID*, versus drain-source voltage, *VDS*. We begin with the top-most curve. This is generated by setting the gate-source voltage, *VGS*, to zero. We then cycle *VDS* from zero to some higher value. Initially, we see a proportional rise in *ID* as *VDS* increases. This is called the *ohmic* or *triode region*.

Eventually, the channel saturates and the current levels out. This is the constant current or saturation region and it occurs for VDS > VP. The breakdown voltage is called BVDGS, or alternately, V(BR)DS. Above this voltage the current increases rapidly. As usual, we do not wish to operate the device in this breakdown region.

If we now repeat the process but this time use a small negative value for VGS, we will trace out a curve of very similar shape. The transition to constant current mode will happen at a slightly lower voltage and the current value will be somewhat lower as well. This process continues in like fashion as we make VGS more and more negative. Eventually, when VGS = VGS(off), the drain current drops to virtually zero (in fact, a small leakage current flows called ID(off)). In contrast, if VGS was allowed to go positive, operation would be lost because the PN junction would become forward- biased and we would lose control of the current via the depletion region. This means that the JFET's current control is entirely in the second quadrant and the largest drain current flows when VGS = 0 V. This current is called IDSS, which stands for the drain current with a shorted gate-source (i.e, if it's shorted, then VGS = 0 V). The JFET cannot produce a continuous current larger than IDSS safely.

The characteristic equation relating drain current and gate-source voltage is shown below. This is valid for the constant current region (i.e., VDS > VP).

$$I_D = I_{DSS}\left(1 - \frac{V_{GS}}{V_{GS(off)}}\right)^2 \qquad (10.1)$$

Where

V_{GS} is the gate-source voltage ($V_{GS(off)} \le V_{GS} \le 0$),

I_D is the drain current,

I_{DSS} is the maximum current,

V_{GS}(off) is the turn-off voltage.

From this we see that the JFET is a square-law device rather than like the BJT which has a logarithmic characteristic.39 In essence, this curve is a portion of a parabola. This means that the JFET's characteristic curve is much more gradual in slope than that of a BJT. This will have important implications when it comes to voltage gain potential and distortion, as we shall see in the following chapter.

It is useful to remember that VGS(off) and IDSS are unique to a given device, rather like β is for a BJT. There can also be a fairly large variation in these parameters. For example, a particular model of JFET might show an IDSS variation between 2 mA and 20 mA, and a VGS(off) variation between -2 V and -8 V. Generally, the most negative VGS(off) values will be associated with the largest IDSS values.

Equation 10.1 is plotted in Figure 10.4. Compare this curve to the curve generated by the Shockley equation for BJTs, Figure 7.1. The graph is shown in normalized form. Instead of plotting for specific values of VGS(off) and IDSS, the axes are presented as fractional portions of the maximums (i.e., the horizontal axis is -VGS/VGS(off) and the vertical axis is ID/IDSS).

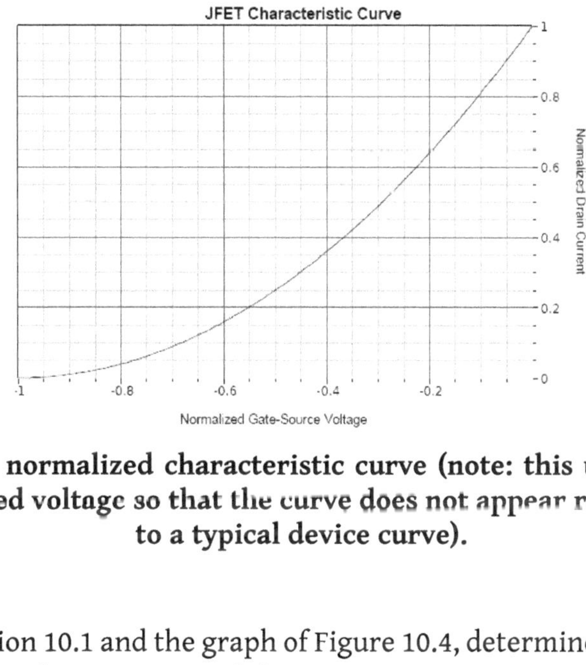

Figure 10.4 JFET normalized characteristic curve (note: this uses-VGS/VGS(off) for the normalized voltage so that the curve does not appear reversed compared to a typical device curve).

Example 10.1

Using both Equation 10.1 and the graph of Figure 10.4, determine the drain current if the gate-source voltage is -1 V and the JFET specs are IDSS = 8 mA and VGS(off) = -2 V.

$$I_D = I_{DSS}\left(1 - \frac{V_{GS}}{V_{GS(off)}}\right)^2$$

$$I_D = 8\,\text{mA}\left(1 - \frac{-1\,\text{V}}{-2\,\text{V}}\right)^2$$

$$I_D = 2\,\text{mA}$$

Using the graph, *VGS/VGS(off)* is 1 V/-2 V, or -0.5. Find this value on the horizontal axis, follow up to the curve and then across to the right vertical axis. The normalized drain current is 0.25, thus *ID* is 0.25 *IDSS*, or 2 mA.

As the characteristic curve plots output current versus input voltage, the slope of this represents the trans-conductance, an important characteristic for biasing and signal analysis. Device trans-conductance is denoted as gm, or alternately as gfs, and given units of siemens. We can derive an equation for trans-conductance by taking the derivative of Equation 10.1.

$$I_D = I_{DSS}\left(1 - \frac{V_{GS}}{V_{GS(off)}}\right)^2$$

$$\frac{dI_D}{dV_{GS}} = -\frac{2I_{DSS}}{V_{GS(off)}}\left(1 - \frac{V_{GS}}{V_{GS(off)}}\right)$$

The coefficient -2 IDSS/VGS(off) is defined as gm0, the trans-conductance when VGS = 0 V. This is the maximum trans-conductance of the device. Substituting, we arrive at

$$g_{m0} = -\frac{2I_{DSS}}{V_{GS(off)}} \tag{10.2}$$

$$g_m = g_{m0}\left(1 - \frac{V_{GS}}{V_{GS(off)}}\right) \tag{10.3}$$

A normalized plot of trans-conductance versus VGS is shown in Figure 10.5. The horizontal axis is -VGS/VGS(off) and the vertical axis is gm/gm0.

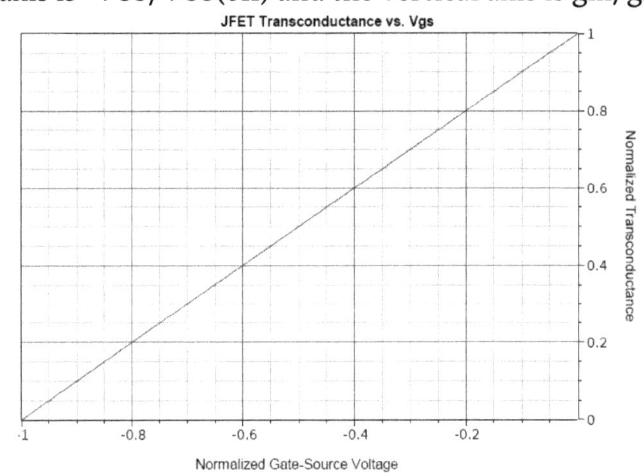

Figure 10.5 Curve of trans-conductance.

From this graph we see that the trans-conductance is a linear function.

Another item of interest regarding these device equations: If we combine Equations 10.1 and 10.3, we generate two equations that will prove useful in upcoming work.

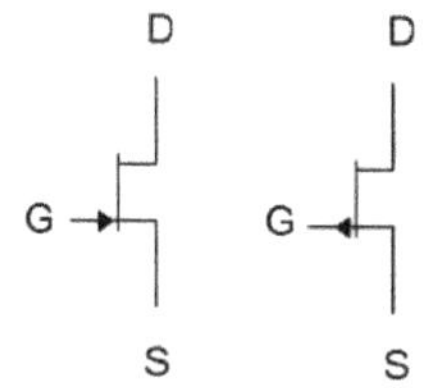

Figure 10.6 JFET schematic symbols: N-Channel (left)P-Channel (right)

$$\frac{g_m}{g_{m0}} = \sqrt{\frac{I_D}{I_{DSS}}} \tag{10.4}$$

$$\frac{I_D}{I_{DSS}} = \left(\frac{g_m}{g_{m0}}\right)^2 \tag{10.5}$$

Before moving on, the schematic symbols for JFETs are shown in Figure 10.6. The middle vertical line represents the channel, and as is usually the case, the arrow points to N material. Sometimes the gate arrow is draw in the middle rather than toward the source. Also, as is the case the BJT, sometimes these symbols are drawn within a circle.

10.3 JFET DATA SHEET INTERPRETATION

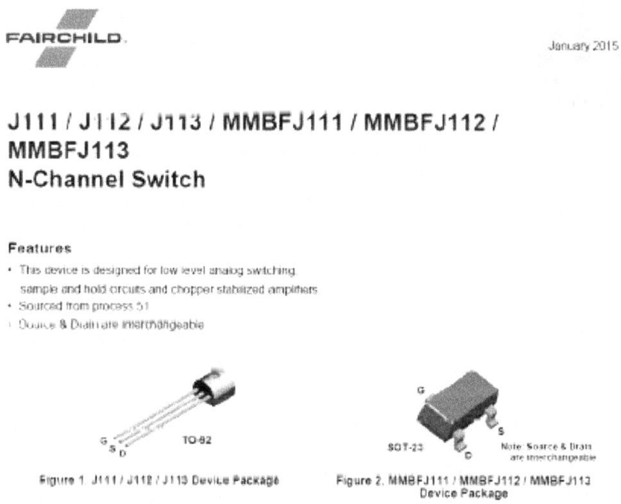
FAIRCHILD

January 2015

J111 / J112 / J113 / MMBFJ111 / MMBFJ112 / MMBFJ113
N-Channel Switch

Features
- This device is designed for low level analog switching sample and hold circuits and chopper stabilized amplifiers
- Sourced from process 51
- Source & Drain are interchangeable

Figure 1. J111 / J112 / J113 Device Package

Figure 2. MMBFJ111 / MMBFJ112 / MMBFJ113 Device Package

Figure 10.7a J111 series N-channel JFET datasheet. Used with permission from SCILLCdba ON Semiconductor.

A data sheet for the J111 series N-channel JFET is shown in Figure 10.7. This is a small signal device designed for audio frequency circuits. It is available in the common TO-92 through-hole package as well as in the surface mount SOT-23 package. Note that the source and drain are interchangeable for this device.

Absolute Maximum Ratings[1,2]

Stresses exceeding the absolute maximum ratings may damage the device. The device may not function or be operable above the recommended operating conditions and stressing the parts to these levels is not recommended. In addition, extended exposure to stresses above the recommended operating conditions may affect device reliability. The absolute maximum ratings are stress ratings only. Values are at T_A = 25°C unless otherwise noted.

Symbol	Parameter	Value	Unit
V_{DG}	Drain-Gate Voltage	35	V
V_{GS}	Gate-Source Voltage	-35	V
I_{GF}	Forward Gate Current	50	mA
T_J, T_{STG}	Operating and Storage Junction Temperature Range	-55 to 150	°C

Notes:

1. These ratings are based on a maximum junction temperature of 150°C.
2. These are steady-state limits. Fairchild Semiconductor should be consulted on applications involving pulsed or low-duty-cycle operations.

Thermal Characteristics

Values are at T_A = 25°C unless otherwise noted.

Symbol	Parameter	Max.		Unit
		J111 / J112 / J113[3]	MMBFJ111 / MMBFJ112 / MMBFJ113[4]	
P_D	Total Device Dissipation	625	350	mW
	Derate Above 25°C	5.0	2.8	mW/°C
$R_{\theta JC}$	Thermal Resistance, Junction-to-Case	125		°C/W
$R_{\theta JA}$	Thermal Resistance, Junction-to-Ambient	200	357	°C/W

Notes:

3. PCB size: FR-4, 76 mm x 114 mm x 1.57 mm (3.0 inch x 4.5 inch x 0.062 inch) with minimum land pattern size.
4. Device mounted on FR-4 PCB 36mm × 18mm × 1.5mm; mounting pad for the collector lead minimum 6cm^2.

Figure 10.7b J111 series N-channel JFET data sheet (cont).

Examining the absolute maximum ratings and thermal characteristics, we find values typical of small signal devices. Maximum drain-gate and gate-source voltages are 35 volts and the maximum power dissipation is 625 milliwatts.

Electrical Characteristics

Values are at T_A = 25°C unless otherwise noted.

Symbol	Parameter	Conditions		Min.	Max.	Unit
Off Characteristics						
$V_{(BR)GSS}$	Gate-Source Breakdown Voltage	I_G = -1.0 μA, V_{DS} = 0		-35		V
I_{GSS}	Gate Reverse Current	V_{GS} = -15 V, V_{DS} = 0			-1.0	nA
V_{GS}(off)	Gate-Source Cut-Off Voltage	V_{DS} = 15 V, I_D = 1.0 μA	111	-3.0	-10.0	V
			112	-1.0	-5.0	
			113	-0.5	-3.0	
I_D(off)	Drain Cutoff Leakage Current	V_{DS} = 5.0 V, V_{GS} = -10 V			1.0	nA
On Characteristics						
I_{DSS}	Zero-Gate Voltage Drain Current[5]	V_{DS} = 15 V, V_{GS} = 0	111	20		mA
			112	5.0		
			113	2.0		
r_{DS}(on)	Drain-Source On Resistance	V_{DS} ≤ 0.1 V, V_{GS} = 0	111		30	Ω
			112		50	
			113		100	

Figure 10.7c J111 series N-channel JFET data sheet (cont).

From the electrical characteristics, note the large variation in *VGS(off)*. For the J111, this runs from a minimum of -3 V to a maximum of -10 V. The J112 and J113 exhibit even wider min/max ratios. Also, note how the larger *VGS(off)* ranges are associated with larger maximums for *IDSS*.

Finally, let's take a look at a series of performance curves shown in Figure 10.7d.

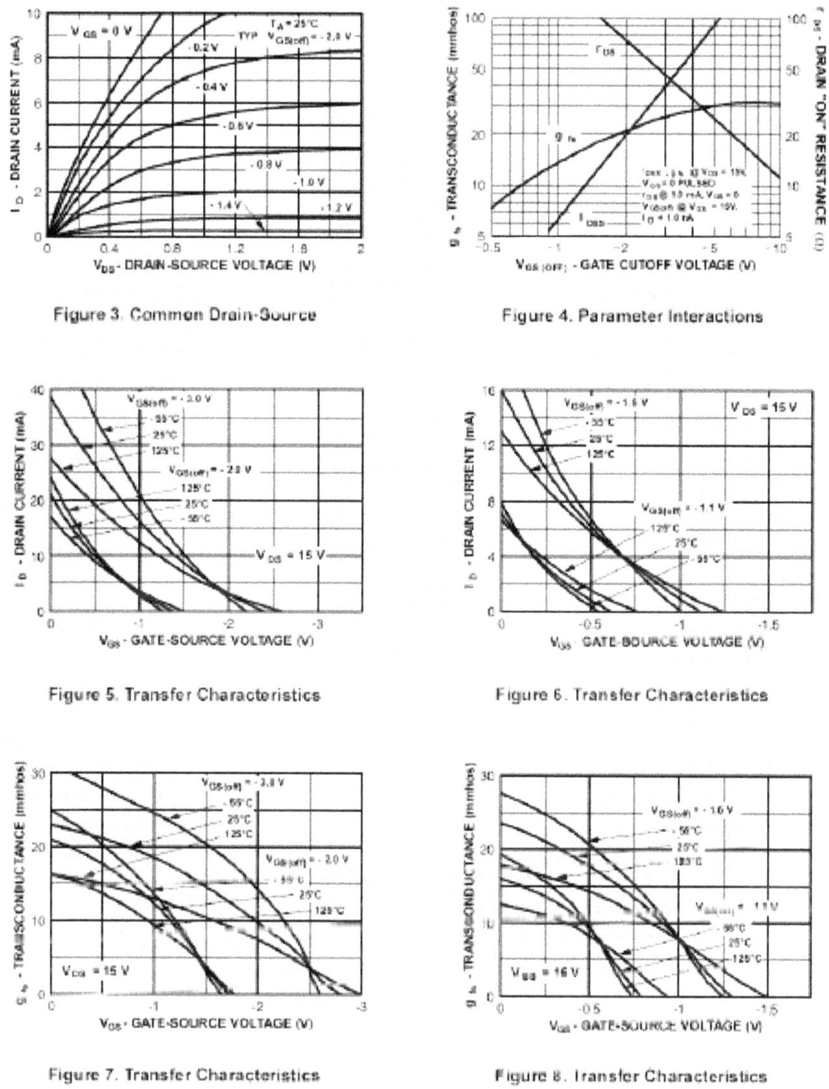

Figure 10.7d J111 series N-channel JFET data sheet (cont).

The upper-left graph is a family of drain curves and corresponds to Figure 10.3, presented earlier. Note that the horizontal scale for VDS only shows the first 2 volts. The point here is to examine the ohmic region. The two middle graphs plot the characteristic curve of the device and correspond to Figure 10.4, although these graphs are drawn rotated around the vertical axis (note that VGS is still shown as a negative value). Two important things may be noted here. First, as already mentioned, large values of IDSS tend to be associated with large values of VGS(off). This graph shows that individual plots tend to scale both horizontally and vertically away from the origin. Second, thermal variations are very much apparent: As the temperature increases, the characteristic curve tends to become less steep.

Finally, the two bottom-most graphs plot the variation of gm with VGS. These correspond to Figure 10.5, although again, the horizontal axis has been rotated around the vertical. Once again we see considerable variation due to temperature. Also, none of the plots exhibit perfect linearity. Further, at lower temperatures, the linearity of the plots decreases even more, warping a relatively straight line into a complex curve.

10.4 JFET BIASING

There are several different ways of biasing a JFET. For many configurations, *IDSS* and $V_{GS(off)}$ will be needed. A simple way to measure these parameters in the lab is shown in Figure 10.8. To measure I_{DSS} we simply ground the gate and source terminals as this forces *VGS* to be 0 V. We insert an ammeter between V_{DD} and the drain, and then set V_{DD} to a value higher than *VP* (+15 VDC generally being sufficient). The resulting ammeter reading is I_{DSS}. Obtaining $V_{GS(off)}$ is only slightly more work. Leaving the ammeter in the drain, unhook the gate from ground and instead connect it to an adjustable negative power supply. Turn the supply more negative until the ammeter reads zero (practically speaking, < 1% of I_{DSS}). At that point the voltage source will be equal to $V_{GS(off)}$.

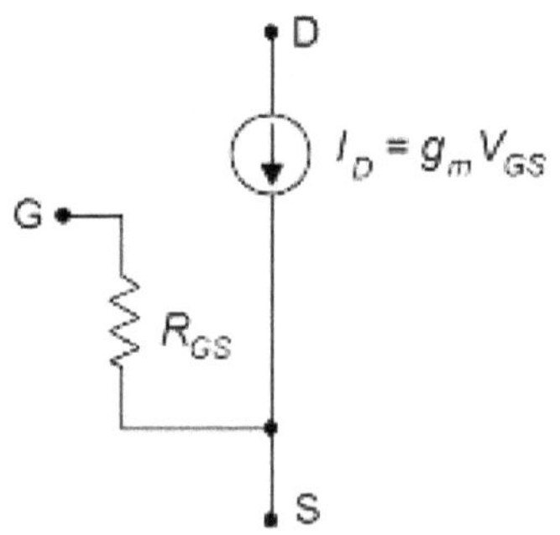

Figure 10.8 Measuring I_{DSS} and $V_{GS(off)}$.

DC Model

Before we begin examining the bias circuits themselves, we need a basic DC model of the JFET. A model sufficient for our analyses is shown in Figure 10.9.

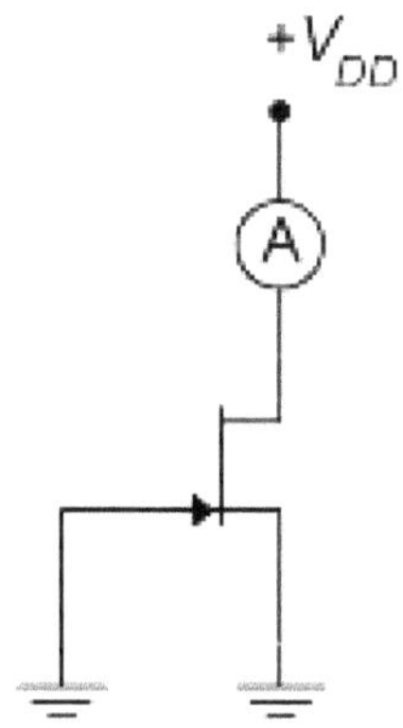

Figure 10.9 DC model of JFET.

The model consists of a voltage-controlled current source, ID, that is equal to the product of the gate-source voltage, VGS, and the trans-conductance, gm. The resistance between the gate and source, RGS, is that of the reverse-biased PN junction, in other words, ideally infinity for DC. As a consequence, in most practical circuits we can assume that gate current, IG, is zero. Therefore, ID = IS.

Constant Voltage Bias

The simplest form of bias is the *constant voltage bias.* The prototype is shown in Figure 10.10 with current directions and voltage polarities shown.

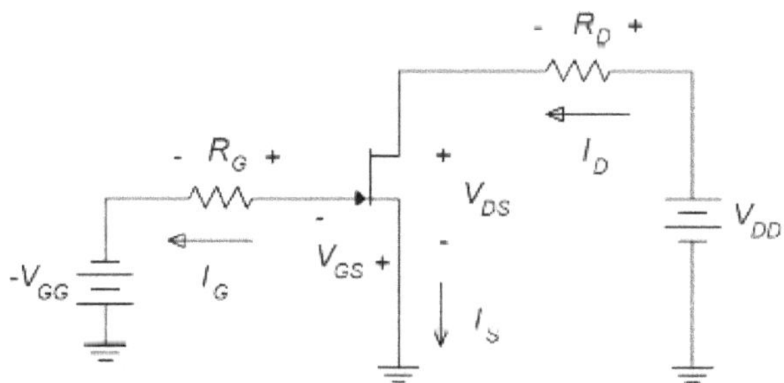

Figure 10.10 Constant voltage bias prototype.

This is a fairly straightforward design using only a couple of resistors and power sources. Figure 10.11 shows the same circuit but with the JFET model inserted, ready for analysis.

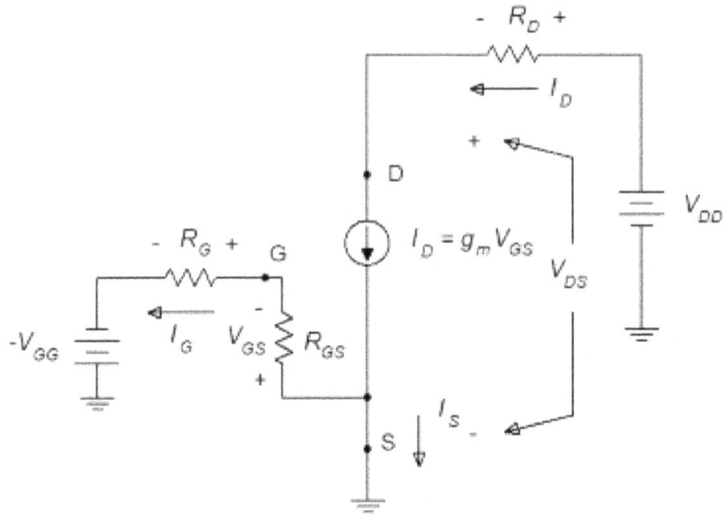

Figure 10.11 Constant voltage bias with model.

Ultimately, the goal here is to determine a means for finding the transistor's drain current and drain-source voltage, along with the potentials across any other components.

To begin, consider the gate-source loop. By KVL, the VGG source must drop across RG and the gate-source junction, $V_{GS.}$

$$V_{GG} = V_{R_G} + V_{GS}$$
$$V_{GG} = I_G R_G + V_{GS}$$

IG is approximately zero so this simplifies to

$$V_{GS} = V_{GG}$$

Given the trans-conductance, gm, we can find ID. Alternately, ID may be found using Equation 10.1 along with the device parameters IDSS and VGS(off). For this circuit, the latter technique tends to be more practical. Once ID is found, the voltage drop across RD may be found, and then VDS is determined from KVL.

Example 10.2

For the circuit of Figure 10.12, determine ID and VDS. Assume IDSS = 10 mA and VGS(off) = -5 V.

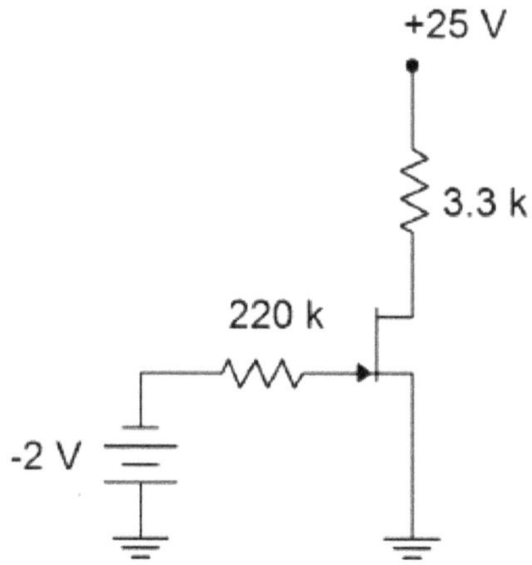

Figure 10.12 Schematic for Example 10.2.

First, because $IG \approx 0$, the drop across RG is ≈ 0 and $VGS = VGG$. Using Equation 10.1

$$I_D = I_{DSS}\left(1 - \frac{V_{GS}}{V_{GS(off)}}\right)^2$$

$$I_D = 10\,\text{mA}\left(1 - \frac{-2\,\text{V}}{-5\,\text{V}}\right)^2$$

$$I_D = 3.6\,\text{mA}$$

Looking at the drain-source loop, KVL shows

$$V_{DD} = I_D R_D + V_{DS}$$
$$V_{DS} = V_{DD} - I_D R_D$$
$$V_{DS} = 25\,\text{V} - 3.6\,\text{mA} \times 3.3\,\text{k}\,\Omega$$
$$V_{DS} = 13.1\,\text{V}$$

While the computation for the constant voltage bias is relatively simple, it does not exhibit a stable Q point. For example, if Example 10.2 is repeated with another JFET, this one with $IDSS = 12$ mA and $VGS(off) = -6$ V, the results are starkly different: ID grows to 5.33 mA and VDS shrinks to 7.4 V. These are considerable changes given the relatively modest shifts in the device parameters. In this regard, the constant voltage bias is reminiscent of the simple base bias configuration used with BJTs.

To get a better understanding of the Q point stability issue, refer to Figure 10.13.

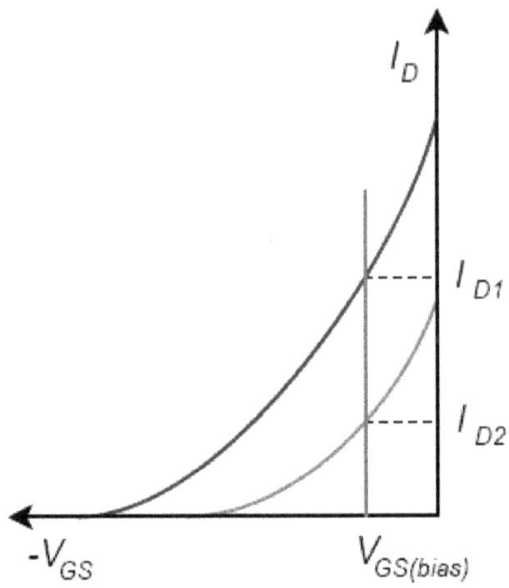

Figure 10.13 Variation for constant voltage bias.

Characteristic curves are plotted here for two different devices, one in green and one in blue. These represent the sort of device parameter variations we might expect to see across a product model. The fixed value of gate bias voltage is shown in red.

From this graph it should be obvious that this form of bias will produce a wide variation in drain current, and thus, is not a good choice for applications that require a stable Q point. If the application does not have this requirement, constant voltage bias offers the advantage of requiring a minimum of components.

Self Bias

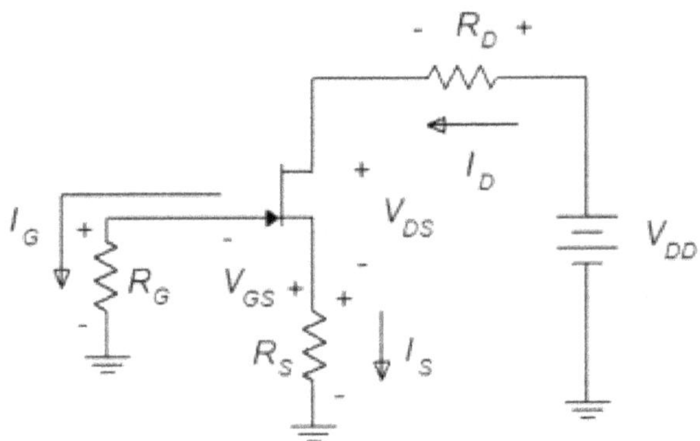

Figure 10.14 Self bias prototype.

Self bias uses a small number of components and only a single power supply, yet it offers better stability than constant voltage bias. The name comes from the fact that the drain current will be used to create a voltage drop that sets up the gate-source, hence the circuit "biases itself". It is also referred to as automatic bias. The self bias prototype is shown in Figure 10.14.

Once again, we may assume that *IG* is 0. As *RG* is connected directly to ground, this means that *VG* ≈ 0 V. This being true, inspection of the schematic reveals that the magnitude of *VGS* must be the same as the voltage across *RS*. Because *ID* = *IS* then

$$V_{GS} = -I_D R_S \qquad (10.6)$$

This value of *VGS* is what generates the drain current. The definition is self-referential. This being the case, how do we analyze the circuit? A proper derivation of the equation for drain current is not trivial. We start with the characteristic equation (Equation 10.1) and expand it.

$$I_D = I_{DSS}\left(1 - \frac{V_{GS}}{V_{GS(off)}}\right)^2$$

$$I_D = I_{DSS}\left(1 - \frac{2V_{GS}}{V_{GS(off)}} + \frac{V_{GS}^2}{V_{GS(off)}^2}\right)$$

$$I_D = I_{DSS} - \frac{2I_{DSS}V_{GS}}{V_{GS(off)}} + \frac{I_{DSS}V_{GS}^2}{V_{GS(off)}^2}$$

+ 2

Substitute using Equation 10.2

$$g_{m0} = -\frac{2I_{DSS}}{V_{GS(off)}}$$

$$I_D = I_{DSS} + g_{m0}V_{GS} + \frac{I_{DSS}V_{GS}^2}{V_{GS(off)}^2}$$

Using Equation 10.6 this can be expanded to

$$I_D = I_{DSS} - g_{m0} I_D R_S + \frac{I_{DSS} I_D^2 R_S^2}{V_{GS(off)}^2}$$

Rearranging yields

$$0 = \frac{I_{DSS} R_S^2}{V_{GS(off)}^2} I_D^2 \quad (1 + g_{m0} R_S) I_D + I_{DSS}$$

This is a quadratic equation in the form ax^2+bx+c and can be solved using the quadratic formula:

$$y = \frac{-b \pm \sqrt{b^2 - 4ac}}{2a}$$

The positive option in the numerator may be ignored as this occurs for *VGS* beyond *VGS(off)*. The result is

$$I_D = 2I_{DSS}\left(\frac{1 + g_{m0}R_S - \sqrt{1 + 2g_{m0}R_S}}{(g_{m0}R_S)^2}\right) \qquad (10.7)$$

Although this is an accurate analytical solution, it's certainly not the sort of equation most people want to memorize or derive as needed. As the gm0 RS term is repeated in this equation multiple times, it is useful to plot this equation in terms of normalized ID versus gm0 RS. This curve is plotted in Figure 10.15.

To use this curve, the first step is to find gm0 RS. The value of RS is determined by inspection and gm0 may be determined by Equation 10.2, repeated below for convenience.

$$g_{m0} = -\frac{2I_{DSS}}{V_{GS(off)}}$$

The value of gm0 RS is found on the horizontal axis, traced up to the curve and then over to the normalized ID ratio. This number is multiplied by IDSS to determine the value of ID.

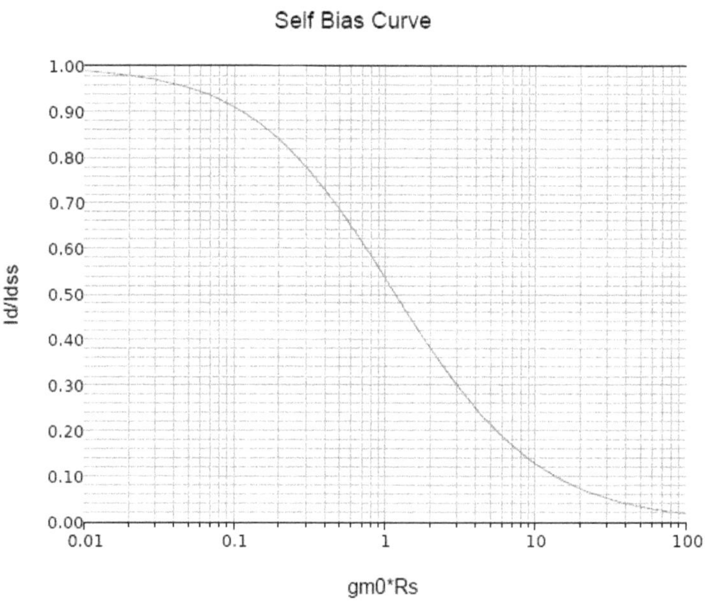

Figure 10.15 Self bias curve.

Example 10.3

Determine *ID* and *VDS* for the circuit shown in Figure 10.16. Assume I_{DSS} = 10 mA and $V_{GS(off)}$ = -4 V.

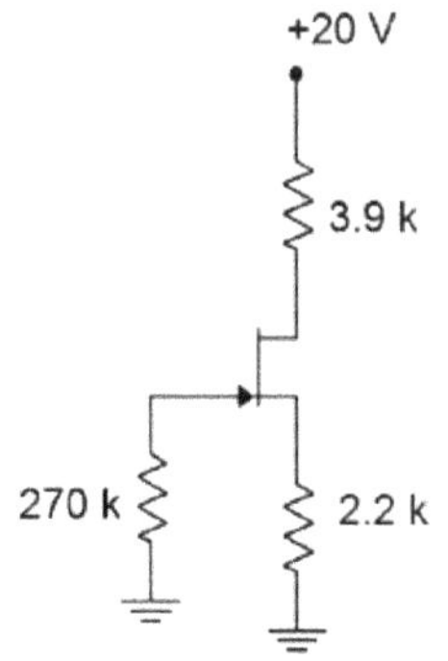

Figure 10.16 Schematic for Example 10.3.

Using the graphical method, first determine g_{m0} R_S.

$$g_{m0} = -\frac{2I_{DSS}}{V_{GS(off)}}$$

$$g_{m0} = -\frac{2 \times 10\,\text{mA}}{-4\,\text{V}}$$

$$g_{m0} = 5\,\text{mS}$$

Therefore *gm0 RS* = 5 mS · 2.2 k Ω = 11. The self bias graph yields approximately 0.12 for the normalized current ratio. Therefore

$$I_D = 0.12 I_{DSS}$$

$$I_D = 0.12 \times 10\,\text{mA}$$

$$I_D = 1.2\,\text{mA}$$

Using Ohm's law and KVL

$$V_D = V_{DD} - I_D R_D$$

$$V_D = 20\,\text{V} - 1.2\,\text{mA} \times 3.9\,\text{k}\Omega$$

$$V_D = 15.32\,\text{V}$$

$$V_S = I_D R_S$$

$$V_S = 1.2\,\text{mA} \times 2.2\,\text{k}\Omega$$

$$V_S = 2.64\,\text{V}$$

$$V_{DS} = V_D - V_S$$

$$V_{DS} = 15.32\,\text{V} - 2.64\,\text{V}$$

$$V_{DS} = 12.68\,\text{V}$$

An alternate technique is to make an initial guess for *VGS*, typically one half of *VGS(off)*. The value of *ID* is then computed from the characteristic equation (Equation 10.1) and compared with the Ohm's law relation, Equation 10.6, rewritten as

ID = -*VGS*/*RS*. Chances are, the two results will not agree so adjust the *VGS* estimate and repeat the process. If done properly, the currents should be closer. Iterate this process until you converge on the answer.

To use this technique for the preceding problem we'd start by assuming *VGS* = -2 V (half of *VGS(off)*). Using this in Equation 10.1 yields *ID* = 2.5 mA, while using Equation10.6 produces *ID* = 910 μA. Obviously the initial estimate was not correct. The second estimate for *VGS* needs to increase negatively as this will decrease the result from Equation 10.1 and increase the result from Equation 10.6, hopefully meeting in the middle. We might try -2.5 volts. This will yield 1.4 mA from Equation 10.1 and1.14 mA from Equation 10.6. As the gap has narrowed, the adjustment for the third estimate will be smaller, so we could try -2.6 volts. This would be relatively close to the value as computed in Example 10.3 (*VGS* = -*VS*).

This approximation technique also offers a clue as to how self bias gains stability over constant voltage bias. If for some reason ID was to increase, this would create a larger voltage drop across RS. Because this voltage is the same magnitude as VGS, this means that VGS grows negatively. A more negative VGS reduces ID, thus opposing the initial change in drain current. This feedback mechanism is similar in function to the BJT collector feedback bias. The stability issue is visualized in Figure 10.17.

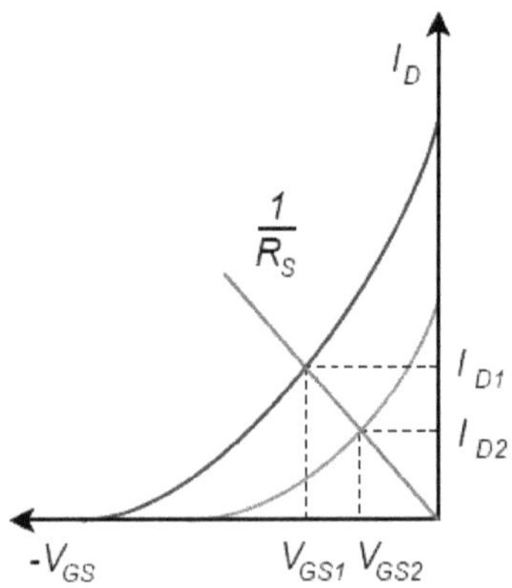

Figure 10.17 Variation for self bias.

Two device curves are plotted to represent parameter variation (green and blue). Equation 10.6 shows the relationship between ID and VGS. If we put this in the form y = mx + b, we find that the line goes through the origin and has a slope of 1/RS. This line is plotted in red. Where the line intersects the device curve yields the drain current and gate-source voltage for that particular device. Unlike constant voltage bias, self bias shifts some variation over to VGS, making ID more stable. In fact, if there is a particular design target for ID or VGS, a rearrangement of Equation 10.6 can be used to find the needed value of RS along with the characteristic curve or equation.

$$R_S = -\frac{V_{GS}}{I_D}$$

For example, if a certain ID is desired, this value could be used with Equation 10.1 to determine the corresponding VGS. These values are then used to find the required RS. Alternately, the normalized values could be obtained via Figure 10.4.

Example10.4

Determine a value for RS to set VGS = -2 V for the circuit shown in Figure 10.18. Assume IDSS = 20 mA and VGS(off) = -4 V.

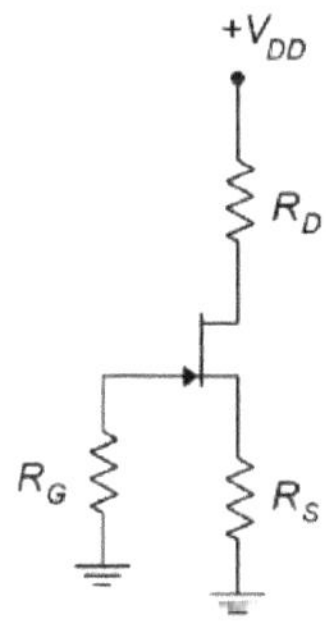

Figure 10.18 Schematic for Example 10.4.

We can determine the drain current using Equation 10.1.

$$I_D = I_{DSS}\left(1 - \frac{V_{GS}}{V_{GS(off)}}\right)^2$$

$$I_D - 20\,\text{mA}\left(1 - \frac{-2\,\text{V}}{-4\,\text{V}}\right)^2$$

$$I_D = 5\,\text{mA}$$

$$R_S = -\frac{V_{GS}}{I_D}$$

$$R_S = -\frac{-2\,\text{V}}{5\,\text{mA}}$$

$$R_S = 400\,\Omega$$

In sum, self bias is a minimal parts count circuit that offers modest stability. The stability can be improved with the addition of other components, as we shall see with the next bias configuration.

Combination Bias

The combination bias configuration (AKA source bias) is based on self bias but adds a negative power supply connected to RS, hence its name. This will enhance the stability of ID, VDS and gm. The combination bias prototype is shown in Figure 10.19.

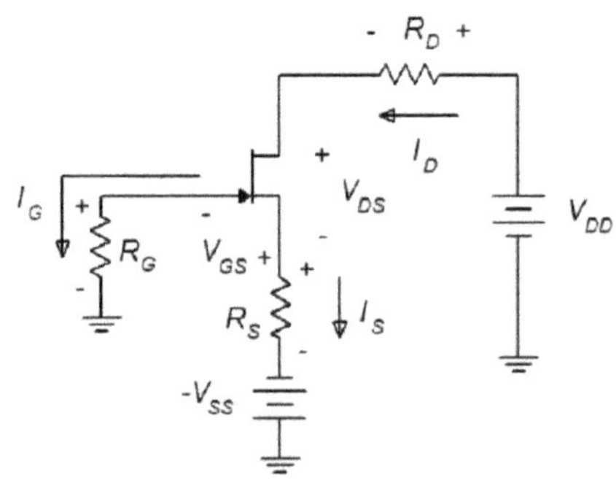

Figure 10.19 Combination bias prototype.

The analysis is similar to that of self bias but with one major twist: the source power supply increases the voltage drop across *RS*. This stabilizes the voltage (and hence, the current) because it is no longer equal to -*VGS*, but rather

$$V_{R_S} = I_D R_S = |V_{GS}| + |V_{SS}| \qquad (10.8)$$

If VSS >> VGS, then we can approximate ID as VSS/RS. As with self bias, an analytical solution for ID is possible. In order to do so, we would begin with the characteristic equation and Equation 10.8. The derivation is left as an exercise.

$$I_D = 2 I_{DSS} \left(\frac{1 + g_{m0} R_S (1+k) - \sqrt{1 + 2 g_{m0} R_S (1+k)}}{(g_{m0} R_S)^2} \right) \qquad (10.9)$$

The formula is very similar to the self bias formula but with the addition of a factor, k. k is a "swamping factor" and is defined as the ratio of VSS to VGS(off). If k = 0, there is no source power supply and the formula reverts back to the simpler self bias formula. On the other hand, if k is very large, ID ≈ VSS/RS.

As was the case with self bias, we can plot Equation 10.9 using the gm0RS factor. A series of three plots for k = 2, 3 and 4 are rendered in Figured 10.2040.

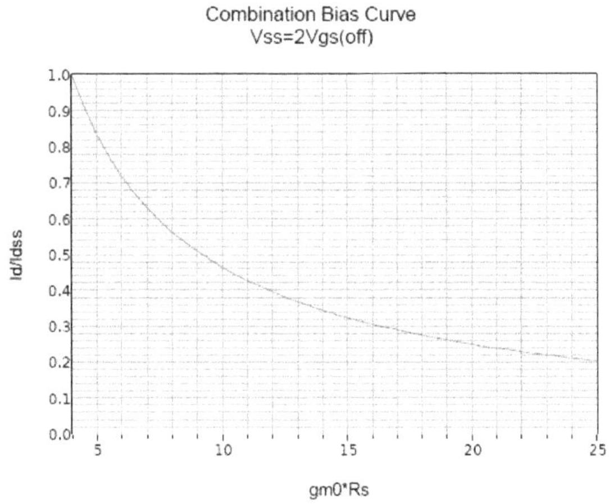

Figure 10.20a Combination bias curve, k = 2.

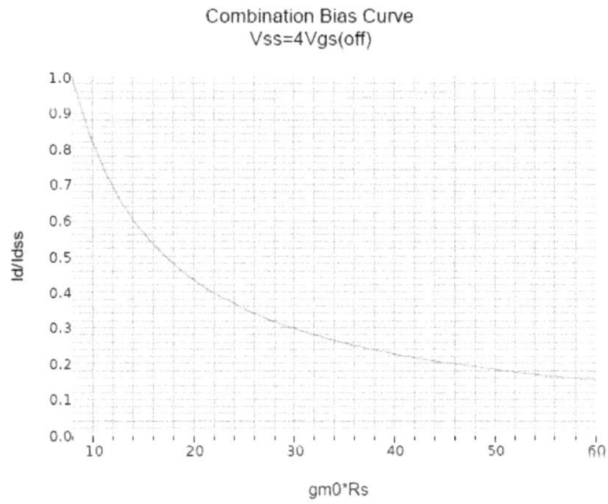

Figure 10.20b Combination bias curve, k = 3.

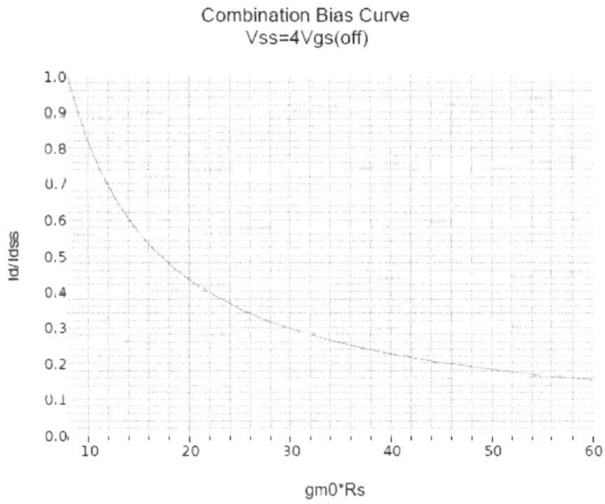

Figure 10.20c Combination bias curve, k = 4.

Example 10.5

Determine I_D and V_{DS} for the circuit shown in Figure 10.21. AssumeI_{DSS} = 12 mA and $V_{GS(off)}$ = -4 V.

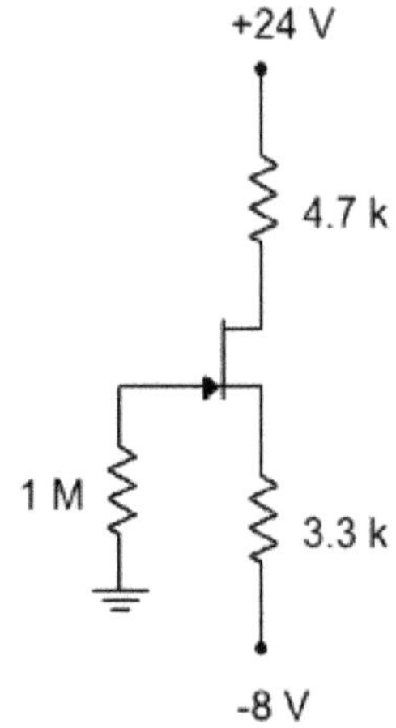

Figure 10.21 Schematic for Example 10.5.

Using the graphical method, first determine g_{m0} RS.

$$g_{m0} = -\frac{2 I_{DSS}}{V_{GS(off)}}$$

$$g_{m0} = -\frac{2 \times 12\,\mathrm{mA}}{-4\,\mathrm{V}}$$

$$g_{m0} = 6\,\mathrm{mS}$$

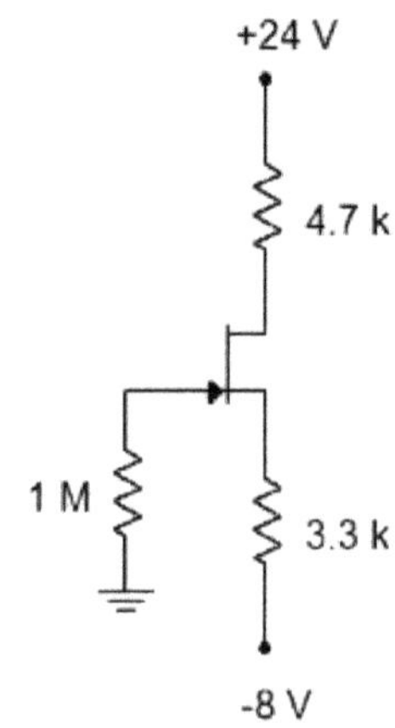

Figure 10.21 Schematic for Example 10.5.

Using the graphical method, first determine gm0 RS.

$$g_{m0} = -\frac{2I_{DSS}}{V_{GS(off)}}$$
$$g_{m0} = -\frac{2\times 12\,\text{mA}}{-4\,\text{V}}$$
$$g_{m0} = 6\,\text{mS}$$

Therefore gm0 RS = 6 mS · 3.3 k Ω = 19.8. The swamping ratio, k, is VSS/VGS(off) = -8/-4 = 2. This requires the graph in Figure 10.20a. This graph yields approximately 0.25 for the normalized current ratio. Therefore

$$I_D = 0.25\,I_{DSS}$$
$$I_D = 0.25\times 12\,\text{mA}$$
$$I_D = 3\,\text{mA}$$

Using Ohm's law and KVL

$$V_D = V_{DD} - I_D R_D$$
$$V_D = 24\,\text{V} - 3\,\text{mA}\times 4.7\,\text{k}\Omega$$
$$V_D = 9.9\,\text{V}$$

$$V_S = V_{SS} + I_D R_S$$
$$V_S = -8\,\text{V} + 3\,\text{mA}\times 3.3\,\text{k}\Omega$$
$$V_S = 1.9\,\text{V}$$

$$V_{DS} = V_D - V_S$$
$$V_{DS} = 9.9\,\text{V} - 1.9\,\text{V}$$
$$V_{DS} = 8\,\text{V}$$

As a crosscheck, using Equation 10.9 yields 3.028 mA for *ID*. The deviation is no doubt due to inaccuracy in reading the graph. In any case, using this value of drain current we find *VS* to be 1.992 volts, a little higher than calculated above. This indicates that *VGS* is -1.992 volts (because *VG* ≈ 0 V). If we plug this value of *VGS* into Equation 10.1, *ID* = 3.024 mA; an excellent match with the deviation being due to accumulated rounding errors.

In order to show the increased Q point stability of the combination bias, we'll repeat the preceding problem using a JFET with a significantly lower *IDSS*.

Example10.6

Determine *ID* for the circuit shown in Figure 10.21. Assume*IDSS* = 8 mA and *VGS(off)* = -4 V.

For this version we'll use Equation 10.9. First determine *gm0 RS*.

$$g_{m0} = -\frac{2I_{DSS}}{V_{GS(off)}}$$

$$g_{m0} = -\frac{2\times 8\,\text{mA}}{-4\,\text{V}}$$

$$g_{m0} = 4\,\text{mS}$$

Therefore *gm0 RS* = 4 mS · 3.3 k Ω = 13.2. The swamping ratio, *k*, is *VSS*/*VGS(off)* = -8/-4 = 2.

$$I_D = 2I_{DSS}\left(\frac{1+g_{m0}R_S(1+k)-\sqrt{1+2g_{m0}R_S(1+k)}}{(g_{m0}R_S)^2}\right)$$

$$I_D = 2\times 8\,\text{mA}\left(\frac{1+13.2(1+2)-\sqrt{1+2\times 13.2(1+2)}}{(13.2)^2}\right)$$

$$I_D = 2.906\,\text{mA}$$

For the graphical method, a reasonable estimate for the normalized ID would be around 0.36, yielding a drain current of 2.88 mA. Stability is apparent because the drain current has dropped only a few percent in spite of the fact that IDSS decreased by 33%.

The graph of Figure 10.22 illustrates nicely the increased stability of the Q point. Once again, we plot two representative device curves in green and blue. As was the case with self bias, a plot line can be drawn, the slope of which is equal to the reciprocal of RS. This plot line does not go though the origin, though. Instead, the x axis intercept is the voltage |VSS|. Thus, the red plot line is shifted along the VGS axis.

As can be seen in the graph, the variation in ID is reduced (although at the expense of variation in VGS). For large values of VSS with correspondingly large values of RS, the bias plot line becomes nearly horizontal, indicating a very stable Q point. With two variables in play, this bias proves to be very flexible. It can also be realized by using a positive voltage divider at the gate and removing VSS (returning RS to ground).

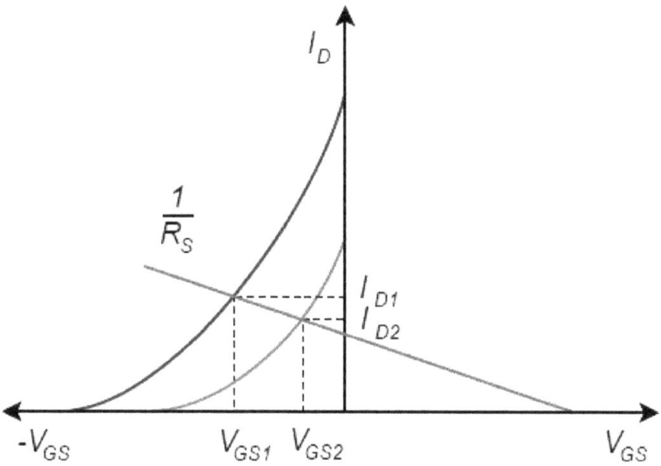

Figure 10.22 Variation for combination bias.

Constant Current Bias

The most stable bias for JFETs relies, oddly enough, on a current source made with a BJT. It is called constant current bias, yet another imaginative tag. Interestingly, although this will keep the Q point very stable, a fixed ID does not guarantee the most stable value of voltage gain. In fact, it might be easier to achieve that goal using combination bias. The prototype constant current bias circuit is shown in Figure 10.23. An NPN BJT is used for an N-channel JFET and a PNP would be used with a P-channel JFET, typically driven from above (i.e., circuit flipped top to bottom).

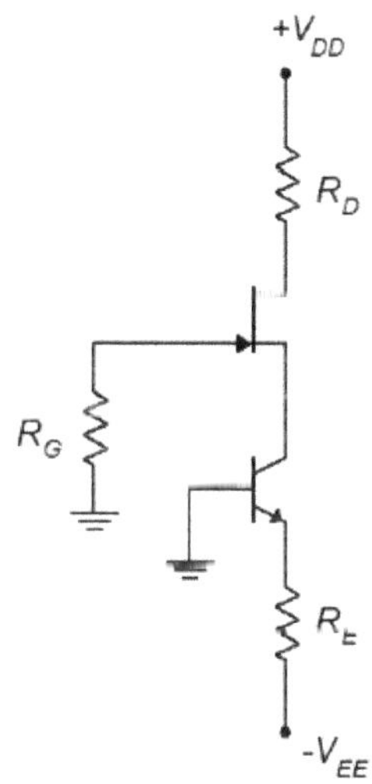

Figure 10.23 Constant current bias prototypes.

Ignoring the JFET for a moment, the BJT is configured as in two-supply emitter bias. In this case the base is tied directly to ground, leaving the emitter at about-0.7 VDC. The remainder of the *VEE* supply drops across *RE*, establishing the emitter current. As the collector is connected directly to the JFET's source terminal, this means that *IS* ≈ *IE*. The source current winds up being just as stable as the emitter current, which we have already seen is very stable. The only requirement is that *IE* should not be programmed to be larger than *IDSS*. This being true, *ID* will set up a corresponding *VGS*. This also establishes *VS* because *VG* ≈ 0 V. Therefore, the source terminal will be a small positive voltage and this is precisely what the BJT needs in order to guarantee that its collector-base junction is reverse-biased.

Computation of circuit currents and voltages is straightforward and does not involve the use of graphical aides. The first step is to examine the BJT's emitter loop and determine *IE*. Once this is found, *IS* and *ID* are known, and all remaining component potentials may be found using Ohm's law and KVL.

OThis technique does not involve the calculation of *VGS*. In fact, because *ID* is very stable, *VGS* will show the widest variation of all biasing circuits when the JFET is changed. If *VGS* is needed, it can be determined via a little algebraic manipulation on Equation 10.1.

Example 10.7

Detemine $I_{D,}$ V_{DS} and V_{GS} in the circuit of Figure 10.24. I_{DSS} = 15 mA and $V_{GS(off)}$ = -3 V.

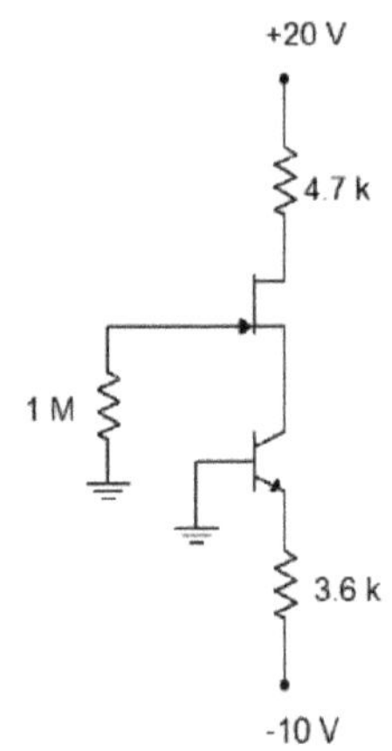

Figure 10.24 Schematic for Example 10.7.

We begin by finding $I_{E.}$

$$I_E = \frac{|V_{EE}| - 0.7\,\text{V}}{R_E}$$
$$I_E = \frac{10\,\text{V} - 0.7\,\text{V}}{3.6\,\text{k}\Omega}$$
$$I_E = 2.58\,\text{mA}$$

I_E is the same as I_S and $_{ID,}$ therefore

$$V_D = V_{DD} - I_D R_D$$
$$V_D = 20\,\text{V} - 2.58\,\text{mA} \times 4.7\,\text{k}\Omega$$
$$V_D = 7.87\,\text{V}$$

To find V_S we note that VS = $-V_{GS}$ and rearrange Equation 10.1.

$$I_D = I_{DSS}\left(1 - \frac{V_{GS}}{V_{GS(off)}}\right)^2$$

$$V_{GS} = V_{GS(off)}\left(1 - \sqrt{\frac{I_D}{I_{DSS}}}\right)$$
$$V_{GS} = -3\,\text{V}\left(1 - \sqrt{\frac{2.58\,\text{mA}}{15\,\text{mA}}}\right)$$
$$V_{GS} = -1.24\,\text{V}$$

Therefore V_S = 1.24 V and

$$V_{DS} = V_D - V_S$$
$$V_{DS} = 7.87\,\text{V} - 1.24\,\text{V}$$
$$V_{DS} = 6{,}63\,\text{V}$$

We turn next to a computer simulation of a similar circuit to validate our methodology.

Computer Simulation

A constant current bias circuit is enter into a simulator as shown in Figure 10.25.

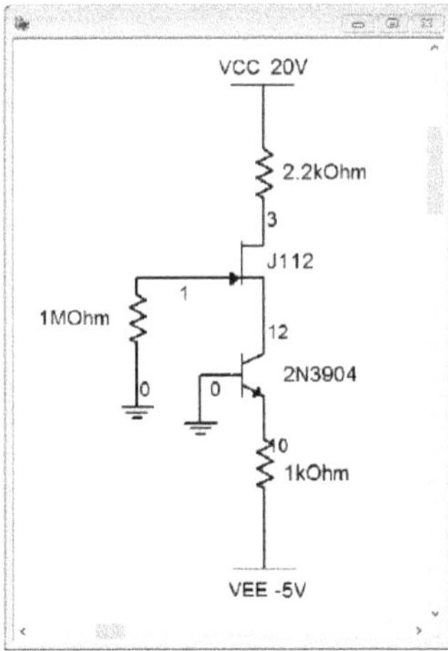

Figure 10.25 Constant current bias circuit in simulator.

A cursory estimate shows that *IE* and *ID* should be around 4.3 mA. Also, *VD* should be approximately 20 V - 4.3 mA · 2.2 kΩ, or about 10.54 volts. The results of a DC operating point analysis are shown in Figure 10.26.

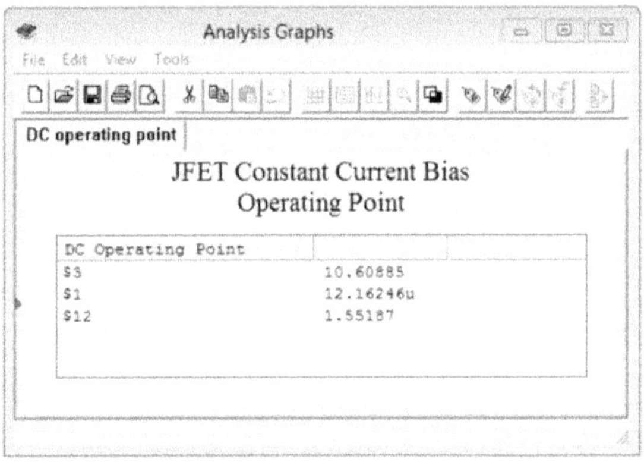

Figure 10.26 Constant current bias DC operating point simulation results.

The drain voltage (node 3) is just over 10.6 volts, agreeing with our estimate. Also, note the minuscule gate voltage (node 1) of 12 μV which verifies our continuing assumption in these circuits that VG ≈ 0 VDC. Finally, we see a modest potential of about 1.5 volts at the source terminal (node 12). This shows the proper reverse-biasing of both the gate-source and collector-base junctions.

Finally, we can examine the Q point variation using Figure 10.27. Here, the plot line is perfectly horizontal and all device variation is manifest in VGS.

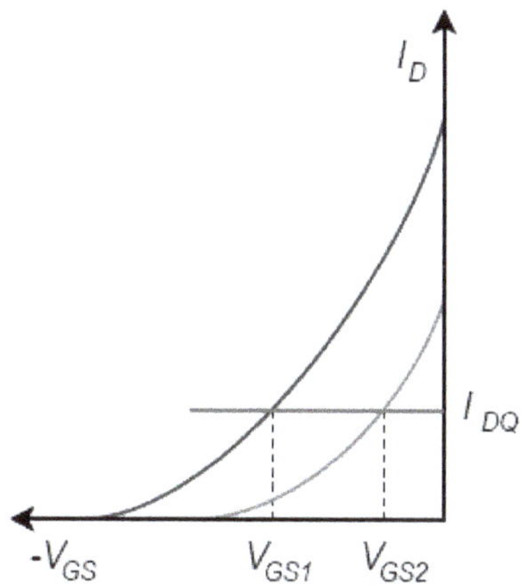

Figure 10.27 Variation for constant current bias.

CONCISE SUMMARY

- The junction field effect transistor is an altogether different device from the bipolar junction transistor. Instead of relying on a forward-biased PN junction to control current, the JFET utilizes a reverse-biased PN junction. Furthermore, the JFET uses voltage control rather than the BJT's current control. In spite of this, a family of JFET drain curves offers similarity to the BJT's collector curves, exhibiting three regions: ohmic, constant current and breakdown.
- The DC model of a JFET includes a voltage-controlled current source in the drain and a very, very large resistance, RGS, from gate to source. This resistance models that of a reverse-biased PN junction. The characteristic equation of the JFET is square-law and is consequently much more gentle in slope than the corresponding equation for a BJT. The maximum current produced by a JFET is IDSS and occurs when VGS = 0 V. VGS must always be negative to ensure proper operation and all negative values will lead to a drain current less than IDSS. Once the gate-source becomes negative enough (at VGS(off)), drain current goes to zero.
- There are several methods to bias JFETs. Perhaps the most simple method is to apply a fixed potential to the gate while grounding the source. This is called constant voltage bias and is the least stable bias in terms of Q point. Self bias uses a minimum of components and offers modest stability. It is a decent general-purpose bias. The addition of a negative power supply to the source resistor leads to the combination bias topology. This circuit offers improvements in stability over self bias. The most stable bias is the constant current bias. This form relies on a BJT to establish a very stable current.

CHAPTER-11

APPLICATIONS OF JFET SMALL SIGNAL AMPLIFIERS

11.1 INTRODUCTION

Junction Field Effect Transistors (JFETs) employed in small-signal amplifiers constitute a vital category within the realm of electronic circuits. Small-signal amplifiers, designed to amplify weak input signals with minimal distortion, find applications in various fields, including audio processing, communication systems, and instrumentation. JFETs, with their inherent voltage-controlled characteristics and high input impedance, are well-suited for these applications. The unique properties of JFETs enable precise control over amplification levels, making them ideal for scenarios where signal fidelity is crucial. As small-signal amplifiers utilizing JFETs play a pivotal role in enhancing the strength of delicate signals, understanding their operational principles and design considerations is essential for engineers involved in the development of electronic systems requiring accurate and efficient signal processing.

The JFET can be used to create both voltage amplifiers and voltage followers. In comparison with the BJT, the JFET tends to have less voltage gain potential. On the other hand, JFET circuits offer the possibility of a much higher input impedance, lower noise and better high frequency performance. There are many other similarities with BJT amplifiers. For example, the possibility of swamping still exists as a means of lowering distortion at the expense of voltage gain. Also, the JFET voltage amplifier inverts the signal, just like the BJT version. When it comes to AC analysis, a key element for the BJT is r'e. For the JFET, the comparable parameter is trans-conductance, gm.

JFET amplifiers and followers can be used with their BJT cousins. Indeed, the combination of the two, each playing to their strengths, has the potential to outperform a design using only one type of device.

Alongside their use in amplifiers and followers, JFETs can also be used in their ohmic region. This includes applications as voltage-controlled resistors and analog switches. In this mode, the device no longer behaves as a constant current source. Instead, the channel resistance becomes a function of the gate-source voltage and can be used as a control element within a voltage divider. As such, it has the capability of changing resistance value much faster than a mechanical potentiometer.

11.2 SIMPLIFIED AC MODEL OF THE JFET

An AC model of the JFET is shown in Figure 11.1. This is essentially the same model as was used for DC analysis. Once again, we have a voltage-controlled current source situated in the drain. The reverse-biased junction shows up as a very large resistance, rGS.

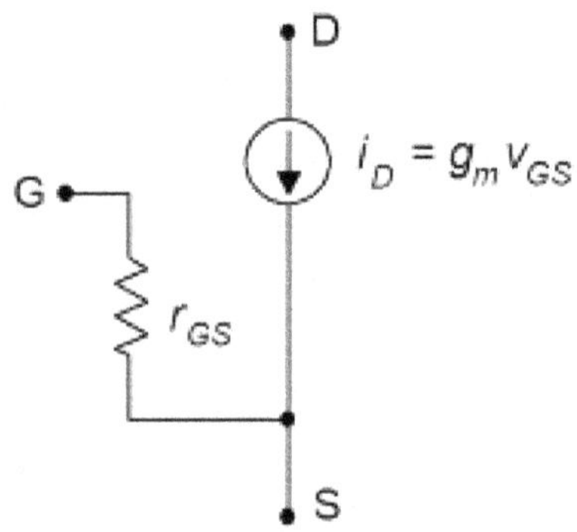

Figure 11.1 AC model of JFET.

It is worth mentioning that this model is suitable only for low frequencies. At higher frequencies, device capacitances can play a major role in the response of the amplifier. There are three device capacitances not shown in the Figure that shunt each pair of terminals: CGS, CDG and CDS. On a data sheet, the "lumped" capacitances are often given. These are Ciss, the capacitance looking into the gate with the source and drain shorted to ground: Ciss = CGS + CDG; and Crss, the capacitance seen from the drain with the gate and source shorted to ground: Crss = CDS + CDG. As we shall see, these capacitances can have a sizable impact on amplifier characteristics such as Zin.

The value of trans-conductance, gm, will prove to be of particular interest. It is roughly of equal importance to r'e in a BJT.41

11.3 COMMON SOURCE AMPLIFIER

The common source amplifier is analogous to the common emitter amplifier. The prototype amplifier circuit with device model is shown in Figure 11.2.

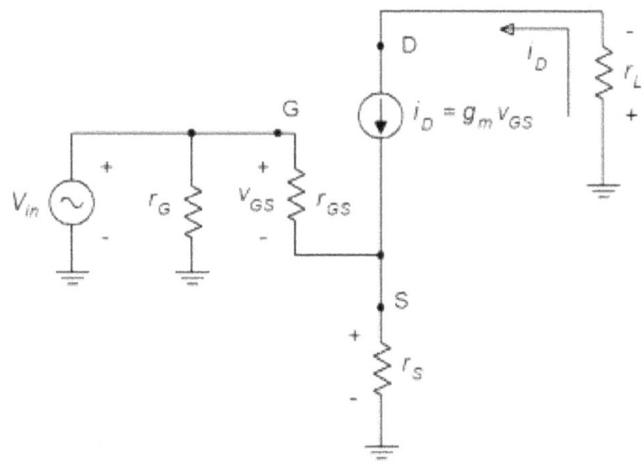

Figure 11.2 Common source amplifier with model.

This circuit includes a swamping resistor, *rS*. The input signal is presented to the gate terminal while the output is taken from the drain.

Voltage Gain

An equation for the voltage gain, Av, is developed as follows. First, we start with the fundamental definition, namely that voltage gain is the ratio of vout to vin, and proceed by expressing these voltages in terms of their Ohm's law equivalents.

$$A_v = \frac{v_{out}}{v_{in}} = \frac{v_D}{v_G}$$

$$A_v = \frac{-i_D r_L}{i_D r_S + v_{GS}}$$

$$A_v = \frac{-g_m v_{GS} r_L}{g_m v_{GS} r_S + v_{GS}}$$

$$A_v = -\frac{g_m r_L}{g_m r_S + 1} \qquad (11.1)$$

If there is no swamping resistor, the first portion of the denominator drops out and the gain simplifies to -gm · rL. The swamping resistor in the source, rS, plays the same role here as it did in the BJT: it helps to stabilize the gain and reduce distortion. It does so at the expense of voltage gain.

Input Impedance

Referring back to Figure 11.2, the input impedance of the amplifier will be rG in parallel with the impedance looking into the gate terminal, Zin(gate). For the non-swamped case, this will be rGS. At low frequencies rGS is very large, well into the megohms. In most practical circuits, rG will be much lower, hence

$$Z_{in} = r_G \,||\, r_{GS} \approx r_G \tag{11.2}$$

Theoretically, for swamped amplifiers *Zin(gate)* will be higher than *rGS* but this is a moot point. In either case, it is relatively easy to obtain a high input impedance, certainly much easier than it is for typical single-device BJT amplifiers.

It might be easy to become complacent and simply assume that *rG* sets the input impedance and that's the end of it. This would be a mistake. As mentioned earlier, with impedances this high, we cannot ignore items such as junction capacitance. For example, for a general purpose device a typical value for *Ciss*, the total input capacitance, may be in the vicinity of 5 to 10 pF. This capacitance appears in parallel with *rG*. If this amplifier is used for ultrasonic signals, the capacitive reactance, *XC*, would be as low as 160 kΩ at 100 kHz. Although this is high compared to typical BJT circuits, it's less than the *RG* values commonly used for biasing. At higher frequencies, the situation is even worse as *XC* decreases with frequency. Also, we are ignoring the Miller effect here which makes the situation even worse than even worse, so perhaps we can say that it's even worser, which is a claim we could also make regarding the grammar of this sentence.

Output Impedance

To investigate the output impedance, we'll refer to Figure 11.3. This circuit is very similar to that of Figure 11.2. The major difference is that the AC load equivalent has been split into its two components, the load itself, *RL*, and the drain biasing resistor, *RD*.

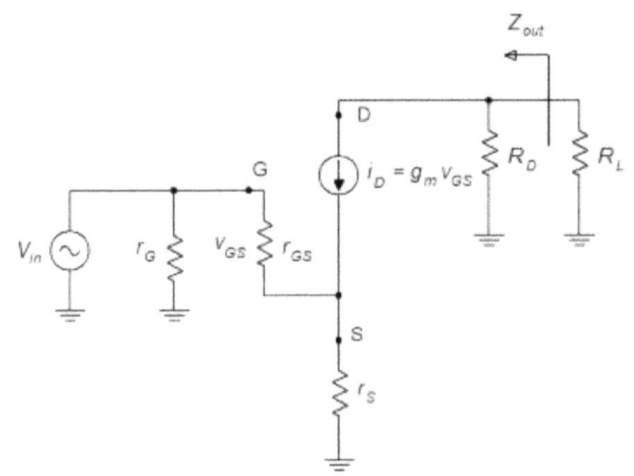

Figure 11.3 Output impedance of common source amplifier.

From the vantage point of RL, peering back into the amplifier we see RD in parallel with the impedance at the drain. At the drain we find the current source, iD. The internal impedance of this equivalent current source is very high compared to typical values for RD (hundreds of kΩ), therefore we can approximate the output impedance as

$$Z_{out} \approx R_D \qquad (11.3)$$

It should be noted that all forms of DC bias discussed in the previous chapter are game here. There are a few limitations to be aware of, though. For example, when using constant voltage bias, swamping is not possible as that bias form does not use a source resistor. In contrast, self bias and combination bias include a source resistor so swamping is a possibility, however, *RS* may need to be split and partially bypassed to achieve the desired results. Finally, constant current bias is not well-positioned to use swamping as that would require some additional work to fit in a new *RS* along with the current source. More typically, the current source will just be bypassed with a capacitor to produce a non-swamped amplifier.

Example 11.1

Determine the voltage gain and input impedance for the circuit shown in Figure 11.4. Assume IDSS = 15 mA and VGS(off) = -3 V.

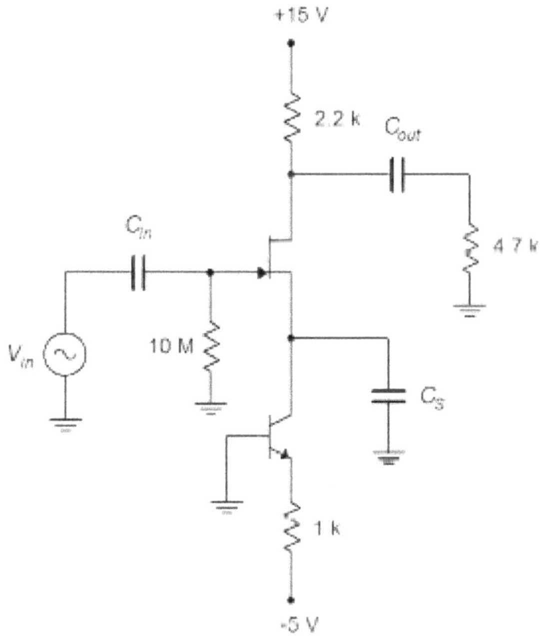

Figure 11.4 Circuit for Example 11.1.

This is an unswamped common source amplifier with constant current bias. We can determine *Zin* via inspection.

$$Z_{in} = Z_{in(gate)} \,||\, R_G$$
$$Z_{in} \approx 10\,\text{M}\Omega$$

To find the voltage gain, we'll first need to find *ID* and *gm0* in order to find *gm*.

$$I_D = \frac{|V_{EE}| - 0.7\,\text{V}}{R_E}$$
$$I_D = \frac{5\,\text{V} - 0.7\,\text{V}}{1\,\text{k}\Omega}$$
$$I_D = 4.3\,\text{mA}$$

$$g_{m0} = -\frac{2I_{DSS}}{V_{GS(off)}}$$
$$g_{m0} = -\frac{30\,\text{mA}}{-3\,\text{V}}$$
$$g_{m0} = 10\,\text{mS}$$

Knowing the current and maximum trans-conductance, we can find gmthrough the use of Equation 10.4.

$$\frac{g_m}{g_{m0}} = \sqrt{\frac{I_D}{I_{DSS}}}$$
$$g_m = g_{m0}\sqrt{\frac{I_D}{I_{DSS}}}$$
$$g_m = 10\,\text{mS}\sqrt{\frac{4.3\,\text{mA}}{15\,\text{mA}}}$$
$$g_m = 5.35\,\text{mS}$$

$$A_v = -\frac{g_m r_L}{g_m r_S + 1}$$
$$A_v = -\frac{5.35\,\text{mS}(2.2\,\text{k}\Omega \parallel 4.7\,\text{k}\Omega)}{5.35\,\text{mS} \times 0\,\Omega + 1}$$
$$A_v = -8.02$$

Example 11.2

Determine the voltage gain and input impedance for the circuit shown in Figure 11.5. Assume *IDSS* = 24 mA and *VGS(off)* = -4 V.

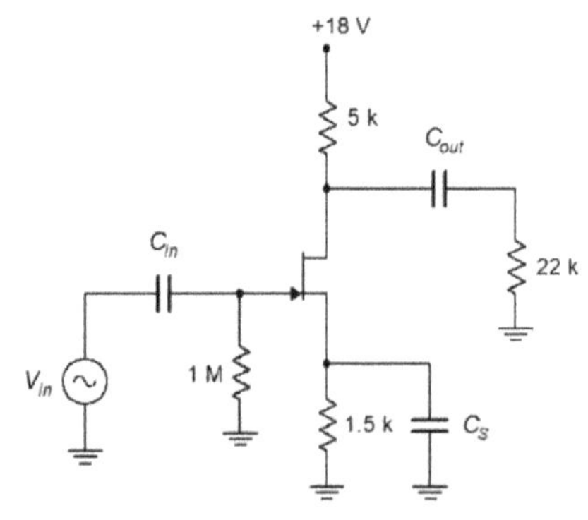

Figure 11.5 Circuit for Example 11.2

This is an unswamped common source amplifier with self bias. Once again, we can determine *Zin* via inspection.

$$Z_{in} = Z_{in(gate)} \parallel R_G$$
$$Z_{in} \approx 1\,\text{M}\Omega$$

To find the voltage gain, we'll first need to find gm0. Also, we can take a shortcut and find the normalized drain current from the self bias graph instead of finding ID itself.

$$g_{m0} = -\frac{2I_{DSS}}{V_{GS(off)}}$$
$$g_{m0} = -\frac{48\,\text{mA}}{-4\,\text{V}}$$
$$g_{m0} = 12\,\text{mS}$$

RS is 1.5 kΩ, therefore gm0RS = 18. From the self bias graph this produces a normalized drain current of 0.08.

$$g_m = g_{m0}\sqrt{\frac{I_D}{I_{DSS}}}$$
$$g_m = 12\,\text{mS}\sqrt{0.08}$$
$$g_m = 3.4\,\text{mS}$$

Again, there is no swamping so *rS* = 0. The gain formula reduces to

$$A_v = -g_m r_L$$
$$A_v = -3.4\,\text{mS}(22\text{k}\,\Omega \parallel 5\text{k}\,\Omega)$$
$$A_v = -13.9$$

We will now turn our attention to the effect of swamping. As in the BJT case, we expect to sacrifice gain and in return, see an improvement in distortion. We shall examine this through the use of a simulation.

Computer Simulation

A common source amplifier using self bias is entered into the simulator as shown in Figure 11.6.

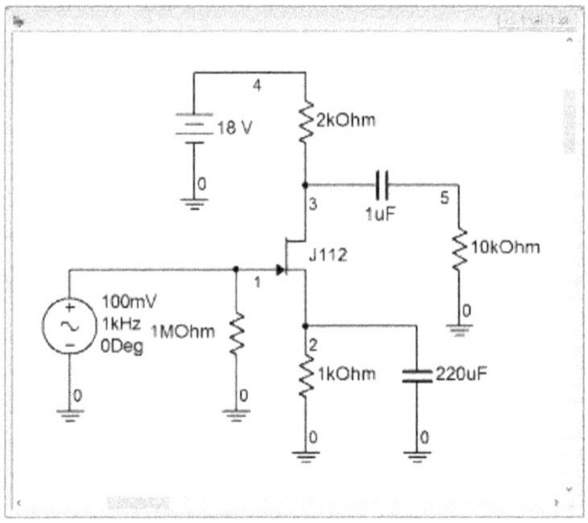

Figure 11.6 Common source amplifier in simulator.

Reasonable device values for this model are *IDSS* = 40 mA and *VGS(off)* = -2.3 V. Based on these we find

$$g_{m0} = -\frac{2I_{DSS}}{V_{GS(off)}}$$
$$g_{m0} = -\frac{80\,\text{mA}}{-2.3\,\text{V}}$$
$$g_{m0} = 34.8\,\text{mS}$$

Given R_S = 1 kΩ, the self-bias equation yields

$$I_D = 2I_{DSS}\left(\frac{1+g_{m0}R_S-\sqrt{1+2g_{m0}R_S}}{(g_{m0}R_S)^2}\right)$$
$$I_D = 2I_{DSS}\left(\frac{1+34.8-\sqrt{1+2\times 34.8}}{(34.8)^2}\right)$$
$$I_D = 1.81\,\text{mA}$$

$$g_m = g_{m0}\sqrt{\frac{I_D}{I_{DSS}}}$$
$$g_m = 34.8\,\text{mS}\sqrt{\frac{1.81\,\text{mA}}{40\,\text{mA}}}$$
$$g_m = 7.4\,\text{mS}$$

$$A_v = -g_m r_L$$
$$A_v = -7.4\,\text{mS}(2\,\text{k}\Omega \parallel 10\,\text{k}\Omega)$$
$$A_v = -12.3$$

The results of a transient analysis are shown in Figure 11.7 for a 100 mV peak input signal.

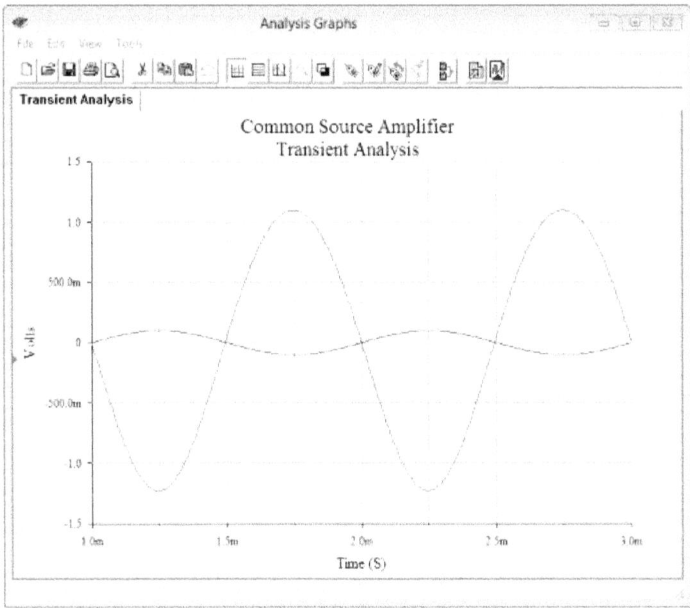

Figure 11.7 Transient analysis of the common source amplifier.

First, the inversion between the input (blue trace) and output (red trace) is obvious. Also, some distortion is evident in the output waveform. Inspecting the positive and negative peaks shows that the positive peaks are a little more broad than the negative ones and don't quite reach the same magnitude. The averaged gain is about-11.75, around 5% low from the estimate and not too bad considering the distortion.

The simulation is run a second time, but this time around the 1 kΩ source resistor is split into a 200 Ω/800 Ω pair with only the 800 Ω being bypassed. The DC bias does not change leaving gm untouched, but the new 200 Ω swamping resistor drops the gain to

$$A_v = -\frac{g_m r_L}{g_m r_S + 1}$$

$$A_v = -\frac{7.4\,\text{mS}(2\,\text{k}\Omega \parallel 10\,\text{k}\Omega)}{7.4\,\text{mS} \times 200\,\Omega + 1}$$

$$A_v = -4.96$$

The output signal from the simulator is shown in Figure 11.8.

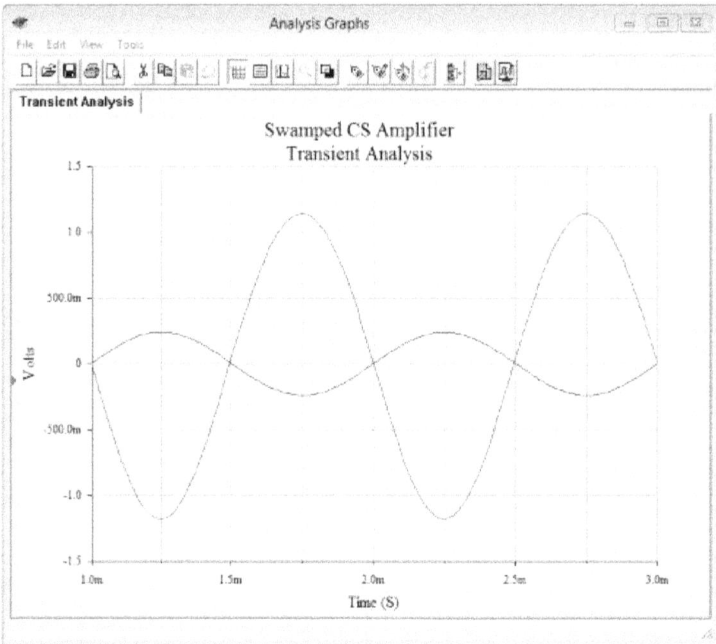

Figure 11.8 Transient analysis of the swamped common source amplifier.

The input signal was raised to 240 mV peak in order to keep the output signals of the two versions at the same amplitude. The symmetry appears to be better here and the gain works out to -4.85, just a few percent low.

Total harmonic distortion (THD) analysis is performed next. The results are shown in Figures 11.9 and 11.10. To keep the comparison fair, the input levels are adjusted to maintain similar output voltages. The non-swamped results are seen in Figure 11.9, and as expected based on the waveform asymmetry, the THD is relatively high at roughly 4%. The swamped version scores better at just over 1.6%, although this is still not stellar performance.

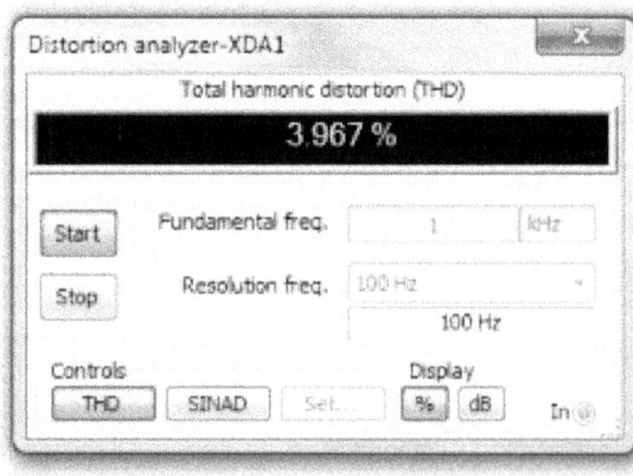

Figure 11.9 THD of common source amplifier.

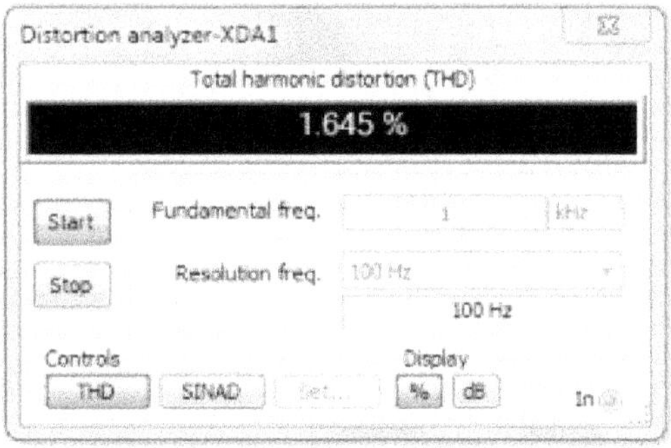

Figure 11.10 THD of swamped common source amplifier.

It is interesting to note that while the voltage gain of the swamped amplifier has dropped to 41% of the non-swamped gain, the THD has dropped to 41% of the non-swamped THD. In fact, given the square-law nature of the characteristic curve, we would expect the distortion to be lower if we used smaller signals. To verify this, the THD simulation is run again for the swamped amplifier, but now using an input signal ten times smaller at only 24 mV peak. The result is shown in Figure 11.11.

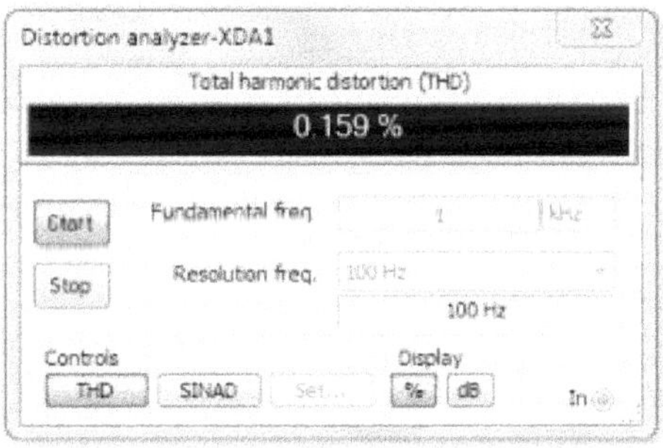

Figure 11.11 THD of swamped common source amplifier with reduced signal level.

The resulting THD is markedly lower for an order of magnitude improvement. We're now at least approaching "hi-fi" territory.

11.4 COMMON DRAIN AMPLIFIER

The common drain amplifier is analogous to the common collector emitter follower. The JFET version is also known as a source follower. The prototype amplifier circuit with device model is shown in Figure 11.12. As with all voltage followers, we expect a non-inverting voltage gain close to unity, a high Zin and low Zout.

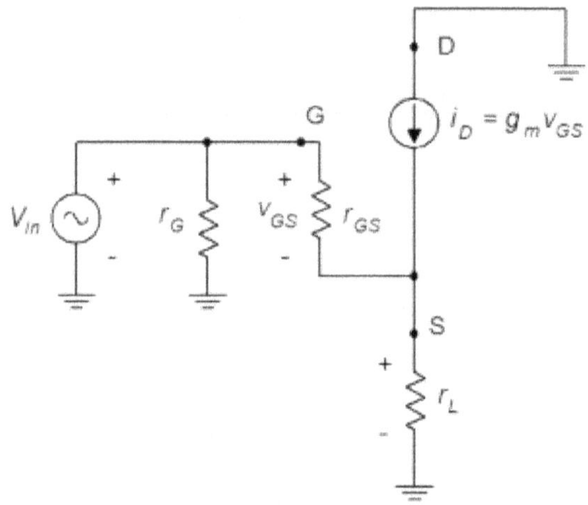

Figure11.12 Common drain (source follower) prototype.

The input signal is presented to the gate terminal while the output is taken from the source. Many bias circuits may be used here as long as they do not have a grounded source terminal such as constant voltage bias.

Voltage Gain

In order to develop an equation for the voltage gain, *Av*, we follow the same path we took with the common source amplifier earlier in this chapter. First, we start with the fundamental definition, namely that voltage gain is the ratio of *vout* to *vin*, and proceed by expressing these voltages in terms of their Ohm's law equivalents.

$$A_v = \frac{v_{out}}{v_{in}} = \frac{v_S}{v_G}$$

$$A_v = \frac{i_D r_L}{i_D r_L + v_{GS}}$$

$$A_v = \frac{g_m v_{GS} r_L}{g_m v_{GS} r_L + v_{GS}}$$

$$A_v = \frac{g_m r_L}{g_m r_L + 1} \qquad (11.4)$$

Equation 11.4 is very similar to the gain equation derived for the swamped common source amplifier; the notable changes being the lack of the minus sign indicating that this circuit does not invert the signal, and *rL* replacing *rS* in the denominator. It is worth remembering that *rL* here is the AC source resistance while in the common source amplifier *rL* is the AC drain resistance. To avoid potential confusion, this equation could also be written as

$$A_v = \frac{g_m r_S}{g_m r_S + 1} \qquad (11.4b)$$

In any event, the goal is to make sure that gmrS >> 1. By doing so, the voltage gain will be very close to unity.

Input Impedance

The analysis for common drain input impedance is virtually identical to that for the swamped common source amplifier. The result is replicated here for convenience.

$$Z_{in} = r_G \parallel r_{GS} \approx r_G \qquad (11.5)$$

Output Impedance

In order to investigate the output impedance, we'll separate the load resistance from the source bias resistor, as shown in Figure 11.13.

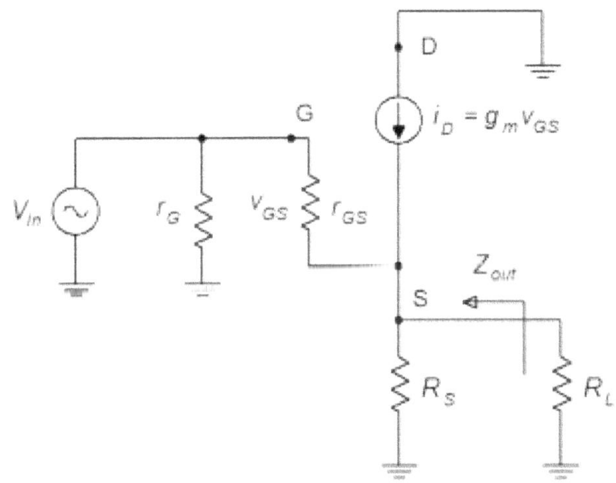

Figure 11.13 Common drain output impedance analysis.

From the position of RL, looking back toward the source we find RS in parallel with the impedance looking back into the source terminal. The voltage at this node is vGS and the current entering this node is iD. The ratio of the two must yield the impedance looking into the source.

$$Z_{source} = \frac{v_{GS}}{i_D}$$

$$Z_{source} = \frac{v_{GS}}{g_m v_{GS}}$$

$$Z_{source} = \frac{1}{g_m} \tag{11.6}$$

Therefore, the output impedance is

$$Z_{out} = R_S \parallel \frac{1}{g_m} \tag{11.7}$$

We can expect this value to be much smaller than the output impedance of typical common source amplifiers.

Example 11.3

For the follower shown in Figure 11.14, determine the input impedance and output voltage. Assume Vin = 100 mV, IDSS = 30 mA, VGS(off) = -2 V.

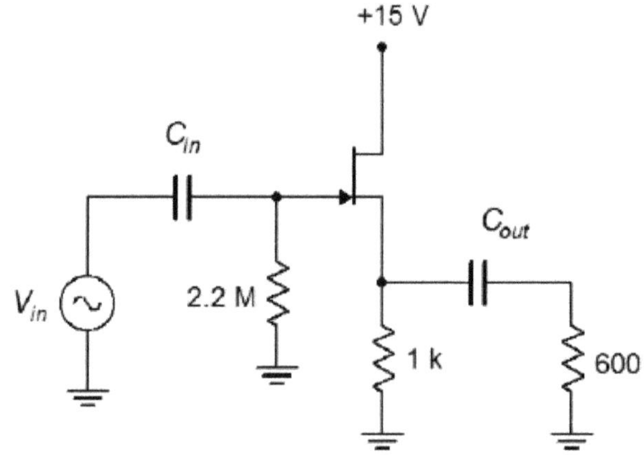

Figure 11.14 Circuit for Example 11.3.

This is a follower using self bias. We'll find gm via the self bias graph.

$$g_{m0} = -\frac{2I_{DSS}}{V_{GS(off)}}$$

$$g_{m0} = -\frac{60\,\text{mA}}{-2\,\text{V}}$$

$$g_{m0} = 30\,\text{mS}$$

RS is 1 kΩ, yielding 30 for gm0 RS. The normalized drain current from the self bias graph is approximately 0.05.

$$g_m = g_{m0}\sqrt{\frac{I_D}{I_{DSS}}}$$
$$g_m = 30\,\text{mS}\sqrt{0.05}$$
$$g_m = 6.71\,\text{mS}$$

$$A_v = \frac{g_m r_S}{g_m r_S + 1}$$
$$A_v = \frac{6.71\,\text{mS}(1\,\text{k}\Omega \parallel 600\,\Omega)}{6.71\,\text{mS}(1\,\text{k}\Omega \parallel 600\,\Omega)+1}$$
$$A_v = 0.716$$

Thus Vout is 71.6 mV. By inspection, Zin may be approximated as 2.2 MΩ.

Example 11.4

For the circuit shown in Figure 11.15, determine the input impedance and output voltage. Assume Vin = 100 mV, IDSS = 36 mA, VGS(off) = 3 V.

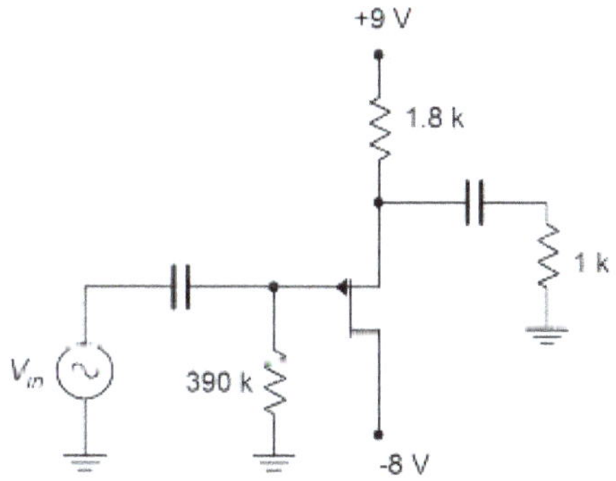

Figure 11.15 Circuit for Example 11.4.

This follower uses combination bias with a P-channel JFET. Note that the source is at the top. We'll find gm via the combination bias graph for k = 3 (k = VSS / VGS(off)).

$$g_{m0} = \frac{2I_{DSS}}{V_{GS(off)}}$$
$$g_{m0} = \frac{72\,\text{mA}}{3\,\text{V}}$$
$$g_{m0} = 24\,\text{mS}$$

RS is 1.8 kΩ, yielding 43.2 for gm0 RS. The normalized drain current from the k = 3 combination bias graph is approximately 0.17.

$$g_m = g_{m0}\sqrt{\frac{I_D}{I_{DSS}}}$$
$$g_m = 24\,\text{mS}\sqrt{0.17}$$
$$g_m = 9.9\,\text{mS}$$

$$A_v = \frac{g_m r_S}{g_m r_S + 1}$$
$$A_v = \frac{9.9\,\text{mS}(1\,\text{k}\Omega \parallel 1.8\,\text{k}\Omega)}{9.9\,\text{mS}(1\,\text{k}\Omega \parallel 1.8\,\text{k}\Omega)+1}$$
$$A_v = 0.864$$

11.5 MULTI-STAGE AND COMBINATION CIRCUITS

The rules for multi-stage circuits utilizing JFETs are the same as those discussed for BJTs: Steps must be taken to ensure that the bias of one stage does not adversely affect the bias of surrounding stages (typically by using coupling capacitors or going to a DC coupled system), the load for a given stage will be the input impedance of the following stage, the input impedance of the system will be the input impedance of the first stage, and the system gain will be the product of the individual stage gains.

Keeping those items in mind, there are no limits concerning mixing BJTs with JFETs, or mixing N-channel with P-channel devices. There are certain practical issues, however, that might dictate where certain devices are used. JFETs, due to their high input impedance and modest gain potential, tend to be used at the front end of amplifying systems. Their comparatively low self-noise is also a bonus at this location. BJTs, on the other hand, have high gain potential and tend to be used in the remaining stages. Their high distortion can be tamed through swamping.

To examine the possibilities, let's walk through the mixed, multi-stage amplifier presented in Figure 11.16.

This amplifier uses a bipolar power supply which gives the designer a lot of flexibility. The first stage consists of a JFET common source amplifier. It utilizes combination bias (notice that RS is connected to the shared negative supply, VEE, that also serves as VSS). CS bypasses the source resistor so this stage does not use swamping. Distortion should not be an issue unless the input signal is fairly large. The load for this stage is RD in parallel with the input impedance to the second stage (coupling capacitor CC will appear ideally as a short for signal frequencies).

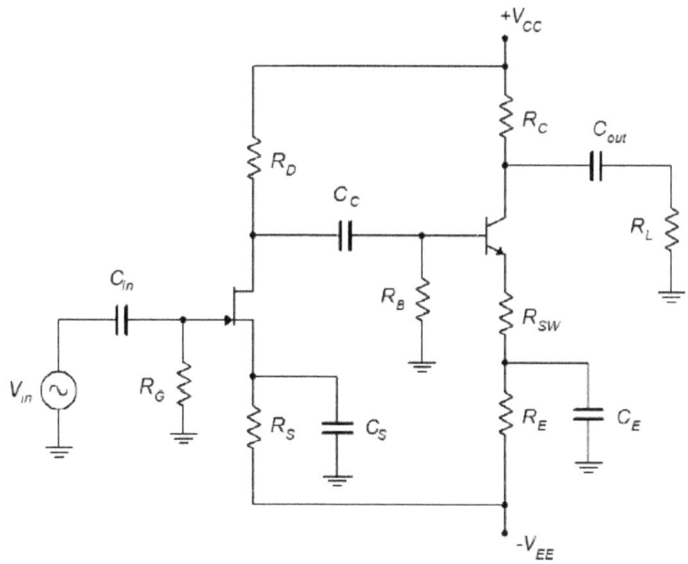

Figure 11.16 Two-stage JFET/BJT amplifier.

The second stage utilizes an NPN BJT configured as a swamped common emitter amplifier. It utilizes two-supply emitter bias. It's input impedance is the parallel combination of RB and Zin(base). The base input impedance, in turn, is a function of β and RSW (r'e will have only a small impact due to the swamping resistor). The load for this stage will be RL in parallel with RC. That value divided by RSW will give the approximate stage gain (again, r'e will have little impact). Although the second stage will be dealing with a larger signal, distortion will be mitigated by the swamping resistor.

The system gain will be the product of the two stage gains. As they both invert the signal, the inversion of the inversion will lead to an output signal that is in phase with the input signal. The system input impedance will depend on the JFET first stage and can be approximated to be equal to *RG*, at least at low frequencies.

11.6 OHMIC REGION OPERATION

As noted in the previous chapter, the JFET's operational curves span three regions. Two have been discussed: the constant current region is where the normal amplifiers and followers are biased, and breakdown is a region to be avoided due to potential damage. The third region is known as the ohmic region, or triode region. It occurs in the area where VDS is less than the pinch-off voltage, VP. In this area, the device behaves more like a resistor than like a current source. If we were to examine a family of drain curves, like those of Figure 10.3, and magnify the area near the origin, we would see something like the plot in Figure 11.17.

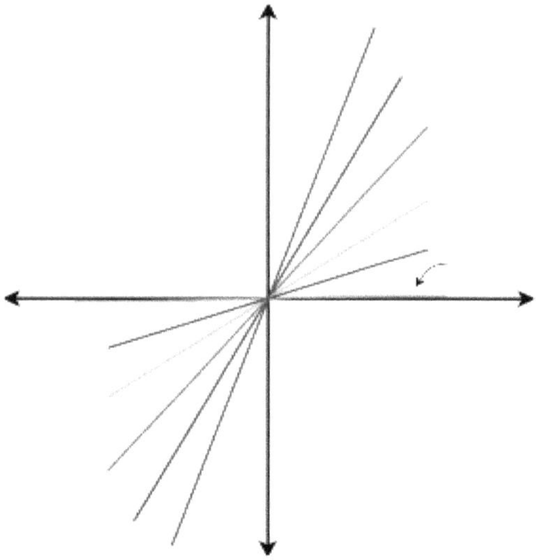

Figure 11.17 FET ohmic region.

If VDS is a small value, typically less than 100 mV or so, each of the curves appears as a straight line. Further, the slope of that line is a function of the gate-source voltage, VGS. The closer VGS is to 0 V, the steeper the slope (violet line) and the closer VGS is to VGS(off), the more shallow the slope (dark red line). Finally, if VGS = VGS(off), the slope is nearly zero (aqua line). Because this is a plot of drain current versus drain-source voltage, the slope indicates the conductance of the channel. In somewhat more useful terms, we can say that the reciprocal of the slope indicates the resistance of the channel. Therefore, if VGS = 0 V, the channel resistance will be at its minimum, and when VGS = VGS(off), the channel resistance will be at its maximum. The maximum channel resistance can be quite high, well into the hundreds of kilo-ohms. The minimum channel resistance varies considerably from device to device. It is found on a data sheet as rDS(on). rDS(on) can be as small as a few ohms for specialized JFETs and as large as hundreds of ohms for general purpose devices.42 For example, the data sheet for the J111 series JFETs found in Figure 10.7 shows maximum values of 30 Ω, 50 Ω and 100 Ω for the J111, J112 and J113, respectively. The channel resistance does not follow a linear relation with VGS.

To be more specific, in this region the drain current no longer follows the characteristic equation we used for biasing (Equation 10.1). The drain current equation in the ohmic region is:

$$I_D = \frac{V_{DS}}{V_P} 2 I_{DSS} \left(\left(1 - \frac{V_{GS}}{V_P} \right) - \frac{V_{DS}}{V_P} \right) \qquad (11.8)$$

Where VP = | VGS(off) | and VGS is to be taken as an absolute value and lies between 0 and VP.

Recalling that, in general, rDS = VDS / ID, we can substitute Equation 11.8 for ID and, after including the definition of gm0, arrive at an expression for rDS:

$$r_{DS} = \frac{V_P}{g_{m0}\left(V_P - V_{GS} - \frac{V_{DS}}{2}\right)}$$

For small values of *VDS*, this reduces to a simple equation:

$$r_{DS} = \frac{V_P}{g_{m0}\left(V_P - V_{GS}\right)} \qquad (11.9)$$

What we have created here is voltage-controlled resistor. Equation 11.9 shows that the resistance of the channel is a function of the gate-source voltage: the channel resistance will be at its minimum (rDS(on)) when VGS = 0 V, and it approaches infinity when VGS equals VGS(off). Generally, there are two applications that make use of the ohmic region: an electronic rheostat/potentiometer and an analog switch. A simple circuit that can be used for either application is shown in Figure 11.18.

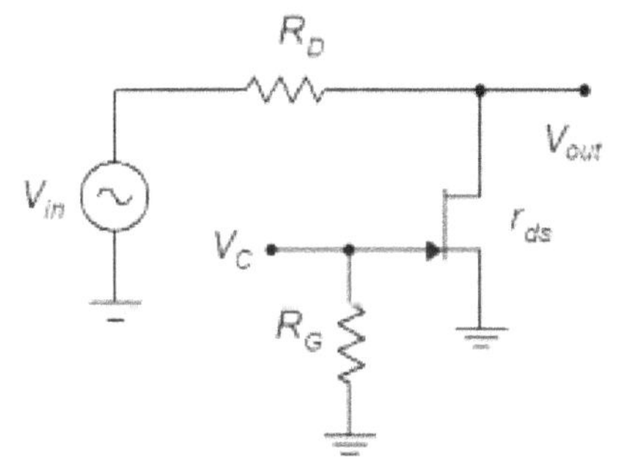

Figure 11.18 Using the JFET as a voltage controlled resistor or switch.

Note that no external bias is applied to the circuit. Instead, a control voltage, *VC*, is applied to the gate and the input signal is applied to a resistor attached to the drain terminal. The output is taken across the JFET's drain-source.

The idea behind this circuit is the basic resistive voltage divider. The JFET's channel resistance, *rDS*, forms a voltage divider along with *RD*.

$$v_{out} = V_{in}\frac{r_{DS}}{r_{DS} + R_D}$$

If rDS >> RD, vout approaches vin. Conversely, if rDS << RD, vout approaches zero. Normally, RD is set somewhere between the maximum and minimum channel resistances in order to obtain the widest range of operation.

As the control voltage VC is VGS, then VC controls the size of vout. If we set VC to 0 V, rDS is very small and thus vout ≈ 0. On the other hand, if VC is set to a large negative potential (beyond VGS(off)), then vout ≈ Vin. If VC is set between these extremes then vout will be somewhere in the middle range. If VC is continuously variable, then the circuit behaves like a solid-state potentiometer. If, in contrast, VC is only set at the limits, then the circuit behaves like a switch, either allowing or preventing the signal from transferring through. This is known as an analog switch.

A single JFET/resistor combination as shown in Figure 11.18 will have limited isolation as an analog switch and only a modest range of adjustment when used as a voltage-controlled potentiometer. To improve performance, multiple circuits can be cascaded or other JFETs can be added to create a "pi" attenuator network.

This voltage-controlled resistor has a huge advantage over traditional electro-mechanical potentiometers and switches: speed. In this circuit, the resistance can be changed at very high rates, essentially, as fast as VC can change. Consequently, it would be no big deal to switch the input signal on and off at rates well over 100,000 times per second. No mechanical switch or potentiometer can hope to perform anywhere near that speed, and any attempt to do so would lead to the devices burning up from the friction. In general, flaming potentiometers are frowned upon during the design and development process, although it would make a decent name for an indie rock band. Another advantage is that a switch can be thrown "remotely", that is, we only need to route the control voltage to the switch operator, not the signal itself. This can reduce system noise. It's also easier to implement if the switch is being "thrown" programmatically, such as via a microcontroller.

Example 11.5

For the circuit shown in Figure 11.19, if the input signal is 50 mV, determine the output voltage for VC = 0 VDC and -6 VDC. Assume VGS(off) = -5 V, rDS(on) = 30 Ω and rDS(off) = 800 kΩ.

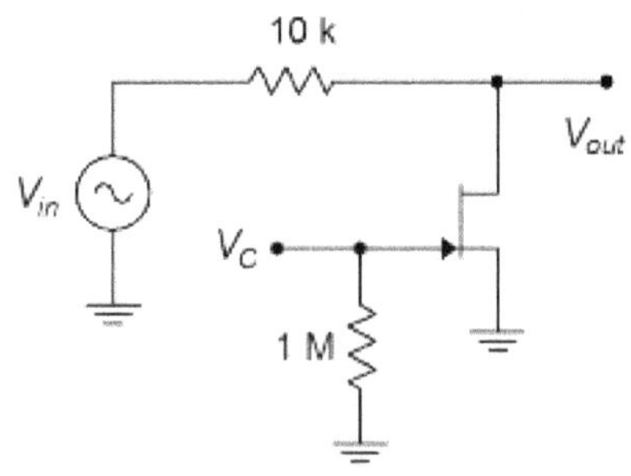

Figure 11.19 Circuit for Example 11.5.

For *VC* = 0 VDC, the channel resistance will be at its minimum of*rDS(on)*.

$$V_{out} = V_{in}\frac{r_{DS(on)}}{R_D + r_{DS(on)}}$$

$$V_{out} = 50\,\text{mV}\frac{30\,\Omega}{10\,\text{k}\Omega + 30\,\Omega}$$

$$V_{out} = 0.15\,\text{mV}$$

The signal has been reduced by a factor of over 330. That's not as good as a mechanical switch but if we cascaded two of these the overall reduction would be more than 100,000:1.

For *VC* = -6 VDC, the channel resistance will be at its maximum of *rDS(off)*.

$$V_{out} = V_{in}\frac{r_{DS(off)}}{R_D + r_{DS(off)}}$$

$$V_{out} = 50\,\text{mV}\frac{800\,\text{k}\Omega}{10\,\text{k}\Omega + 800\,\text{k}\Omega}$$

$$V_{out} = 49.4\,\text{mV}$$

This represents nearly 99% of the input signal, so the signal is passed through cleanly.

Computer Simulation

To verify the results of the preceding example, the circuit is entered into a simulator as shown in Figure 11.20. A J111 JFET model is used which has parameters similar to those used in the example.

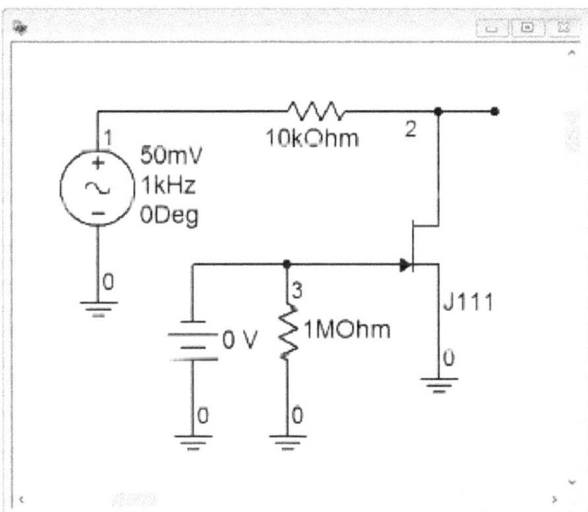

Figure 11.20 Analog switch in simulator.

A transient analysis is run twice; the first time is with a control voltage of 0 V and the second with a control voltage of -6 V. In the first case, the output should be just a small residual and in the second, we should see the full input signal. The results of the first trial are shown in Figure 11.21 while the second is shown in Figure 11.22.

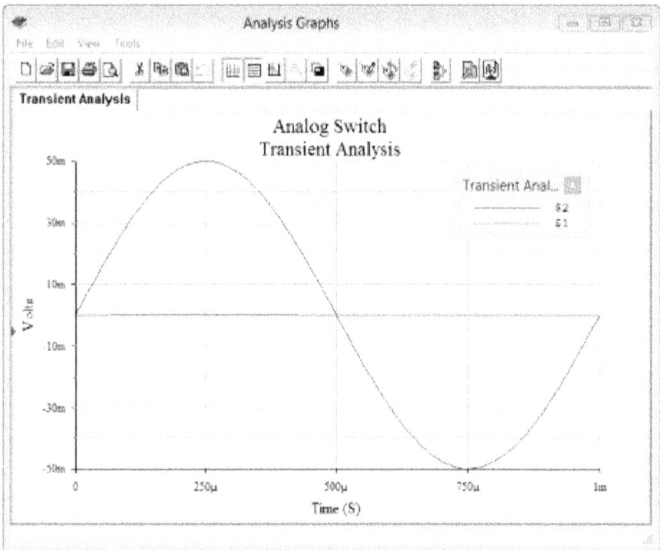

Figure 11.21 Transient analysis using V_C = 0 V.

With *VC* at 0 V, the output trace (red, at node 2) is nearly flat. The precise value of its peak is 0.167 mV, not far from the value calculated in Example 11.5.

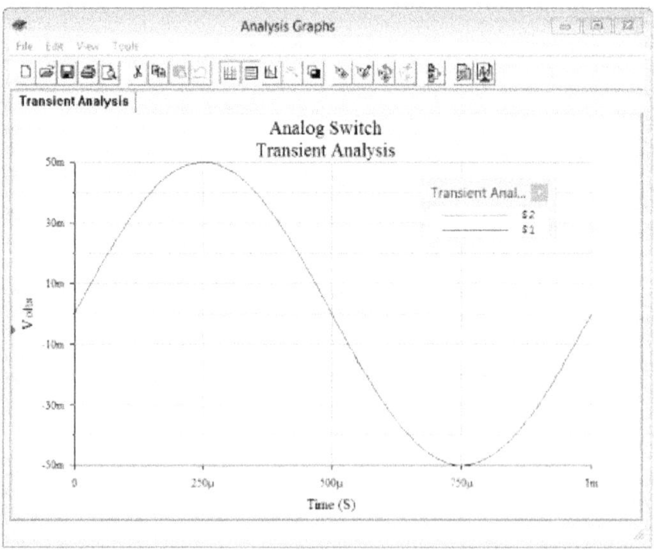

Figure 11.22 Transient analysis using V_C = -6 V.

In contrast, when VC = -6 V the JFET is off, offering a high impedance and no loss of signal. At first glance, it may appears as though the output trace is missing but what has happened is that it is hidden behind the input trace (blue, node 1). The amplitudes are virtually identical so the blue trace completely obscures the red trace.

SUMMARY

JFETs can be used to create both voltage amplifiers and voltage followers. The common source configuration is similar to the BJT's common emitter configuration. It offers voltage gain with signal inversion. The amplifier can be built upon any of the bias schemes presented in the preceding chapter. Biasing circuits that made use of a source resistor, such as self bias and combination bias, may also use swamping. Swamping will decrease available voltage gain but reduce distortion.

The JFET voltage follower, or source follower, is similar to the BJT's emitter follower. It offers a voltage of gain of nearly unity without inversion, a high input impedance and a low output impedance.

In general, JFETs do not offer as high of a gain as BJTs. The parameter comparable to the BJT's *r'e* is the trans-conductance, *gm*. Further, they tend to offer very high input impedance values compared to BJTs. This is due to using a reverse biased junction instead of a forward biased junction.

JFETs can also be used in their ohmic region to create voltage-controlled resistances and analog switches. A key parameter in these applications is the minimum channel resistance, *rDS(on)*.

Problems

AnalysisProblems

- For the amplifier of Figure 11.23, determine *Zin* and *Av*. *IDSS* = 12 mA, *VGS(off)* = -2 V, *VDD* = 15 V, *RG* = 220 kΩ, *RD* = 2 kΩ, *RL* = 3.3 kΩ, *RS* – 330 Ω.
- For the amplifier of Figure 11.23, determine *Zin* and *Vout*. *Vin* = 50 mV, *IDSS* = 15 mA, *VGS(off)* = -3 V, *VDD* = 20 V, *RG* = 270 kΩ, *RD* = 2 kΩ, *RL* = 6.8 kΩ, *RS* = 270 Ω.

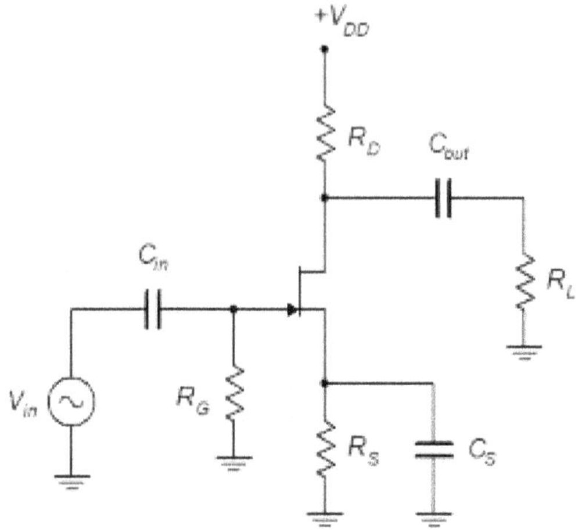

Figure 11.23

- For the amplifier of Figure 11.24, determine *Zin* and *Vout*. *Vin* = 60 mV, *IDSS* = 10 mA, *VGS(off)* = -3 V, *VDD* = 20 V, *VSS* = -6 V, *RG* = 270 kΩ, *RD* = 2 kΩ, *RL* = 4 kΩ, *RS* = 1.8 kΩ, *RSW* = 200 Ω.

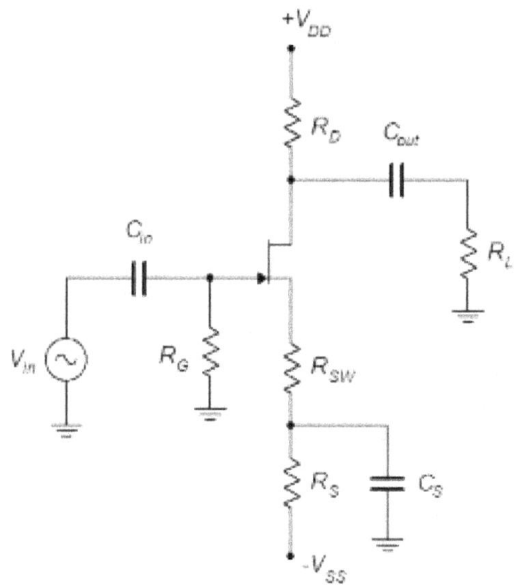

Figure 11.24

- For the amplifier of Figure 11.24, determine *Zin* and *Av*. *IDSS* = 12 mA, *VGS(off)* = -2 V, *VDD* = 18 V, *VSS* = -4 V, *RG* = 330 kΩ, *RD* = 2.2 kΩ, *RL* = 10 kΩ, *RS* = 3 kΩ, *RSW* = 100 Ω.

- For the amplifier of Figure 11.25, determine *Zin* and *Av*. *IDSS* = 12 mA, *VGS(off)* = -2 V, *VDD* = 18 V, *VEE* = -4 V, *RG* = 390 kΩ, *RD* = 2.2 kΩ, *RE* = 1 kΩ, *RL* = 20 kΩ.

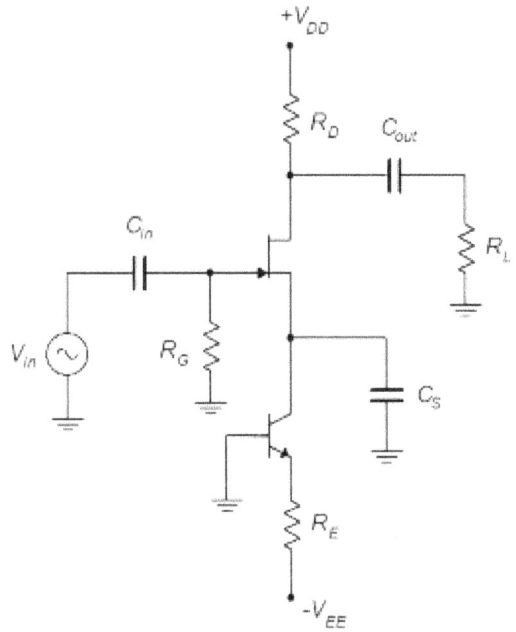

Figure 11.25

- For the amplifier of Figure 11.25, determine *Zin* and *Vout*. *Vin* = 70 mV, *IDSS* = 12 mA, *VGS(off)* = -2 V, *VDD* = 18 V, *VEE* = -4 V, *RG* = 390 kΩ, *RD* = 2.2 kΩ, *RL* = 20 kΩ.

- For the circuit of Figure 11.26, determine *Zin* and *Av*. *IDSS* = 12 mA, *VGS(off)* = -2 V, *VDD* = 10 V, *RG* = 220 kΩ, *RL* = 3.3 kΩ, *RS* = 330 Ω.

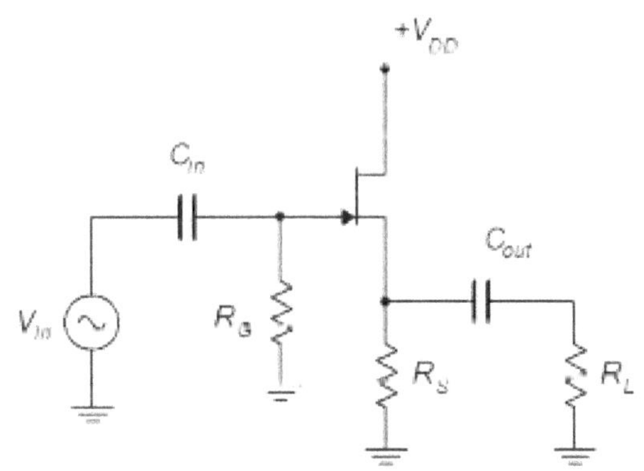

Figure 11.26

- For the circuit of Figure 11.26, determine *Zin* and *Vout*. *Vin* = 200 mV, *IDSS* = 15 mA, *VGS(off)* = -3 V, *VDD* = 12 V, *RG* = 270 kΩ, *RL* = 1.8 kΩ, *RS* = 270 Ω.
- For the circuit of Figure 11.27, determine *Zin* and *Vout*. *Vin* = 100 mV, *IDSS* = 10 mA, *VGS(off)* = -3 V, *VDD* = 15 V, *VSS* = -6 V, *RG* = 470 kΩ, *RL* = 4 kΩ, *RS* = 1.8 kΩ.

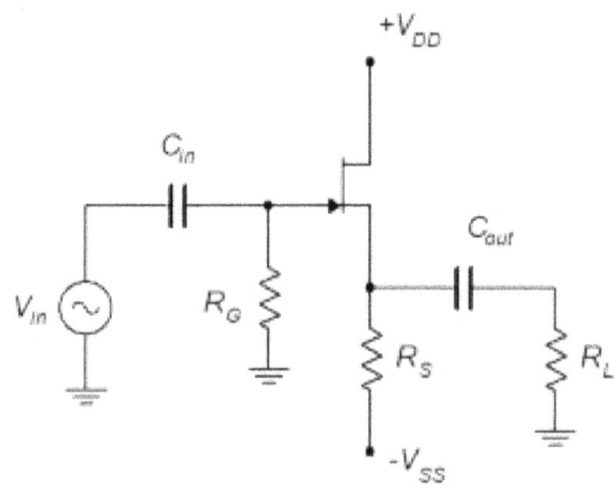

Figure 11.27

- For the circuit of Figure 11.27, determine *Zin* and *Av*. *IDSS* = 18 mA,*VGS(off)* = -2 V, *VDD* = 14 V, *VSS* = -6 V, *RG* = 360 kΩ, *RL* = 10 kΩ, *RS* = 1 kΩ.
- For the circuit of Figure 11.28, determine *Vout*. *Vin* = 100 mV, *rDS(on)* = 50 Ω,*rDS(off)* = 1 MΩ, *VGS(off)* = -3 V, *VC* = -6 V, *RG* = 270 kΩ, *RD* = 6.8 kΩ.
- For the circuit of Figure 11.28, determine *Vout*. *Vin* = 100 mV, *rDS(on)* = 75 Ω,*rDS(off)* = 750 kΩ, *VGS(off)* = -3 V, *VC* = 0 V, *RG* = 180 kΩ, *RD* = 5.1 kΩ.

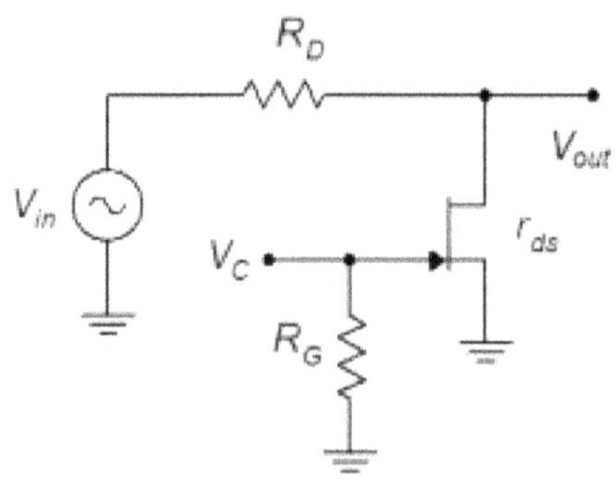

Figure 11.28

Design Challenge Problems

- Following the circuit of Figure 11.24, design an amplifier with a gain of at least 4 and an input impedance of at least 300 kΩ. *RL* = 10 kΩ. The JFET has the following parameters: *VGS(off)* = -2 V, *IDSS* = 15 mA. Try to use standard resistor values.
- Using the circuit of Figure 11.26, design a follower with a gain of at least0.7 and an input impedance of at least 500 kΩ. *RL* = 1 kΩ. The JFET has the following parameters: *VGS(off)* = -3 V, *IDSS* = 20 mA. Try to use standard resistor values.

Computer Simulation Problems

- Utilizing manufacturer's data sheets, find devices with the following specifications (typical) and verify them using the measurement techniques presented in the prior chapter. Device 1: *VGS(off)* = -2 V, *IDSS* = 15 mA. Device 2: *VGS(off)* = -3 V, *IDSS* = 20 mA.
- Using the device model from the preceding problem, verify the design of Problem 13.
- Using the device model from Problem 15, verify the design of Problem

CHAPTER-12

METAL OXIDE SEMICONDUCTOR FETS (MOSFETS)

12.1 INTRODUCTION: AN OVERVIEW

Metal-Oxide-Semiconductor Field-Effect Transistors (MOSFETs) represent a cornerstone in modern electronics, serving as essential components in a myriad of applications. MOSFETs operate based on the modulation of the conductivity of a semiconductor channel by an electric field. Consisting of a metal gate separated from the semiconductor material by a thin oxide layer, MOSFETs come in two main types: N-channel and P-channel. One of the distinguishing features of MOSFETs is their ability to function as voltage-controlled devices, making them integral in digital logic circuits, memory devices, and power electronics. Due to their high input impedance and low power consumption, MOSFETs are particularly well-suited for applications requiring efficient signal amplification and switching. As foundational elements in integrated circuits and microelectronics, MOSFETs play a pivotal role in shaping the landscape of contemporary electronic technologies.

The MOSFET shares many similarities with the JFET including very low gate current and being modeled as a voltage-controlled current source. It is also available in N- and P-channel varieties. Unlike the JFET, it has two variations: the depletion-enhancement mode variant, or DE-MOSFET; and the enhancement-only mode variant, or E-MOSFET. All of the bias types discussed for JFETs will work for DE-MOSFETs, plus a few others. E- MOSFETs, on the other hand, require new biasing prototypes.

For AC analysis, both common source and common drain amplifier topologies may be realized with DE- and E- MOSFETs. The equations for input impedance, voltage gain and the like are generally unchanged from the JFET. E-MOSFETs are also available as power devices. They have certain advantages over power BJTs, including higher speed and a negative temperature coefficient of trans-conductance which means they are less likely to suffer from thermal instabilities such as current hogging.

One item of practical importance is that MOSFETs are very susceptible to ESD (electrostatic discharge) and special precautions must be taken to prevent accidental damage to the device. Unlike both the JFET and the BJT, the MOSFET does not rely on a PN junction for its operation. Instead, it uses a charge-based system not unlike a capacitor. The gate is, in fact, insulated from the channel. For this reason it is sometimes referred to as an IGFET, which stands for Insulated Gate FET. This insulation layer will lead to very, very high input resistance due to extremely low gate current but also leads to the issue of ESD susceptibility.

12.2 THE DE-MOSFET

Like the JFET, the DE-MOSFET is based around the idea of modulating current flow through the drain-source channel by generating a depletion layer from a gate-source voltage. It achieves this through an entirely different process, though. To understand how the device is constructed, a simplified functional drawing of an N-channel DE- MOSFET is shown in Figure 12.1.

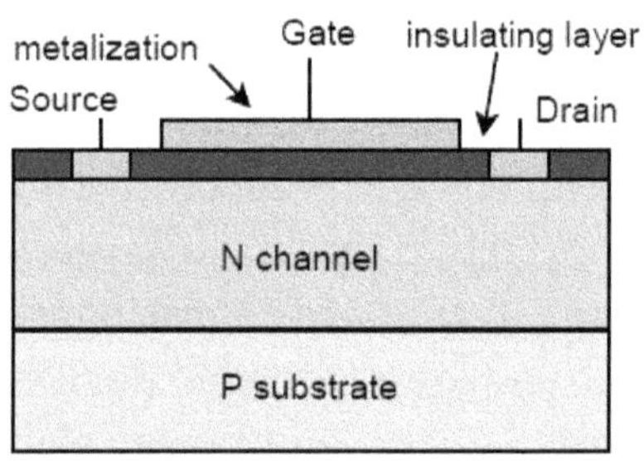

Figure 12.1 DE-MOSFET internal structure.

Here we see the N-type channel sitting on a P substrate. Drain and source leads are attached to either end. Above the channel is a very thin insulating layer (silicon dioxide is one possibility). Above this we have a metallization to which the gate terminal is attached. Note that there is no PN junction involved with the gate. To this we shall add external bias sources and limiting resistors, as shown in Figure 12.2.

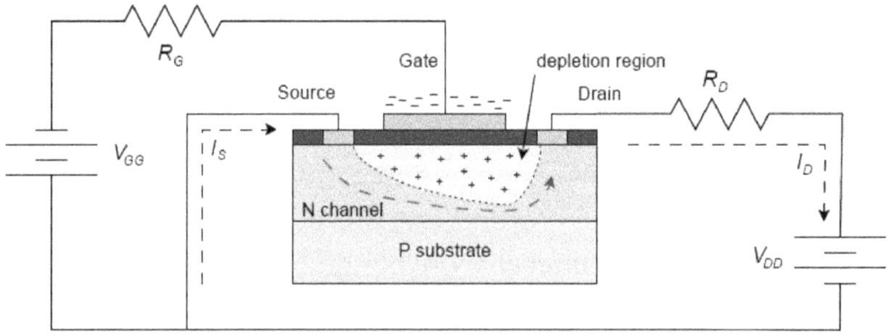

Figure 12.2 DE-MOSFET bias with electron flow.

The dashed lines represent electron current flow as in our previous device analyses. A positive supply, VDD, is attached to the drain via a limiting resistor. A second supply, VGG, is attached to the gate. Gate current can be approximated as zero, so VGS = VGG. If VGS is zero, a certain amount of current will flow through the channel based on the channel's physical parameters and the applied drain-source potential. For relatively low values of VDS, the channel will behave somewhat like a resistance. This is the same ohmic region as seen with the JFET. As VDS increases, the channel will saturate and begin to behave like a constant current source. If VDS is brought too high, the drain current increases sharply as the device enters the breakdown region. The general behavior mimics that of a JFET. Note that the current moves laterally, across the device, so this type of construction is referred to as a lateral MOSFET.

If VGS is set to a modest negative voltage, a depletion region will develop inside the channel. Basically, the gate is acting like one plate of a capacitor, the channel like the other plate, and the insulating layer is the dielectric. Just like a capacitor, the negative charge on the gate "plate" leads to an equivalent positive charge on the channel "plate". As the channel is made of N-type material, this action creates a region devoid of free charges, hence a depletion region. This depletion region will lead to pinch-off sooner, and thus a lower current in the saturation region. The more negative VGS is made, the greater the depletion region and the lower the corresponding drain current. Eventually, if VGS is brought negative enough, the channel will be blocked and no drain current will flow. This voltage is referred to as VGS(off) (again). The current produced when VGS = 0 V is likewise referred to as IDSS. This mode of operation is referred to as depletion mode because of the depletion region that is created.

What makes the DE-MOSFET distinct from the JFET is what happens when VGS > 0 volts. In a JFET, this would forward bias the junction and control would be lost.

Here, however, a positive VGS simply reverses the polarities associated with the gate and channel "plates". Thus, a positive VGS enhances channel conductivity and drain current increases as VGS is brought more positive. This mode of operation is called enhancement mode. This also means that IDSS is no longer the maximum drain current of which the device is capable. A characteristic curve is shown in Figure 12.3, below.

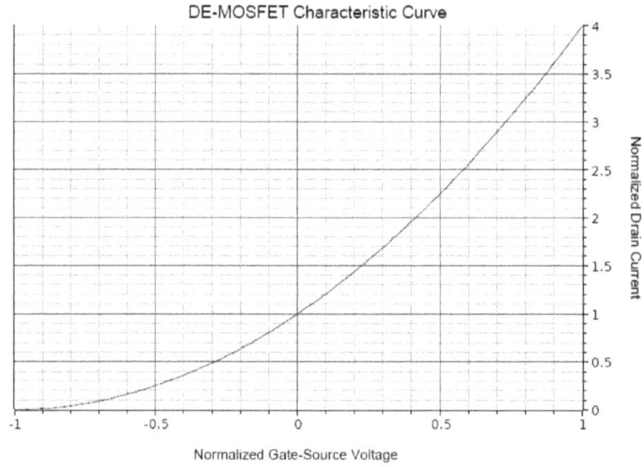

Figure 12.3 Characteristic curve for DEMOSFET (note: this uses-VGS/VGS(off) for the normalized voltage so that the quadrants do not appear reversed compared to a typical device curve).

This curve is essentially the same curve as presented for the JFET with the exception that it extends into the first quadrant. This makes the DE-MOSFET a unique device in that it can operate in two different quadrants.

$$I_D = I_{DSS}\left(1 - \frac{V_{GS}}{V_{GS(off)}}\right)^2 \quad (12.1)$$

The device equation for operation when *VDS* > *VP* is also the same, but with an extended range for *VGS*:

Where

V_{GS} is the gate-source voltage ($V_{GS(off)} \leq V_{GS} \leq V_{GS(max)}$),

I_D is the drain current,

I_{DSS} is the maximum current,

$V_{GS(off)}$ is the turn-off voltage.

VGS(max) may be found on a data sheet. Above this voltage the insulating layer will be damaged and the device will cease to function properly. A typical value for this might be in the range of 20 to 30 volts. The trick is that given the very small gate current, even a simple electrostatic discharge can damage the device. It is very easy to develop hundreds of volts static on the human body. In fact, it is generally not noticeable until the potential reaches a few thousand volts (as in body hair standing up). The consequence of this is that simply picking up the device could destroy it.[43]

There are a couple different ways of dealing with this issue. The first way is to add back-to-back Zener diodes across the device during its manufacture. The problem with this is that the diode leakage current will be greater than the gate current and this degrades performance. The other technique is to prevent the charge from getting to the device in the first place. For example, the MOSFET can be shipped in conductive plastic (not to be confused with ordinary plastic or polystyrene foam).

Some devices are shipped with a metal shorting that encompasses all of the leads. Also, during manufacture or prototyping, environmental controls are established to minimize the creation of static charges, optimal humidity being important as one example. Workers who handle devices may work on special conductive mats or wear wrist straps that are attached to ground. These items are only mildly conductive, that is, of high resistance, as it would not be safe to electrically ground a human working in an electrical lab. The devices are conductive enough to bleed off static charge but not so conductive as to present a shock hazard. Once installed on the circuit board, normal ESD precautions apply.

As the device's characteristic equation has not changed, many of the items derived for the JFET still apply to the DE-MOSFET. This includes the trans-conductance equation plot found in Chapter 10.

As the trans-conductance equation is unchanged with the exception of an extended range for *VGS*, the definition for *gm0* is also unchanged.

$$g_{m0} = -\frac{2I_{DSS}}{V_{GS(off)}} \tag{12.2}$$

$$g_m = g_{m0}\left(1 - \frac{V_{GS}}{V_{GS(off)}}\right) \tag{12.3}$$

It is worth noting that *gm0* no longer represents the maximum device trans-conductance because *IDSS* no longer represents the maximum drain current as seen in Figure 12.3. To illustrate it another way,

$$g_m = g_{m0}\sqrt{\frac{I_D}{I_{DSS}}} \tag{12.4}$$

It is very important to watch the sign of *VGS* in Equation 12.3. In enhancement mode, a positive *VGS* will lead to a *gm* greater than *gm0* due to the double negative.

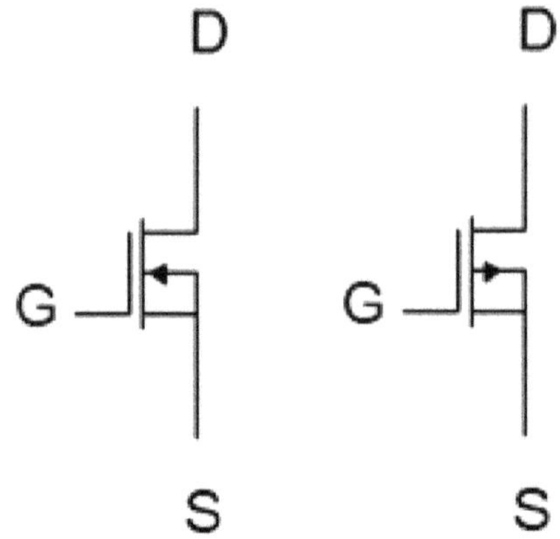

Figure 12.4 DE-MOSFET schematic symbols. N-channel (left) and P-channel (right).

The schematic symbols for the DE-MOSFET are shown in Figure 12.4. As is the norm, the arrow points in the direction of N material, with the central vertical bar representing the channel. The arrow is attached to the substrate. In some devices this is brought out of the packaging as a fourth lead although in many it is simply tied back to the source terminal as shown here. Finally, note how the gate terminal is not drawn connected to the body of the device, emphasizing its isolated nature.

12.3 DE-MOSFET BIASING

As the characteristic equations of the JFET and DE-MOSFET are the same, the DC biasing model is the same. Consequently, the DE-MOSFET can be biased using any of the techniques used with the JFET including self bias, combination bias and current source bias as these are all second quadrant biasing schemes (i.e., have a negative *VGS*). The self bias and combination bias equations and plots from Chapter 10 may be used without modification. The DE-MOSFET also allows first quadrant operation so a couple of new biasing forms become available: zero bias and voltage divider bias. In reality, both are variations on constant voltage bias but which utilize the first quadrant.

ZeroBias

Zero bias is unique. In some ways it can be thought of as a cross between self bias and constant voltage bias. Like self bias, it does not require a second DC source for the gate or source terminal. Like constant voltage bias, there is no need for a source resistor, *RS*. A prototype of zero bias is shown in Figure 12.5. There is no question that it is a minimal parts-count circuit.

Zero bias is so named because it operates at *VGS* = 0 V. Recall that the gate current is ideally zero, thus there is no drop across *RG* and *VG* = 0 V as a consequence. The source is tied directly to ground, therefore *VGS* must equal 0 V. As *VGS* doesn't change, this can be thought of as a form of constant voltage bias. The interesting bit is that when an AC signal is applied to the gate, its negative portion will pull the MOSFET down into depletion mode and the positive portion will push the operation into enhancement mode. Because the device can operate in this fashion, conducting current while straddling zero, so to speak, DE-MOSFETs are sometimes referred to as *normally on* devices.

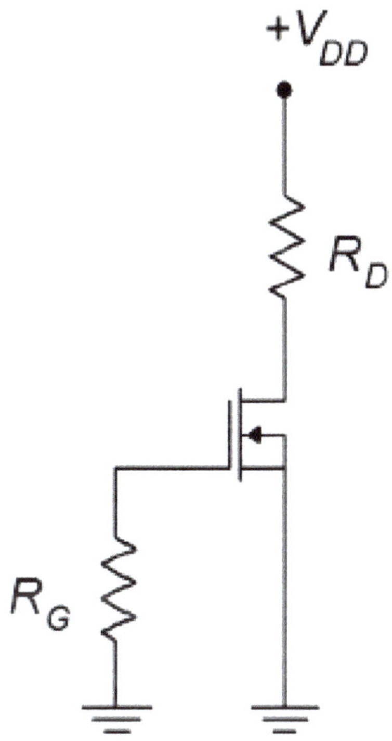

Figure 12.5 Zero bias prototype.

Determining the operating point for zero bias is startlingly easy. Because VGS = 0 V, ID must equal IDSS and gm must equal gm0. Like all constant voltage biasing schemes, though, Q point stability is not very good. Another point to notice is that, as there is no source resistor, this bias is only applicable to non-swamped common source amplifiers. It cannot be used with a source follower or swamped amplifier (if a small swamping resistor is inserted into the source, technically the circuit can be classified as self bias, although the AC signal may still push operation into enhancement mode).

Example 12.1

Determine *ID*, *VD* and *gm0* for the circuit shown in Figure 12.6. Assume *IDSS* = 12 mA and *VGS(off)* = -3 V.

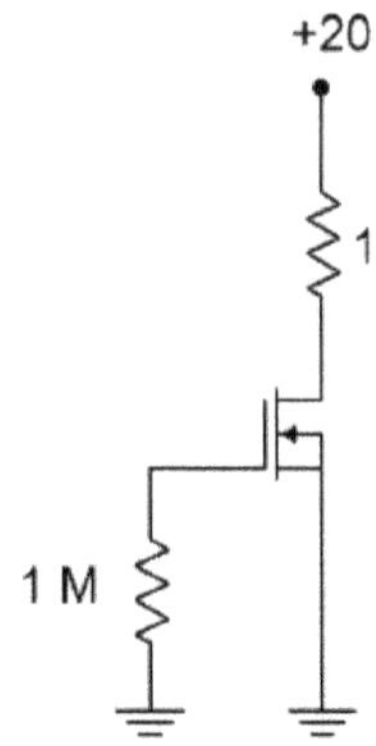

Figure 12.6 Circuit for Example 12.1.

By inspection, as this is zero bias *ID* = *IDSS*, and therefore *ID* = 12 mA. Using KVL and Ohm's law we can find *VD*.

$$V_D = V_{DD} - I_D R_D$$
$$V_D = 20\,\text{V} - 12\,\text{mA} \times 1\,\text{k}\Omega$$
$$V_D = 8\,\text{V}$$

$$g_{m0} = -\frac{2I_{DSS}}{V_{GS(off)}}$$
$$g_{m0} = -\frac{2 \times 12\,\text{mA}}{-3\,\text{V}}$$
$$g_{m0} = 8\,\text{mS}$$

Voltage Divider Bias

Voltage divider bias is a form of constant voltage bias that operates in enhancement mode. A prototype circuit is shown in Figure 12.7. Note that the source terminal is connected directly to ground. This is important. If this was not the case, this would be a form of combination bias (basically shifting the VSS supply up to ground and then shifting the gate voltage from ground up to a positive VSS to maintain the same differential voltage). As such, it would be operating in depletion mode.

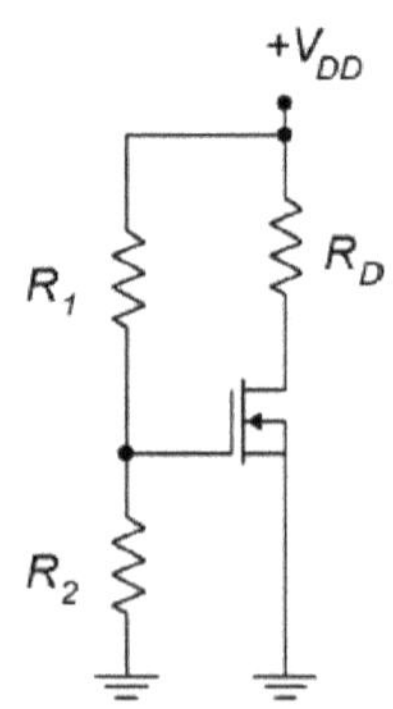

Figure 12.7 Voltage divider bias prototype.

The voltage divider comprised of *R1* and *R2* will establish a DC bias potential on the gate. As the source is at ground, *VGS* = *VG* = *VR2*. Given that *VDD* must be positive, then *VGS* must be positive, and enhancement mode operation is a given.

The most direct way to handle this is to determine the voltage divider potential and use either the characteristic equation (Equation 10.1) or associated graph to determine the drain current. Once *ID* is found, the drain-source voltage may be found via the standard Ohm's law/KVL route.

Before continuing, note that the values of the divider resistors can be very high without creating biasing problems (unlike the BJT version of voltage divider bias). This is because the gate current is so small that even when using megohm values for the divider, the loading caused by the gate will not be noticeable.

Example 12.2

For the circuit of Figure 12.8, determine *ID* and *VD*. Assume *IDSS* = 2 mA and*VGS(off)* = -6 V.

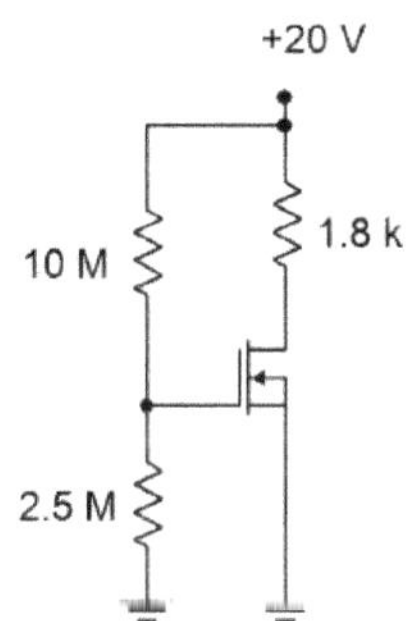

Figure 12.8 Circuit for Example 12.2.

The voltage divider will yield *VGS*.

$$V_{GS} = V_{DD}\left(\frac{R_2}{R_1 + R_2}\right)$$

$$V_{GS} = 20\,\text{V}\left(\frac{2.5\,\text{M}\Omega}{10\,\text{M}\Omega + 2.5\,\text{M}\Omega}\right)$$

$$V_{GS} = 4\,\text{V}$$

Use Equation 10.1 to find *ID*.

$$I_D - I_{DSS}\left(1 - \frac{V_{GS}}{V_{GS(off)}}\right)^2$$

$$I_D = 2\,\text{mA}\left(1 - \frac{4\,\text{V}}{-6\,\text{V}}\right)^2$$

$$I_D = 5.56\,\text{mA}$$

Use KVL and Ohm's law to find *VD.*

$$V_D = V_{DD} - I_D R_D$$
$$V_D = 20\,\text{V} - 5.56\,\text{mA} \times 1.8\,\text{k}\Omega$$
$$V_D = 9.99\,\text{V}$$

Alternately, using the curve of Figure 12.3, we would first find the normalized gate-source voltage which is 4 V/6 V or 0.667 (note that the curve plots *-VGS/VGS(off)* so that the quadrants do not appear reversed). From this the normalized drain current, *ID/IDSS*, may be determined to be approximately 2.8, yielding a drain current of 5.6 mA.

12.4 THE E-MOSFET

The E-MOSFET is available in both low power and high power versions. It operates in enhancement mode (first quadrant) only. The construction of the low power version is similar to that of the DE-MOSFET but with one important distinction. A simplified cross-section of an N-channel E-MOSFET is shown in Figure 12.9.

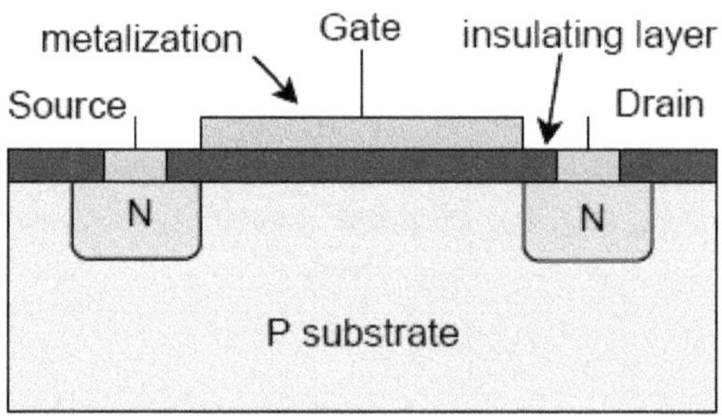

Figure 12.9 E-MOSFET internal structure.

In the E-MOSFET, the P material extends up through the channel and to the gate insulating layer. This has the effect of preventing current flow with negative gate- source voltages applied. Consequently, E-MOSFETs are sometimes referred to as *normally off* devices. In fact, the E-MOSFET will not conduct if *VGS* is zero, or even for small positive values of *VGS*. The P and N materials functionally create energy hills or barriers that prevent current flow through the channel. This can be compared to an NPN BJT that has an open base terminal: no collector current would flow (unless the collector-emitter voltage exceeded the breakdown limit).

To understand how the E-MOSFET functions, refer to Figure 12.10. This diagram shows the device with positive drain and gate supplies attached to it through limiting resistors. The dashed lines indicate electron current flow. As with the DE-MOSFET, the gate can be seen as one plate of a capacitor while the P material serves as the other plate. A positive voltage on the gate will lead to a negative charge on the P material side. If the charge is large enough, all of the holes in the P material can be filled leaving the portion of the material situated near the isolation layer neutral(neither P nor N). Any further increase in gate voltage injects more negative charge into this region, this making it behave like N material. This is called an *N-type inversion layer* and it allows a path for current to flow. The more positive we make the gate voltage, the greater the effect, and the greater the current.

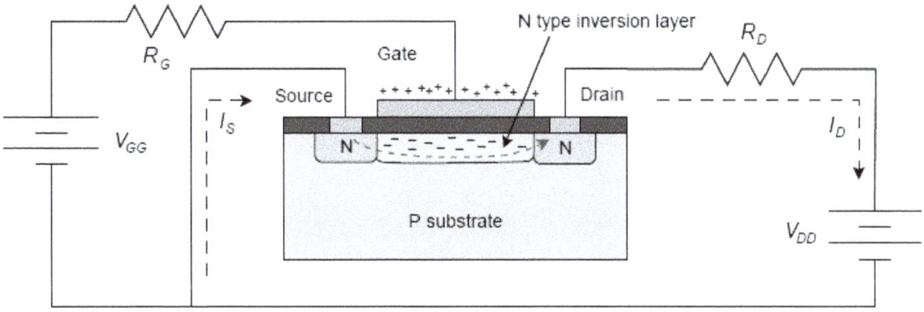

Figure 12.10 E-MOSFET bias with electron flow.

The voltage where the current begins to flow is called the threshold voltage and is usually denoted as *VGS(th)*, although it is sometimes shortened to *Vth* or just *Vt*. Like both the JFET and DE-MOSFET, the E-MOSFET drain curve family exhibits three characteristic regions: the *ohmic* or *triode* region, the *constant current* or *saturation* region, and the *breakdown* region.

The characteristic equation for the E-MOSFET operating in its constant current region is given below. Like the other FETs examined, this is a square-law device.

$$I_D = k(V_{GS} - V_{GS(th)})^2 \qquad (12.5)$$

Where

I_D is the drain current,

V_{GS} is the gate-source voltage ($V_{GS(th)} \leq V_{GS} \leq BV_{GS}$),

$V_{GS(th)}$ is the the threshold voltage,

k is a device parameter (a constant, units of amps/volt^2 or siemens/volt).

The derivative of Equation 12.5 yields the trans-conductance.

$$g_m = \frac{dI_D}{dV_{GS}} = 2k(V_{GS} - V_{GS(th)}) \qquad (12.6)$$

Equation 12.5 is plotted in Figure 12.11. The normalized gate-source voltage is *VGS / VGS(th)* and the normalized drain current is the ratio of *ID* to the current generated when *VGS* is twice *VGS(th)*. This curve is reminiscent of the characteristic curve of a BJT. First, they are both in the first quadrant. Second, both curves exhibit an increasing positive slope. Finally, the curves don't begin to "take off" until some specific turn-on voltage is reached. In the case of the BJT, that voltage is approximately 0.7 V for a silicon device. For the E-MOSFET, that voltage is *VGS(th)*. Obviously though, the MOSFET curve does not increase as rapidly as the BJT curve.

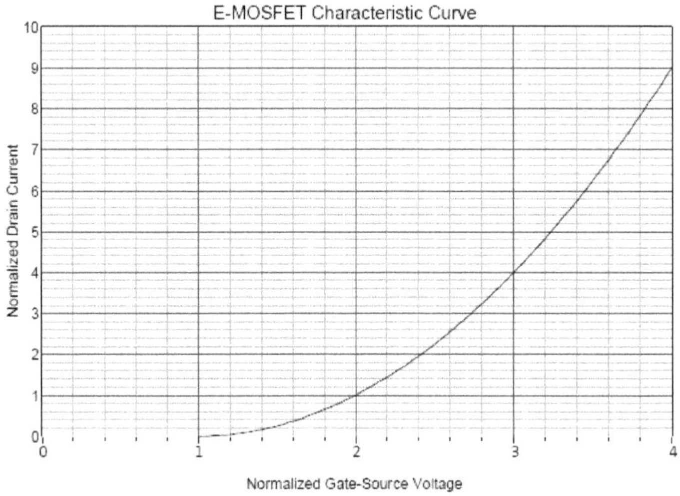

Figure 12.11 Characteristic curve for E-MOSFET.

The power E-MOSFET utilizes a different construction from low power MOSFETs and offers certain advantages over power BJTs including very fast switching speed and lower drive current demands. Consequently, they tend to be favored over BJTs in high power, high speed switching applications such as switching power supply regulators, DC-to-DC converters and class D amplifiers (Chapter 14). These devices also exhibit extremely low *rDS(on)* values, in some cases just a few milliohms. There are different methods of construction, the most recent being the trench style. A cutaway view is shown in Figure 12.12.

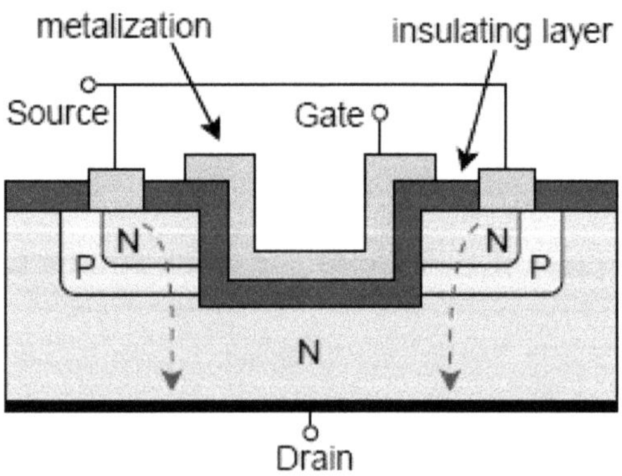

Figure 12.12 Construction of powerTrench E-MOSFET.

A power E-MOSFET is made of a large number of cells, each featuring the U-shaped gate "trench" (an earlier style used a V-shaped trench). Note the location of the drain, now opposite of the gate and source. The advantage here is that the current flows *vertically* rather than laterally. This results in a much lower *rDS(on)* and considerably greater current capacity. The characteristic curve still echoes that of Figure 12.11 although it tends to be steeper when compared to low power devices.

12.5 E-MOSFET DATA SHEET INTERPRETATION

A data sheet for an E-MOSFET, the FDMS86180, is shown in Figure 12.13. This is an N-channel, high power device using trench construction.

FAIRCHILD.

August 2016

FDMS86180

N-Channel Shielded Gate PowerTrench® MOSFET
100 V, 151 A, 3.2 mΩ

Features

- Shielded Gate MOSFET Technology
- Max $r_{DS(on)}$ = 3.2 mΩ at V_{GS} = 10 V, I_D = 67 A
- Max $r_{DS(on)}$ = 7.9 mΩ at V_{GS} = 8 V, I_D = 33 A
- 50% Lower Qrr than Other MOSFET Suppliers
- Lowers Switching Noise/EMI
- MSL1 Robust Package Design
- 100% UIL Tested
- RoHS Compliant

General Description

This N-Channel MV MOSFET is produced using Fairchild Semiconductor's advanced PowerTrench® process that incorporates Shielded Gate technology. This process has been optimized to minimize on-state resistance and yet maintain superior switching performance with best in class soft body diode.

Applications

- Primary DC-DC MOSFET
- Synchronous Rectifier in DC-DC and AC-DC
- Motor Drive
- Solar

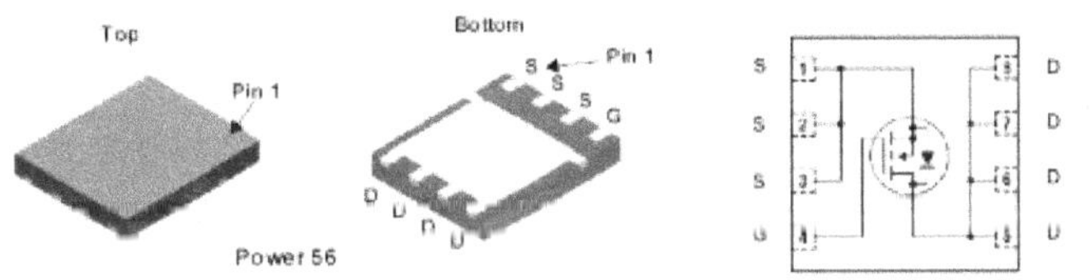

MOSFET Maximum Ratings T_A = 25 °C unless otherwise noted

Symbol	Parameter			Ratings	Units
V_{DS}	Drain to Source Voltage			100	V
V_{GS}	Gate to Source Voltage			±20	V
I_D	Drain Current -Continuous	T_C = 25 °C	(Note 5)	151	A
	-Continuous	T_C = 100 °C	(Note 5)	95	
	-Continuous	T_A = 25 °C	(Note 1a)	21	
	-Pulsed		(Note 4)	775	
E_{AS}	Single Pulse Avalanche Energy		(Note 3)	486	mJ
P_D	Power Dissipation	T_C = 25 °C		138	W
	Power Dissipation	T_A = 25 °C	(Note 1a)	2.7	
T_J, T_{STG}	Operating and Storage Junction Temperature Range			-55 to +150	°C

Thermal Characteristics

$R_{\theta JC}$	Thermal Resistance, Junction to Case		0.9	°C/W
$R_{\theta JA}$	Thermal Resistance, Junction to Ambient	(Note 1a)	45	

Package Marking and Ordering Information

Device Marking	Device	Package	Reel Size	Tape Width	Quantity
FDMS86180	FDMS86180	Power 56	13 "	12 mm	3000 units

Figure 12.13a FDMS86180 data sheet. Used with permission from SCILLC dbaON Semiconductor

One of the first things that might jump out is the "100% RoHS Compliant" green leaf logo in the upper center, meaning that the device meets the Restriction of Hazardous Substances directive. The device comes in the flat, multi-pin Power

56 package and features an *rDS(on)* of just a few milliohms. Continuous current capability at room temperature is 151 amps with a pulsed current maximum of 775 amps. In Figure 12.13b we find a breakdown voltage of 100 volts and an *IDSS* of only 1 μA. Recall that this is a normally off device, and thus *IDSS* represents a leakage current. Continuing, *VGS(th)* varies between 2.0 and 4.0 volts, with 3.2 volts being typical. The forward trans-conductance, *gm* (here referred to as *gFS*) is 144 siemens at a drain current of 67 amps. This is orders of magnitude greater than what we might see with small signal devices. Turn-on and turn-off times are measured in the tens of nanoseconds, verifying the high speed switching ability of the device.

Electrical Characteristics T_J = 25 °C unless otherwise noted

Symbol	Parameter	Test Conditions	Min.	Typ.	Max.	Units
Off Characteristics						
BV_{DSS}	Drain to Source Breakdown Voltage	I_D = 250 μA, V_{GS} = 0 V	100			V
$\frac{\Delta BV_{DSS}}{\Delta T_J}$	Breakdown Voltage Temperature Coefficient	I_D = 250 μA, referenced to 25 °C		73		mV/°C
I_{DSS}	Zero Gate Voltage Drain Current	V_{DS} = 80 V, V_{GS} = 0 V			1	μA
I_{GSS}	Gate to Source Leakage Current	V_{GS} = ±20 V, V_{DS} = 0 V			100	nA
On Characteristics						
$V_{GS(th)}$	Gate to Source Threshold Voltage	V_{GS} = V_{DS}, I_D = 370 μA	2.0	3.2	4.0	V
$\frac{\Delta V_{GS(th)}}{\Delta T_J}$	Gate to Source Threshold Voltage Temperature Coefficient	I_D = 370 μA, referenced to 25 °C		-8		mV/°C
$r_{DS(on)}$	Static Drain to Source On Resistance	V_{GS} = 10 V, I_D = 67 A		2.4	3.2	mΩ
		V_{GS} = 6 V, I_D = 33 A		3.8	7.9	
		V_{GS} = 10 V, I_D = 67 A, T_J = 125 °C		4.0	5.4	
g_{FS}	Forward Transconductance	V_{DS} = 5 V, I_D = 67 A		144		S
Dynamic Characteristics						
C_{iss}	Input Capacitance	V_{DS} = 50 V, V_{GS} = 0 V, f = 1 MHz		4439	6215	pF
C_{oss}	Output Capacitance			2663	3730	pF
C_{rss}	Reverse Transfer Capacitance			24	55	pF
R_g	Gate Resistance		0.1	0.8	1.6	Ω
Switching Characteristics						
$t_{d(on)}$	Turn-On Delay Time	V_{DD} = 50 V, I_D = 67 A, V_{GS} = 10 V, R_{GEN} = 6 Ω		24	39	ns
t_r	Rise Time			12	22	ns
$t_{d(off)}$	Turn-Off Delay Time			30	48	ns
t_f	Fall Time			7	14	ns
Q_g	Total Gate Charge	V_{GS} = 0 V to 10 V; V_{DD} = 50 V, I_D = 67 A		60	84	nC
Q_g	Total Gate Charge	V_{GS} = 0 V to 6 V; V_{DD} = 50 V, I_D = 67 A		38	54	nC
Q_{gs}	Gate to Source Charge			20		nC
Q_{gd}	Gate to Drain "Miller" Charge			12		nC
Q_{oss}	Output Charge	V_{DD} = 50 V, V_{GS} = 0 V		175		nC
Drain-Source Diode Characteristics						
V_{SD}	Source to Drain Diode Forward Voltage	V_{GS} = 0 V, I_S = 2.1 A (Note 2)		0.7	1.2	V
		V_{GS} = 0 V, I_S = 67 A (Note 2)		0.8	1.3	
t_{rr}	Reverse Recovery Time	I_F = 33 A, di/dt = 300 A/μs		44	71	ns
Q_{rr}	Reverse Recovery Charge			109	207	nC
t_{rr}	Reverse Recovery Time	I_F = 33 A, di/dt = 1000 A/μs		33	53	ns
Q_{rr}	Reverse Recovery Charge			235	376	nC

Figure 12.13b FDMS86180 data sheet (cont).

A series of performance graphs are found in Figure 12.13c. In the upper left is a section of drain curves showing the ohmic region through *VDS* = 5 V. The plot directly below this shows the increase in *rDS(on)* as temperature rises. There is about a three-fold variation across the temperature range. At lower left is the characteristic curve variation. Note that the curves are less steep as temperature increases, showing a decrease in *gm* and thus, verifying a negative temperature coefficient of trans-conductance.

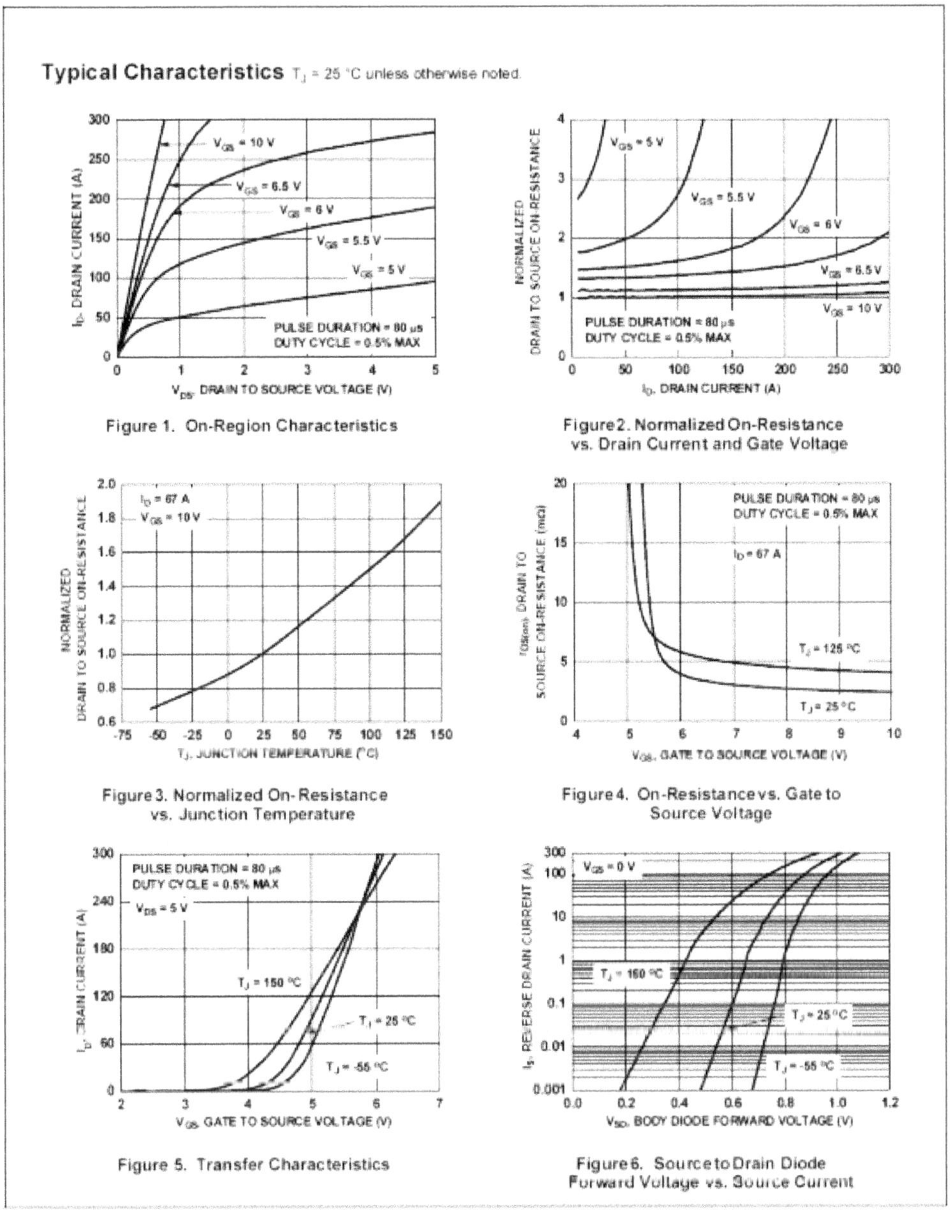

Figure 12.13c FDMS86180 data sheet (cont).

12.6 E-MOSFET BIASING

As the E-MOSFET operates only in the first quadrant, none of the biasing schemes used with JFETs will work with it. First, it should be noted that for large signal switching applications biasing is not much of an issue as we simply need to confirm that there is sufficient drive signal to turn the device on. For linear amplifiers we can use variations on constant voltage bias such as voltage divider bias, or on drain feedback bias.

Voltage Divider Bias

Voltage divider bias is reminiscent of the divider circuit used with BJTs. Indeed, the N-channel E-MOSFET requires that its gate be higher than its source, just as the NPN BJT requires a base voltage higher than its emitter. The major differences between the two are that the E-MOSFET's input gate current is negligible compared to base current and that the gate-source voltage will be most likely higher than the 0.7 volt drop seen across the base-emitter junction. Also, the gate-source voltage will not be locked to a specific voltage but will vary depending on the remainder of the circuit.

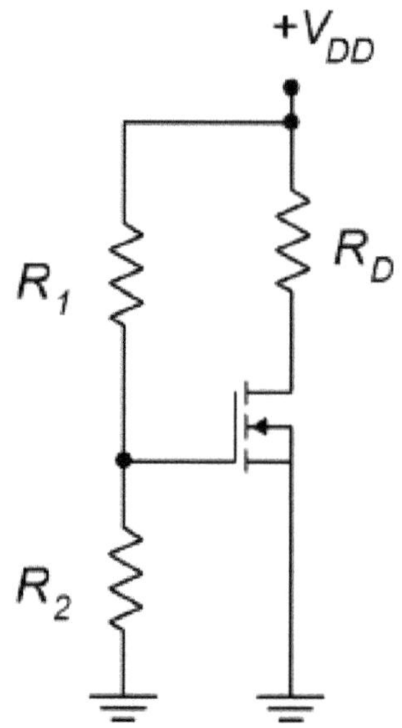

Figure 12.14 Voltage divider bias for E-MOSFET.

The prototype for the voltage divider bias is shown in Figure 12.14. In general, the layout it is the same as the voltage divider bias used with the DE-MOSFET. The resistors *R1* and *R2* set up the divider to establish the gate voltage. As the source terminal is tied directly to ground, this means that *VGS* = *VG*. The potential across *R2* needs to be set above *VGS(th)* for proper operation in accordance with Equation 12.5. Knowing the value of *VG*, either the characteristic equation or the corresponding normalized drain current plot can be used to determine the drain current. The only factor missing is the device constant, *k*. This can be computed for any particular device based on the *ID(on)*, *VGS(on)* coordinate pair specified in the data sheet (or measured in lab). An example is shown in Figure 12.15.

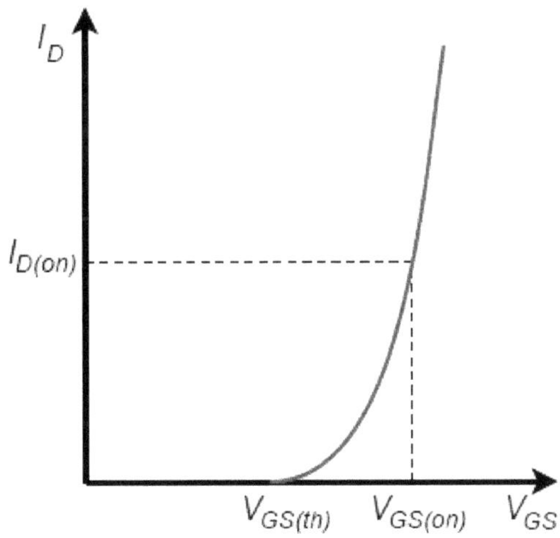

Figure 12.15 Coordinate pair on E-MOSFET curve.

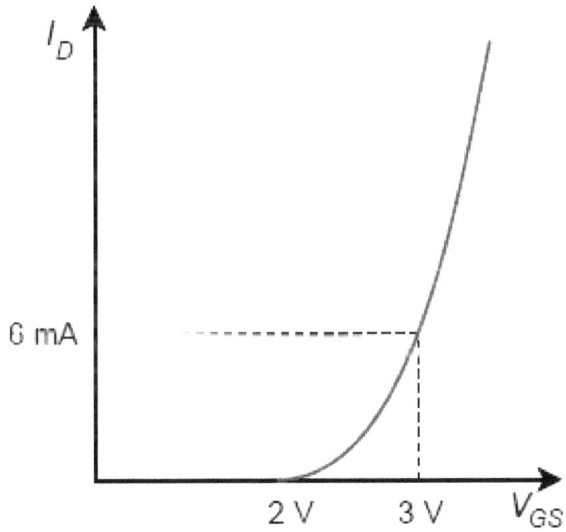

Figure 12.16b Device curve for Example 12.3.

The constant k is found via a rearrangement of Equation 12.5:

$$k = \frac{I_{D(on)}}{\left(V_{GS(on)} - V_{GS(th)}\right)^2}$$

This value can then be used for other biasing points.

Example 12.3

For the circuit and matching device curve of Figure 12.16, find *ID* and *VDS*.

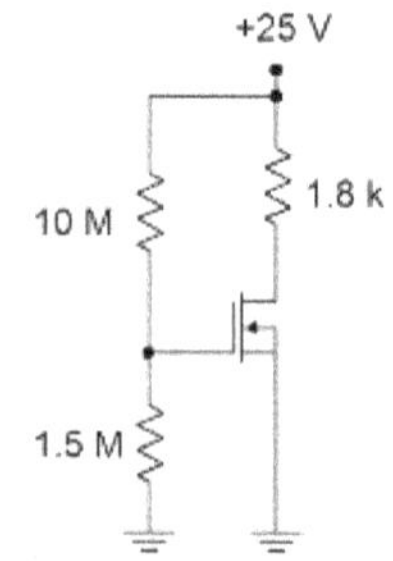

Figure 12.16a Circuit for Example 12.3.

First find the value of *k*:

$$k = \frac{I_{D(on)}}{(V_{GS(on)} - V_{GS(th)})^2}$$
$$k = \frac{6\text{ mA}}{(3\text{ V} - 2\text{ V})^2}$$
$$k = 6\text{ mA/V}^2$$

Now determine the gate voltage:

$$V_G = V_{DD}\frac{R_2}{R_1 + R_2}$$
$$V_G = 25\text{ V}\frac{1.5\text{ M}\Omega}{10\text{ M}\Omega + 1.5\text{ M}\Omega}$$
$$V_G = 3.26\text{ V}$$

The source is grounded so *VGS* = *VG*.

$$I_D = k(V_{GS} - V_{GS(th)})^2$$
$$I_D = 6\text{mA/V}^2(3.26\text{ V} - 2\text{ V})^2$$
$$I_D = 9.54\text{ mA}$$

$$V_{DS} = V_{DD} - I_D R_D$$
$$V_{DS} = 25\text{V} - 9.54\text{ mA} \times 1.8\text{ k}\Omega$$
$$V_{DS} = 7.83\text{ V}$$

In closing, note that it is possible to decouple the voltage divider using the same method employed with BJTs in Figure 7.13. Very large value resistors are available in only a limited variety of sizes so this technique has an added benefit. The divider resistors can use more convenient sizes because *R1* and *R2* will not set the input impedance; it will be set by the decoupling resistor.

6 mA

Drain Feedback Biass

Drain feedback bias utilizes the aforementioned "on" operating point from the characteristic curve. The idea is to establish a drain current via an appropriate selection of the drain resistor and power supply. The prototype of the drain feedback circuit is shown in Figure 12.17.

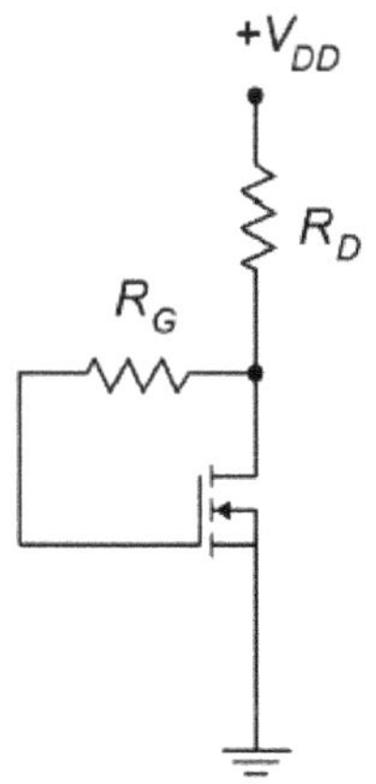

Figure 12.17 Drain feedback bias prototype.

This is relatively simple layout using few components. The key to understanding its operation is the KVL summation:

$$V_{DD} = V_{R_D} + V_{R_G} + V_{GS}$$
$$V_{DD} = I_D R_D + I_G R_G + V_{GS}$$

Gate current is negligible which means that

$$V_{DD} = I_D R_D + V_{GS}$$

and also

$$V_{DS} = V_{GS}$$

Therefore,

$$V_{GS} = V_{DS} = V_{DD} - I_D R_D \qquad (12.7)$$

Equation 12.7 can be used as the basis for the design of the bias circuit.

Example 12.4

Utilizing the prototype of Figure 12.17, determine values for *RD* and *RG* such that the drain current is 8 mA. Assume *VDD* = 20 V, *ID(on)* = 5 mA at *VGS(on)* = 4 V, and *VGS(th)* = 2.5 V.

First find the value of *k*:

$$k = \frac{I_{D(on)}}{\left(V_{GS(on)} - V_{GS(th)}\right)^2}$$

$$k = \frac{5\,\text{mA}}{\left(4\,\text{V} - 2.5\,\text{V}\right)^2}$$

$$k = 2.22\,\text{mA/V}^2$$

CONCISE SUMMARY

- There are two types of MOSFETs: the depletion-enhancement or DE-MOSFET and the enhancement-only or E-MOSFET. Both devices are constructed using an insulated gate instead of a PN junction and both devices exhibit a square-law characteristic curve. Like the JFET, MOSFETs are modeled as voltage-controlled current sources. Both devices show very, very small gate currents due to the insulated gate. They are static sensitive and precautions must be taken when handling them to avoid damage from ESD.
- The DE-MOSFET exhibits the same characteristic curve as the JFET, however, the curve extends into the first quadrant (enhancement mode). Consequently, *IDSS* is no longer the largest drain current possible, but rather, represents a middle ground. The DE-MOSFET can utilize all of the bias prototypes that are used with JFETs, including self bias, constant current bias and combination bias. Due to its dual quadrant capability, other biasing types are also possible including zero bias and voltage divider, both of which are variations on constant voltage bias.
- The E-MOSFET operates in the first quadrant only (enhancement mode). Compared to the DE-MOSFET, its characteristic curve is shifted positive such that *VGS(off)* is now *VGS(th)*, and *IDSS* signifies the off-state leakage current. E-MOSFETs are available in both low power and high power variants. The high power versions utilize an alternate internal structure that allows drain current to flow vertically rather than horizontally. This results in very high current carrying ability and very low values for *rDS(on)*.

CHAPTER-13

MOSFET SMALL SIGNAL AMPLIFIERS: COMMON SOURCE OF AMPLIFIERS

13.1 INTRODUCTION: AN OVERVIEW

Metal-Oxide-Semiconductor Field-Effect Transistors (MOSFETs) operating as small-signal amplifiers in the common-source configuration form a crucial category within electronic circuitry. In this configuration, the MOSFET serves as a voltage amplifier, enhancing weak input signals with minimal distortion. Common source amplifiers are prevalent in applications such as audio processing, communication systems, and integrated circuits due to their favorable characteristics, including high input impedance and voltage gain. MOSFETs, being voltage-controlled devices, offer precise control over amplification levels, making them well-suited for scenarios where signal fidelity is paramount. As small-signal amplifiers using MOSFETs continue to play a pivotal role in electronic systems, understanding their operational principles, design considerations, and advantages is essential for engineers and designers working on circuits that demand efficient and accurate signal processing.

MOSFETs can be used to create both common source voltage amplifiers and common drain voltage followers (i.e., source followers). Both circuits offer the potential for very high input impedance due to the extremely low gate current MOSFETs provide. As with JFET amplifiers, at higher frequencies input capacitance dominates and reduces the input impedance. Not all bias prototypes lend themselves to all possible AC circuits. For example, zero bias for a DE-MOSFET is not suitable for followers or swamped amplifiers as it lacks a source resistor. The same is true for voltage divider biasing used with both DE- and E-MOSFETs. These biasing schemes are suitable for non-swamped amplifiers, though.

In general, MOSFET amplifiers tend to have good high frequency performance, offer low noise and exhibit low distortion with modestly sized input signals. Compared to BJTs, their voltage gain magnitude is lower.

A key parameter in determining gain is the device's trans-conductance, gm. Trans-conductance varies widely depending on the kind of MOSFET used. A small signal DE-MOSFET may exhibit a trans-conductance of just a few millisiemens. In contrast, a high power E-MOSFET may exhibit a trans-conductance of over 100 siemens.

13.2 MOSFET SOURCE AMPLIFIERS

Before we can examine the common source amplifier, an AC model is needed for both the DE- and E-MOSFET. A simplified model consists of a voltage-controlled current source and an input resistance, rGS. This model is shown in Figure 13.1. The model is essentially the same as that used for the JFET. Technically, the gate-source resistance is higher in the MOSFET due to the insulated gate, and this is useful in specific applications such as in the design of electrometers, but for general purpose work it is a minor distinction. The impedance associated with the current source is not shown as it is typically large enough to ignore. Similarly, the device capacitances are not shown. It is worth noting that the capacitances associated with small signal devices might be just a few picofarads, however, a power device might exhibit values of a few nanofarads.

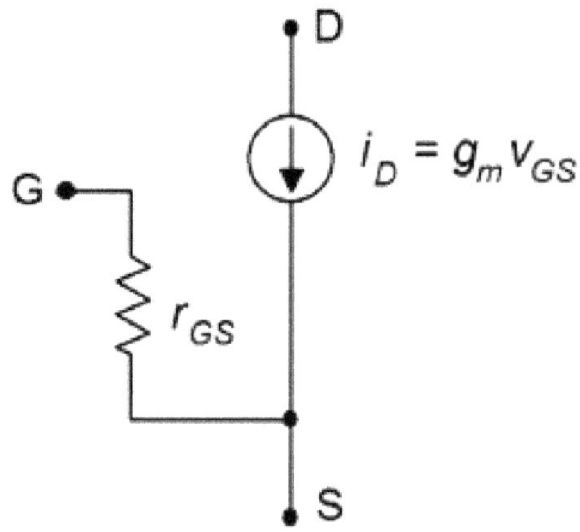

Figure 13.1 AC device model for MOSFETs.

As the device model is the same for both DE- and E-MOSFETs, the analysis of voltage gain, input impedance and output impedance will apply to both devices. The only practical differences will be how the trans-conductance is determined, and circuit variations due to the differing biasing requirements which will affect the input impedance. In fact, there will be a great uniformity between JFET-based circuits and DE-MOSFET circuits operating in depletion mode.

An AC equivalent of a swamped common source amplifier is shown in Figure 13.2. This is a generic prototype and is suitable for any variation on device and bias type. Ultimately, all of the amplifiers can be reduced down to this equivalent, occasionally with some resistance values left out (either opened or shorted). For example, if the amplifier is not swamped then rS = 0. Similarly, rG might correspond to a single gate biasing resistor or it might represent the equivalent of a pair of resistors that set up a gate voltage divider.

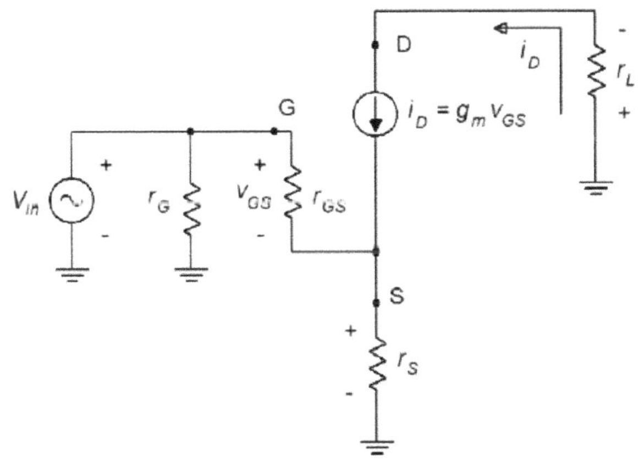

Figure 13.2 Generic common source amplifier equivalent.

In order to derive an equation for the voltage gain, we start with its definition, namely that voltage gain is the ratio of vout to vin. We then proceed by expressing these voltages in terms of their Ohm's law equivalents. Note that rL can also be called rD.

Voltage Gain

This is the general equation for voltage gain. If the amplifier is not swamped then the first portion of the denominator drops out and the gain simplifies to

$$A_v = \frac{v_{out}}{v_{in}} = \frac{v_L}{v_G} = \frac{v_D}{v_G}$$

$$A_v = \frac{-i_D r_L}{i_D r_S + v_{GS}}$$

$$A_v = \frac{-g_m v_{GS} r_L}{g_m v_{GS} r_S + v_{GS}}$$

$$A_v = -\frac{g_m r_L}{g_m r_S + 1} \qquad (13.1)$$

or, if preferred

$$A_v = -\frac{g_m r_D}{g_m r_S + 1} \qquad (13.1b)$$

This is the general equation for voltage gain. If the amplifier is not swamped then the first portion of the denominator drops out and the gain simplifies to

$$A_v = -g_m r_L \qquad (13.2)$$

or alternately

$$A_v = -g_m r_D \qquad (13.2b)$$

The swamping resistor, *rS*, plays the same role here as it did with both the BJT and JFET. Swamping helps to stabilize the gain and reduce distortion, but at the expense of voltage gain.

Input Impedance

Referring back to Figure 13.2, the input impedance of the amplifier will be *rG* in parallel with the impedance looking into the gate terminal, *Zin(gate)*. At a minimum this will be *rGS* (it is somewhat higher when swamped but this can be ignored in most cases). At low frequencies *rGS* is very large, perhaps as high as 10^{12} ohms. In most practical circuits, *rG* will be much lower, hence

$$Z_{in} = r_G \,||\, r_{GS} \approx r_G \qquad (13.3)$$

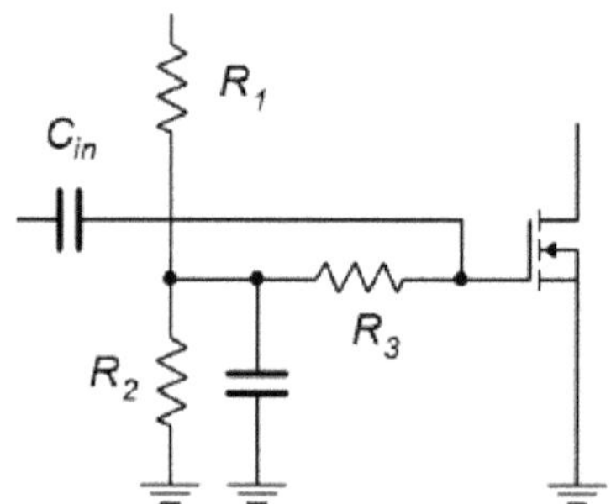

Figure 13.3 Decoupled voltage divider.

It is important to reiterate that *rG* is the equivalent resistance seen prior to the gate terminal that is seen from the vantage point of *Vin*. In the case of self bias, combination bias, zero bias and constant current bias, this will be the single biasing resistor *RG*. For simple voltage divider biasing, *rG* will be the parallel combination of the two divider resistors (i.e., *R1* || *R2*). For decoupled voltage divider biasing, as shown in Figure 13.3, *rG* will be the decoupling resistor (i.e., *R3*) that is connected between the divider and the gate. This is because the divider node is bypassed to ground via a capacitor. Finally, for drain feedback biasing, *rG* is the Millerized *RG* that bridges the drain and gate.

Output Impedance

The derivation of output impedance is unchanged from the JFET case. From the perspective of the load, the output impedance will be the drain biasing resistor, RD, in parallel with the internal impedance of the current source within the device model. RD tends to be much lower than this, and thus, the output impedance can be approximated as RD.

Therefore we may state

$$Z_{out} = r_{model} \,||\, R_D \approx R_D \tag{13.4}$$

Zout = *rmodel* || *RD* ≈ *RD* (13.4)

At this point, a variety of examples are in order to illustrate some of the myriad combinations.

Example 13.1

For the amplifier in Figure 13.4, determine the input impedance and load voltage. Vin = 20 mV, VDD = 20 V, RG = 1 MΩ, RD = 1.8 kΩ, RSW = 20 Ω, RS = 400 Ω, RL = 12 kΩ, IDSS = 40 mA, VGS(off) = −1 V.

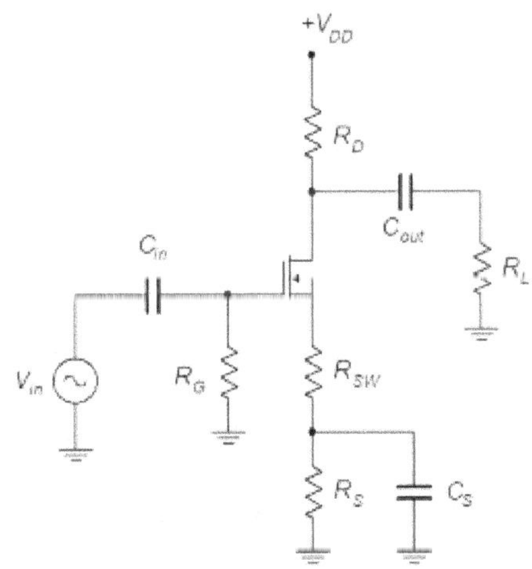

Figure 13.4 Circuit for Example 13.1.

This is a swamped common drain amplifier utilizing self bias. *Zin*can be determined via inspection.

$$Z_{in} = Z_{in(gate)} \,||\, R_G$$
$$Z_{in} \approx 1\,\text{M}\Omega$$

To find the load voltage we'll need the voltage gain, and to find the gain we'll first need to find *gm0*.

$$g_{m0} = -\frac{2I_{DSS}}{V_{GS(off)}}$$
$$g_{m0} = -\frac{80\,\text{mA}}{-1\,\text{V}}$$
$$g_{m0} = 80\,\text{mS}$$

The combined DC value of *RS* is 420 Ω, therefore *gm0RS* = 33.6. From the self bias equation or graph this produces a drain current of 1.867 mA.

$$g_m = g_{m0}\sqrt{\frac{I_D}{I_{DSS}}}$$
$$g_m = 80\,\text{mS}\sqrt{\frac{1.867\,\text{mA}}{40\,\text{mA}}}$$
$$g_m = 17.3\,\text{mS}$$

The swamping resistor, *rS*, is 20 Ω. The voltage gain is

$$A_v = -\frac{g_m r_L}{g_m r_S + 1}$$
$$A_v = -\frac{17.3\,\text{mS}(1.8\,\text{k}\Omega \,||\, 12\,\text{k}\Omega)}{17.3\,\text{mS} \times 20\,\Omega + 1}$$
$$A_v = -20.1$$

And finally

$$V_{load} = A_v V_{in}$$
$$V_{load} = -20.1 \times 20\,\text{mV}$$
$$V_{load} = 402\,\text{mV}$$

Computer Simulation

The amplifier of Example 13.1 is simulated to verify the results. The circuit is entered into the simulator as shown in Figure 13.5. One issue is finding an appropriate DE-MOS device to match the parameters used in the example. The BSS229 proves to be reasonably close. This device model was tested for IDSS by applying a 20 volt source to the drain and shorting the source and gate terminals to ground in the simulator. The current was just under the 40 mA target. Similarly, a negative voltage was attached to the gate and adjusted until the drain current dropped to nearly zero in order to determine VGS(off). The model's value was just under the desired -1 volt. Consequently, we can expect the simulation results to be close to those predicted, although not identical.

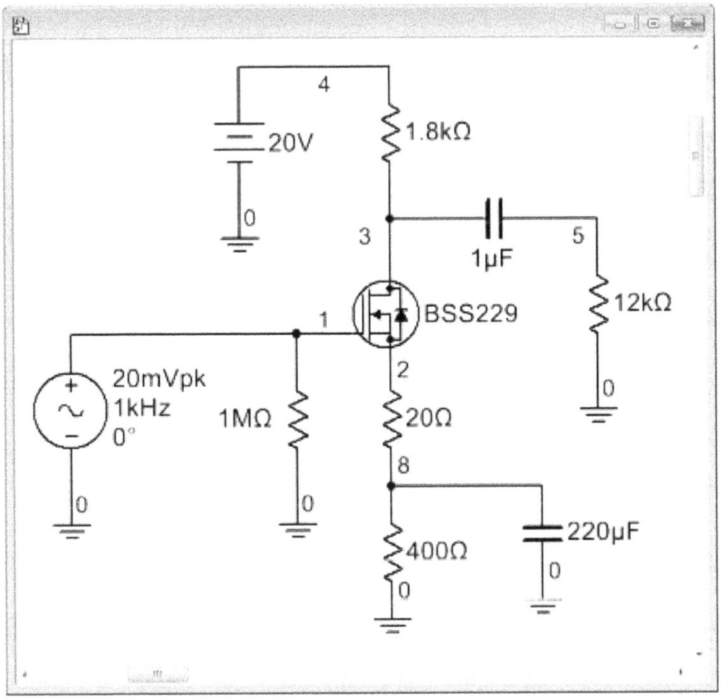

Figure 13.5 The circuit of Example 13.1 in the simulator.

The transient analysis is run next and is shown in Figure 13.6. The expected signal inversion is obvious. The peak amplitude is 417 mV, just a few percent higher than the calculated value. At least some of this deviation is due to the model's variation from the assumed device parameter values.

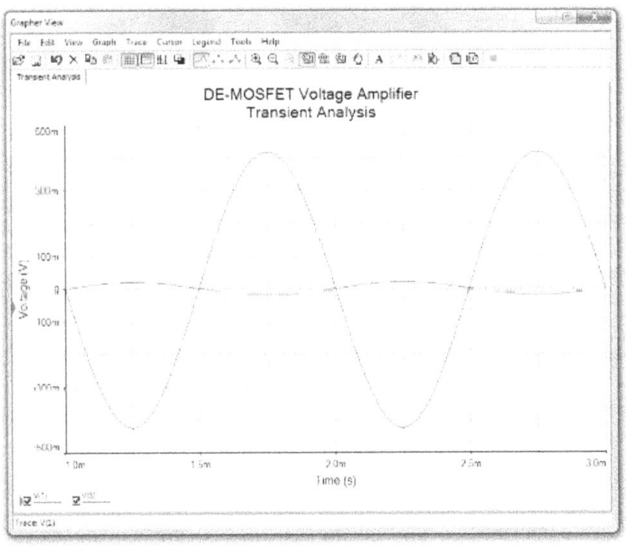

Figure 13.6 Transient analysis simulation for the circuit of Example 13.1.

A DC bias check is also performed. The drain current was calculated to be 1.867 mA. This yields an RD voltage of a little over 3 volts, thus we expect to see a drain voltage of about 17 volts. Similarly, we would expect the source terminal to be sitting at around 700 to 800 mV and the gate at about 0 V.

The results of the DC operating point simulation are shown in Figure 13.7. The agreement with the predicted values is quite good, especially considering that the device model is not a perfect match.

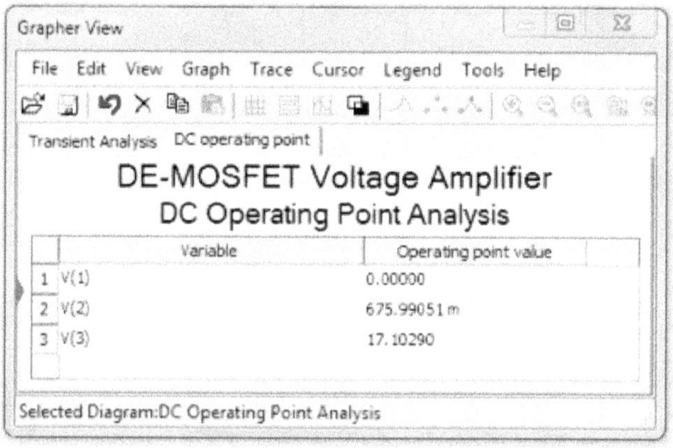

Figure 13.7 DC bias simulation for the circuit of Example 13.1.

Example 13.2

For the circuit of Figure 13.8, determine the voltage gain and input impedance. Assume *VGS(th)* = 2 V, *ID(on)* = 50 mA at *VGS(on)* = 5 V.

First find the value of *k*:

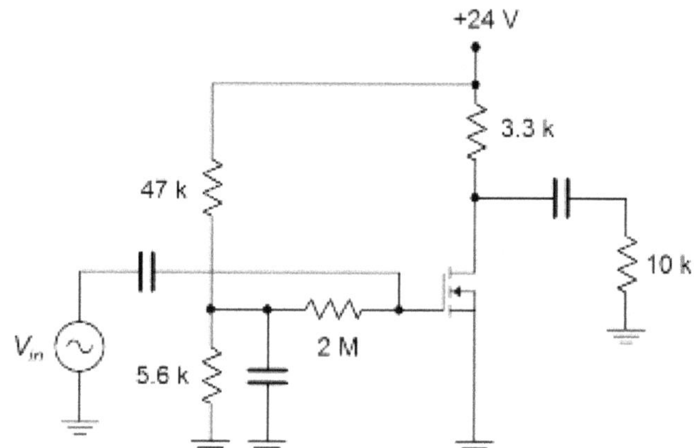

Figure 13.8 Circuit for Example 13.2.

This circuit uses power supply decoupling. The voltage drop across the 2 MΩ resistor is small enough to ignore as the current passing through it is gate current.

Therefore the gate voltage is determined by the divider. Also, as the left end of the 2 MΩ resistor is tied to an AC ground due to the bypass capacitor, it represents the input impedance.

$$Z_{in} = 2\ \text{M}\Omega \parallel Z_{in(gate)} \approx 2\ \text{M}\Omega$$

$$V_G = V_{DD}\frac{R_2}{R_1+R_2}$$
$$V_G = 24\,\text{V}\frac{5.6\,\text{k}\Omega}{47\text{k}\Omega+5.6\,\text{k}\Omega}$$
$$V_G = 2.56\,\text{V}$$

The source is grounded so VGS = VG.

$$I_D = k(V_{GS} - V_{GS(th)})^2$$
$$I_D = 5.56\,\text{mA/V}^2(2.56\,\text{V} - 2\,\text{V})^2$$
$$I_D = 1.74\,\text{mA}$$

$$g_m = 2k(V_{GS} - V_{GS(th)})$$
$$g_m = 2\times 5.56\,\text{mA/V}^2(2.56\,\text{V} - 2\,\text{V})$$
$$g_m = 6.23\,\text{mS}$$

This amplifier is not swamped so the simplified gain equation may be used.

$$A_v = -g_m r_D$$
$$A_v - -6.23\,\text{mS}(3.3\,\text{k}\Omega \parallel 10\,\text{k}\Omega)$$
$$A_v = -15.5$$

Example 13.3

For the circuit of Figure 13.9, determine the voltage gain and input impedance. Assume VGS(off) = -0.75 V and IDSS = 6 mA.

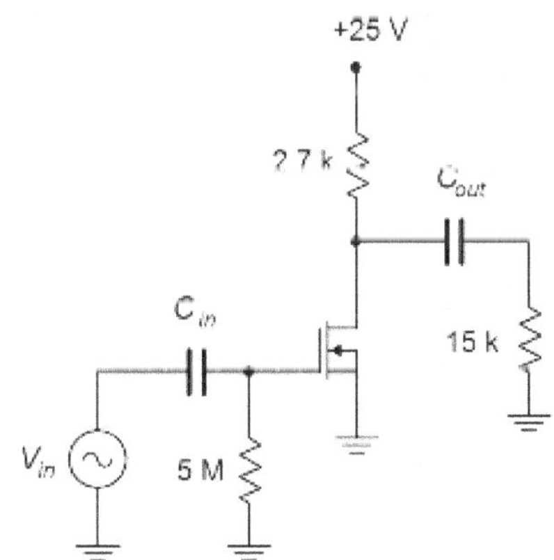

Figure 13.9 Circuit for Example 13.3.

This amplifier uses zero bias, therefore ID = IDSS and gm = gm0.

$$g_{m0} = -\frac{2\,I_{DSS}}{V_{GS(off)}}$$
$$g_{m0} = -\frac{2\times 6\,\text{mA}}{-0.75\,\text{V}}$$
$$g_{m0} = 16\text{mS}$$

This amplifier is not swamped so we may use the simplified equation for voltage gain.

$$A_v = -g_m r_D$$
$$A_v = -16\,\text{mS}\,(2.7\text{k}\,\Omega \;||\; 15\text{k}\,\Omega)$$
$$A_v = -36.6$$

Finally, for the input impedance we have

$$Z_{in} = 5\ \text{M}\Omega \;||\; Z_{in(gate)} \approx 5\ \text{M}\Omega$$

13.3 MOSFET COMMON DRAIN FOLLOWERS

As discussed under the section on JFETs, the common drain amplifier is also known as the source follower. The prototype amplifier circuit with device model is shown in Figure 13.10. As with all voltage followers, we expect a non-inverting voltage gain close to unity with a high Zin and a low Zout.

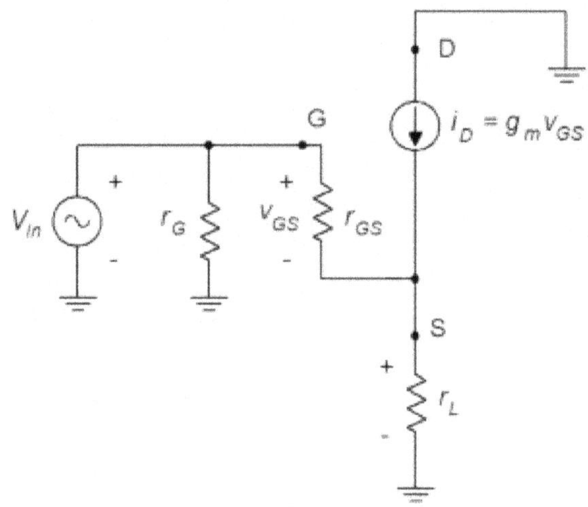

Figure 13.10 Common drain (source follower) prototype.

As is usual, the input signal is applied to the gate terminal and the output is taken from the source. Because the output is at the source, biasing schemes that have the source terminal grounded, such as zero bias and voltage divider bias, cannot be used.

Voltage Gain

The voltage gain equation for the common drain follower is developed as follows: We begin with the fundamental definition that voltage gain is the ratio of vout to vin, and proceed by expressing these voltages in terms of their Ohm's law equivalents. The load is now located at the MOSFET's source, and thus can be referred to as either rL or rS.

$$A_v = \frac{v_{out}}{v_{in}} = \frac{v_S}{v_G} = \frac{v_L}{v_G}$$

$$A_v = \frac{i_D r_L}{i_D r_L + v_{GS}}$$

$$A_v = \frac{g_m v_{GS} r_L}{g_m v_{GS} r_L + v_{GS}}$$

$$A_v = \frac{g_m r_L}{g_m r_L + 1} \qquad (13.5)$$

or, if preferred

$$A_v = \frac{g_m r_S}{g_m r_S + 1} \qquad (13.5b)$$

If g_{mrS}>> 1, the voltage gain will be very close to unity; a desired outcome.

Input Impedance

The analysis for source follower's input impedance is virtually identical to that for the common source amplifier. The same commentary applies regarding the simplification of gate biasing resistors to arrive at the value of r_G.

$$Z_{in} = r_G \,||\, r_{GS} \approx r_G \qquad (13.6)$$

Output Impedance

In order to determine the output impedance, we modify the circuit of Figure 13.10 by separating the load resistance from the source bias resistor. This is shown in Figure 13.11.

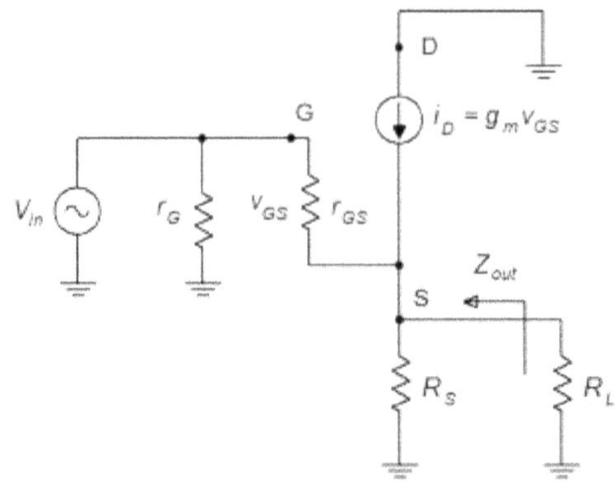

Figure 13.11 Source follower output impedance analysis.

Looking back into the source from the perspective of the load we find that the source biasing resistor, RS, is in parallel with the impedance looking back into the source terminal.

$$Z_{out} = R_S \parallel Z_{source}$$

To find Zsource, note that the voltage at the source is vGS and the current entering this node is iD. The ratio of the two will yield the impedance looking back into the source.

$$Z_{source} = \frac{v_{GS}}{i_D}$$

$$Z_{source} = \frac{v_{GS}}{g_m v_{GS}}$$

$$Z_{source} = \frac{1}{g_m} \qquad (13.7)$$

Therefore, the output impedance is

$$Z_{out} = R_S \parallel \frac{1}{g_m} \qquad (13.8)$$

Looking at Equation 13.8 it is obvious that the higher the trans-conductance, the lower the output impedance. As noted earlier, a large trans-conductance also means that the voltage gain will be close to unity. As a general rule then, a large trans-conductance is desired for the source follower.

Time for a few illustrative examples.

Example 13.4

For the circuit of Figure 13.12, determine the voltage gain and input impedance. Assume VGS(off) = -0.8 V and IDSS = 30 mA.

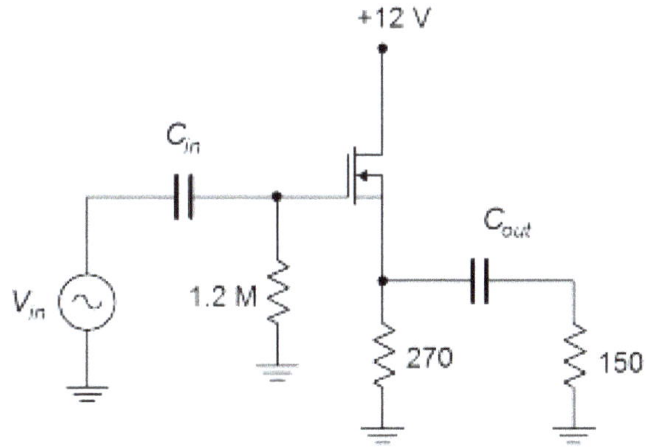

Figure 13.12 Circuit for Example 13.4.

This amplifier uses self bias so we need to determine *gm0RS*.

$$g_{m0} = -\frac{2I_{DSS}}{V_{GS(off)}}$$
$$g_{m0} = -\frac{2 \times 30\,\text{mA}}{-0.8\,\text{V}}$$
$$g_{m0} = 75\,\text{mS}$$

The DC source resistance is the 270 Ω biasing resistor resulting in *gm0 RS* = 16.2. From the self bias equation or graph this produces a drain current of 2.61 mA.

$$g_m = g_{m0}\sqrt{\frac{I_D}{I_{DSS}}}$$
$$g_m = 75\,\text{mS}\sqrt{\frac{2.61\,\text{mA}}{30\,\text{mA}}}$$
$$g_m = 22.1\,\text{mS}$$

The voltage gain is

$$A_v = \frac{g_m r_S}{g_m r_S + 1}$$
$$A_v = \frac{22.1\,\text{mS}(270\,\Omega \,||\, 150\,\Omega)}{22.1\,\text{mS} \times (270\,\Omega \,||\, 150\,\Omega) + 1}$$
$$A_v = 0.68$$

Finally, for the input impedance we have

$$Z_{in} = 1.2\,\text{M}\Omega \,||\, Z_{in(gate)} \approx 1.2\,\text{M}\Omega$$

Example 13.5

For the circuit of Figure 13.13, determine the voltage gain and input impedance. Assume *VGS(off)* = -2.5 V and *IDSS* = 80 mA.

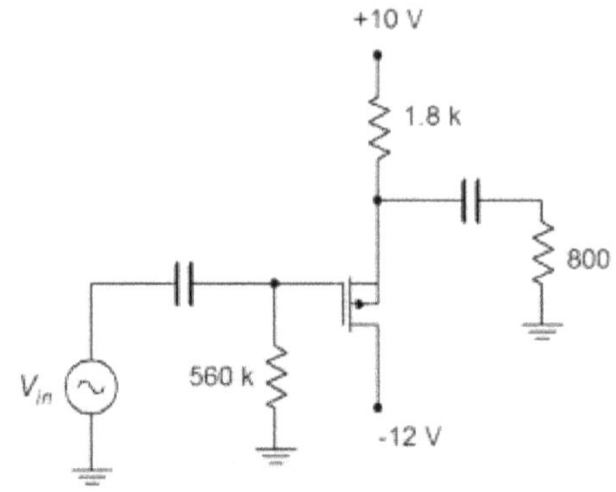

Figure 13.13 Circuit for Example 13.5.

This follower uses a P-channel device with combination bias. Note that the source terminal is toward the top of the schematic. First, determine *gm0RS* and the bias factor, *k*. Then the combination bias equation can be used to determine the drain current.

$$g_{m0} = -\frac{2I_{DSS}}{V_{GS(off)}}$$

$$g_{m0} = -\frac{2\times 80\,\text{mA}}{-2.5\,\text{V}}$$

$$g_{m0} = 64\,\text{mS}$$

The DC source resistance is the 1.8 kΩ biasing resistor resulting in *gm0 RS* = 115.2. The bias factor is *VSS/VGS(off)*, or 4. The combination bias equation (Equation 10.9) yields *ID*= 6.67 mA.

We can now find the trans-conductance and voltage gain.

$$g_m = g_{m0}\sqrt{\frac{I_D}{I_{DSS}}}$$

$$g_m = 64\,\text{mS}\sqrt{\frac{6.67\,\text{mA}}{80\,\text{mA}}}$$

$$g_m = 18.5\,\text{mS}$$

The voltage gain is

$$A_v = \frac{g_m r_S}{g_m r_S + 1}$$

$$A_v = \frac{18.5\,\text{mS}\,(1.8\,\text{k}\Omega \,||\, 800\,\Omega)}{18.5\,\text{mS}\times(1.8\,\text{k}\Omega \,||\, 800\,\Omega) + 1}$$

$$A_v = 0.91$$

Lastly, the input impedance is

$$Z_{in} = 560\ \text{k}\Omega \parallel Z_{in(gate)} \approx 560\ \text{k}\Omega$$

CONCISE SUMMARY

- DE- and E-MOSFET devices may be used to create both common source voltage amplifiers and common drain voltage followers. The common source amplifiers may be swamped or non-swamped, depending on the bias form used. If the bias type does not utilize a source resistor, swamping is not available. This includes zero bias for the DE-MOSFET and voltage divider bias for both the DE- and E-MOSFET.
- Like their JFET counterparts, MOSFET common source amplifiers exhibit moderate inverting voltage gain, very high input impedance and moderate output impedance. The input impedance is a function of the biasing resistor configuration situated in front of the gate as the impedance looking into the gate itself is very, very high at low frequencies.
- The MOSFET common drain followers also behave similarly to the JFET version. Again we see a non-inverting voltage gain approaching unity, a very high input impedance and a low output impedance. The higher the trans-conductance is, the closer the gain will be to unity and the lower the output impedance will be.

CHAPTER-14

BASICS OF CLASS D POWER AMPLIFIERS

14.1 INTRODUCTION: AN OVERVIEW

Class D power amplifiers represent a significant advancement in audio amplifier technology, offering high efficiency and compact design for various applications. Unlike traditional linear amplifiers, Class D amplifiers operate by rapidly switching the input signal using pulse-width modulation (PWM) techniques. This on-off switching allows the amplifiers to deliver power to the load efficiently, minimizing heat dissipation. Class D amplifiers have become popular choices in audio systems, including portable speakers, car audio systems, and home theater setups, due to their ability to provide high-quality sound reproduction while consuming less power. Understanding the basics of Class D power amplifiers involves delving into the principles of PWM, the advantages of high efficiency, and the trade-offs in terms of harmonic distortion. As these amplifiers continue to gain prominence in the audio industry, grasping their fundamental concepts is essential for engineers and audio enthusiasts alike.

Much has been written in this text regarding the efficiency of various power amplifier topologies. While class A is known for its circuit layout simplicity, it is also known for its very low efficiency. Class B and class AB, while more complex than class A, present serious improvements in efficiency. In spite of these improvements, the family of class B amplifiers can hardly be considered as exhibiting high efficiency. Although not explicitly covered in this text, class G and H topologies are variations on class B and attempt to increase efficiency through the use of multiple sets of power supply rails or output devices, and in the process, tick the complexity up to another level.

The class D amplifier is perhaps the last word in amplifier efficiency. Theoretically with ideal devices, the efficiency of the output stage approaches 100%. Unlike the other amplifier forms, the transistors used in class D amplifiers never operate in the linear region; the output devices only operate as a switch, in either saturation or cutoff. High switching speed turns out to be a huge plus as it plays a major role in efficiency.

The increase in efficiency comes at a considerable increase in circuit complexity, however, for some applications this turns out to be a very good trade-off. As odd as it might at first seem, the two areas where class D topologies have taken root are at the opposite ends of the power output spectrum. The first area is perhaps the most obvious, mainly, very high output power amplifiers. An example might be an amplifier used as part of a large public address system and capable of delivering in excess of 1000 watts into a loudspeaker. High efficiency does two things here: First, it reduces the waste heat in the amplifier itself, and second, it reduces the current draw from the AC mains. Both of these are serious issues in a PA system used to fill a stadium or large concert hall as there may be dozens of such amplifiers comprising the system. As a bonus, improved efficiency also leads to a lighter and small enclosure because the need for heat sink area and mass will be reduced, as will the size of the AC power supply transformer. These traits will also reduce production costs and help offset the design complexity cost. The advantages have become so great that, in recent years, class D designs dominate the high end professional audio power amplifier market as well as the very high power automotive audio market (here there is another system limitation working in favor of class D, and that's the limited current capacity of the vehicle's alternator to deliver current).

The second area where class D has found acceptance is for low power portable devices. Examples include personal music devices, cell phones and hearing aids. Output powers for these applications might range from tens of milliwatts up to a few watts, so excess heat is generally not a big problem except in the most compact of enclosures. What is a problem, though, is the energy budget. Unlike a large PA amplifier that might pump out in excess of two horsepower, these portable devices do not have the luxury of running off of the AC mains with tens or even hundreds of amps of current capacity. Instead, these devices are restricted to battery power and batteries can only store so much energy. For a given battery capacity, higher efficiency directly translates into longer battery life. Another way of thinking about this is that, given a higher efficiency, a smaller battery can be used to achieve the same battery life, and this means that the unit can be both smaller and less expensive. Of course, nothing says that we can't opt for a little of each.

14.2 CLASS D BASICS

The key to the high efficiency of class D operation is to only operate the output devices as switches. That is, they are operated at the extreme ends of the load line, in either cutoff or saturation. The only exception to this rule is when the output device is transitioning from one state to another. Either BJTs or E-MOSFETs can be used, although for reasons that we shall examine, E-MOSFETs tend to be preferred in many applications and therefore we shall use them here as a general rule.

To understand the power advantage of class D switching, consider the circuit of Figure 14.1. Here we have an E-MOSFET being driven by a square wave at its gate terminal. The square wave runs from zero to some voltage well above VGS(th), sufficient to fully turn-on the MOSFET.

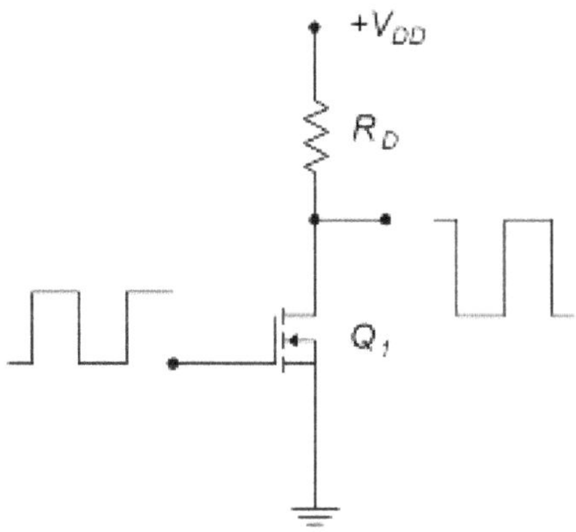

Figure 14.1 A simple MOSFET switch.

At the drain, an inverted square wave appears that ideally runs from +VDD down to zero volts. When the MOSFET is off, no drain current flows. Therefore, the drop across the drain resistor, RD, is zero. Consequently, by KVL, VDS must equal VDD. When the gate signal goes high, it turns on the MOSFET causing a large current. Ideally, this current equals VDD/RD and VDS = 0 V. In reality, the MOSFET will present some resistance (rDS(on)) and this essentially creates a voltage divider between the device and RD. Obviously, if rDS(on) << RD then we can approximate the low state as zero volts.

In the ideal case, the transistor dissipates no power. Here's why: When the gate is low, the device is off so no current flows. Although the device voltage is very high, the product of the device's voltage and current is zero. When the gate is high, the transistor turns on and conducts maximum current, however, the voltage across the device is zero, and the resulting product is zero once again. Therefore, the device dissipates no power.

The reality of the situation reveals that some power is indeed wasted by the output device. There are two chief culprits: non-zero turn-on voltage (caused by rDS(on) for example) and state transitions that are not instantaneous (i.e., the rise and fall times of the waveforms are not zero). These effects are illustrated in Figure 14.2.

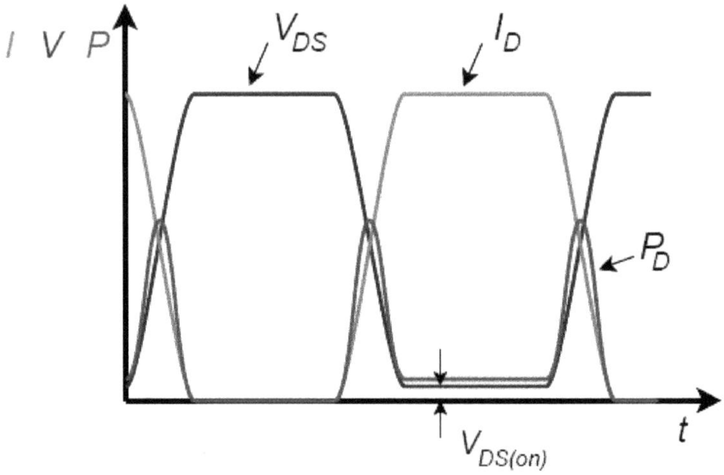

Figure 14.2 Switching losses.

The rise and fall times are exaggerated for clarity as is the voltage drop across the transistor. The green curve represents the device current while the blue curve indicates the device voltage. Note that during the on-state the transistor voltage is not zero, although it is very small. The product of these two curves is the power dissipated by the transistor and is shown in red. In the off-state, no power is dissipated. In the on-state, a small amount of power is dissipated due to the device's on-voltage. Power spikes also occur at the state transitions as neither the current nor the voltage are at zero.

The area under the red curve represents the total power dissipated by the output transistor. In order to minimize this loss and maximize the efficiency, we would like as low of an rDS(on) as possible along with very fast switching speeds. As power E- MOSFETs tend to switch faster than their BJT counterparts, this presents a compelling argument for their use. Further, large power E-MOSFETs may exhibit rDS(on) values of just a few milliohms, guaranteeing relatively small on-state losses (compared to load power) even with drain currents of several tens of amps.

14.3 PULSE WIDTH MODULATION

Clearly, class D presents the possibility of minimal wasted power and high efficiency. We are now left with the problem of how to turn a series of pulses into a continuous, smoothly varying waveform, such as a voice or music signal. There are a few ways to accomplish this; it's a matter of encoding the amplitude of the original signal into the pulse train that drives the output devices. Theoretically, as long as the "area under the curve" for a segment of input signal is identical to the area represented by the pulse train, we will have successfully encoded the signal and we then should be able to decode it, turning it back into a smoothly varying output signal. For this to work properly, the pulse train will have to be at much higher frequency than the input signal in order to follow its changes over time. One way to do this is through *pulse density modulation*, or PDM. The idea is to produce a number of narrow pulses to represent the area. If the input amplitude is large, we create a large number of pulses and if the amplitude is small, we produce a small number of pulses. While this technique can work, it is somewhat challenging to turn this pulse train back into the desired signal at the load.

Another technique to encode the input is *pulse width modulation*, or PWM. Instead of altering the number of pulses in a given period of time, we keep the frequency constant and adjust the width of the pulses. If the input amplitude is high, the width of the corresponding pulse will be wide and if the amplitude is low, the pulse width will be narrow. The decoding of PWM is easier than that of PDM and is generally the preferred route.

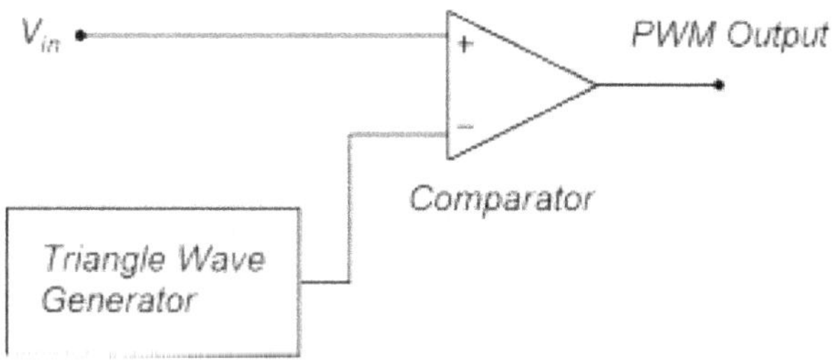

Figure 14.3 PWM encoder.

Generating PWM is a relatively straightforward affair. All we need is a triangle wave and a comparator. The comparator has two input terminals: the signal to be encoded (input signal) and the reference wave (triangle wave). It has a two-state, logical output. The output will be high if the signal is more positive than the reference and it will be low if the reference is more positive than the signal. This is shown in block diagram form in Figure 14.3.

As mentioned, the triangle wave needs to be at a much higher frequency than the signal that is being encoded. As a general rule, this frequency should be at least ten times the highest input signal frequency.

$$f_{triangle} \geq 10 f_2 \qquad (14.1)$$

A simulation showing the PWM waveforms is presented in Figure 14.4.

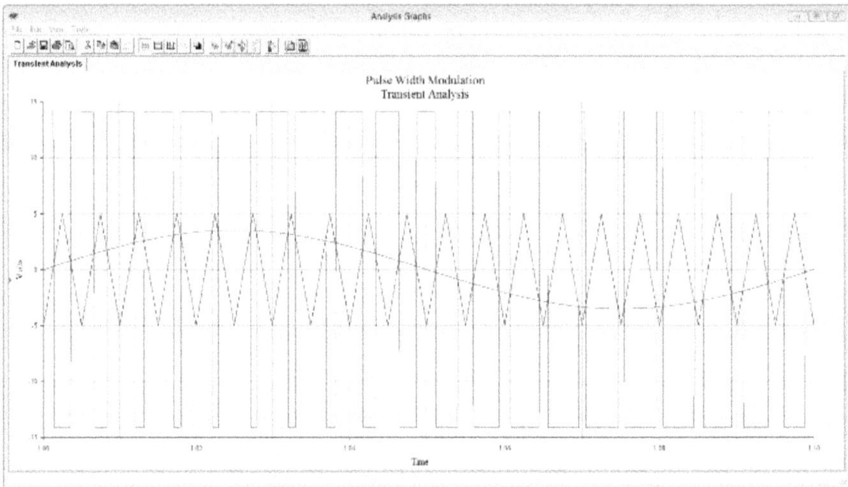

Figure 14.4 PWM waveforms.

The input signal is the red sine wave. The blue triangle wave is the reference and is approximately 20 times higher in frequency. The green wave is the PWM output.

Note that when the red input signal climbs above the blue reference triangle, the green output goes high, otherwise, the output is low. Thus, the duty cycle of the pulses correlates with the input signal amplitude. The input signal should not exceed the amplitude of the triangle wave otherwise accuracy will impaired. Also, the accuracy of the encoding process is dependent on the linearity of the triangle wave, so a high quality triangle wave generator is needed. Lastly, for accuracy and ease of decoding, the output pulses should not be allowed to become too thin, so the input signal should be limited to about 75% of the amplitude of the triangle wave.

Reconstituting the Output

Although we have successfully encoded the input signal into a series of pulses, we still need to decode them, that is, reconstitute the original continuously variable input signal. Mathematically, the PWM signal contains all of the original input signal frequency components and amplitudes, it just has added a large of number of new frequency components. These new components are multiples (harmonics) of the fundamental PWM frequency, and consequently, they are all higher than the input signal frequencies. As such, they may be removed with a low-pass filter. This will effectively reconstitute the original signal (but at a much higher amplitude, of course). A simple passive LC filter would be appropriate in this instance as it must pass large currents and voltages. An example is shown in Figure 14.5.

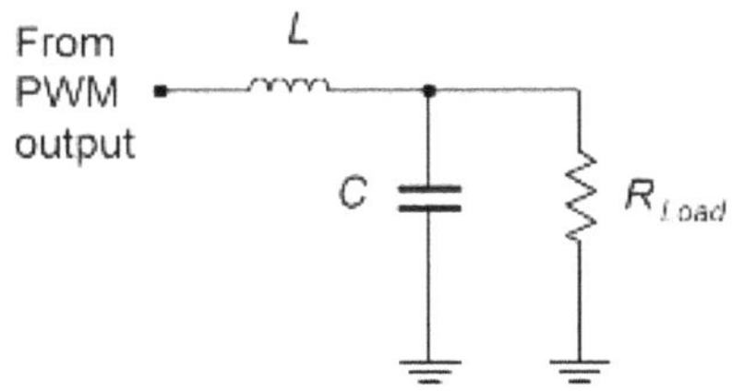

Figure 14.5 Passive low-pass filter.

At low frequencies, *XL* will be very small while *XC* will be very high, thus virtually all of the input signal frequencies will pass on to the load. At high frequencies, such as the harmonics of the PWM pulses, the situation is reversed: *XL* is large and *XC* is small, creating a large loss so that these components do not reach the load.[44] The critical frequency of the network is set to the highest input signal frequency (e.g., for high fidelity audio, slightly above 20 kHz).

We now have a complete outline for the class D amplifier, as shown in Figure 14.6.

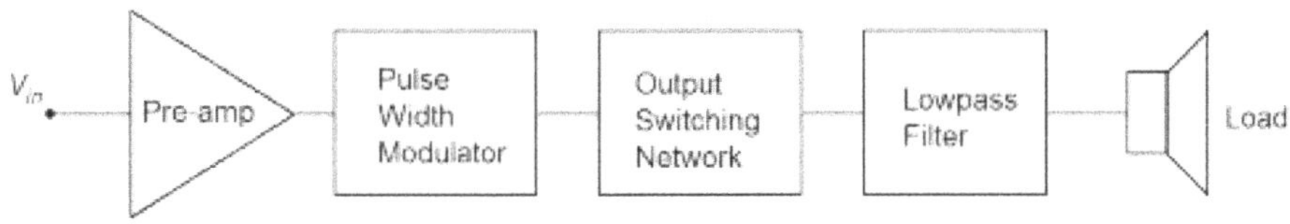

Figure 14.6: Class D amplifier block diagram.

The pre-amp can be comprised of any of the linear amplifier outlines presented in earlier chapters, whether they use BJTs or FETs. What remains then, is further investigation into the output switching network.

14.4 OUTPUT CONFIGURATIONS

If we apply the switching concept to a dual supply, push-pull topology, we arrive at the generic circuit of Figure 14.7.

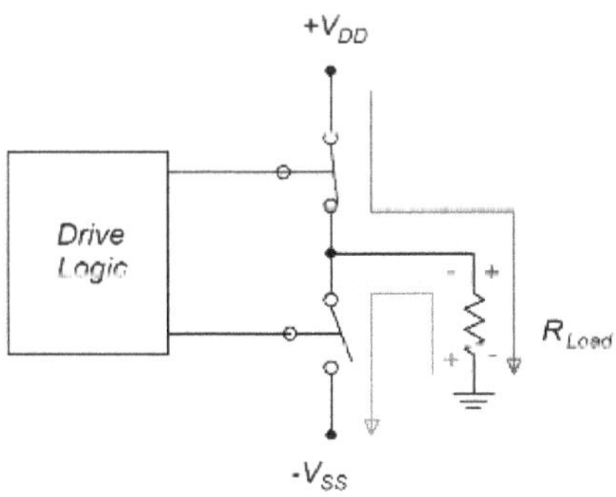

Figure 14.7 Generic push-pull switching.

The two output devices are alternately switched on and off. When the upper device is on, the lower device is off, and current flows from the positive supply to the load (blue path). Alternately, when the lower device is on, the upper device is off, and current flows through the load via VSS (red path). We could use either BJTs or E- MOSFETs for these devices.

Two obvious variations exist of the generic output circuit. The first version, shown in Figure 14.8, appears to be a direct take-off of a class B output. It is shown with E- MOSFETs but could be made with BJTs. Biasing details are not shown, instead a generic "driver" circuit block will prove sufficient for our discussion.

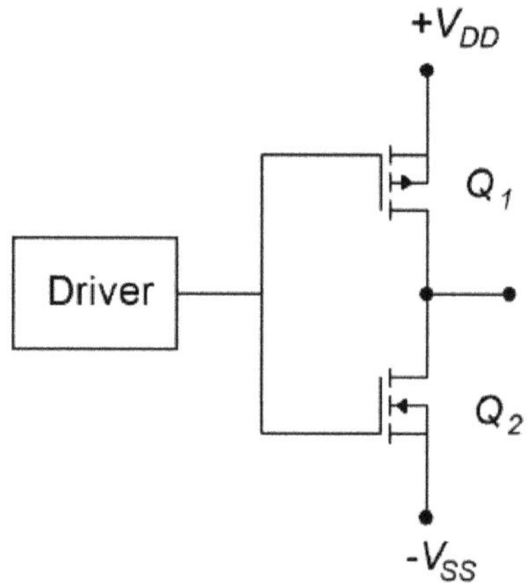

Figure 14.8 Basic push-pull switching.

In this circuit, the driver produces a bipolar pulse train that swings from negative to positive rather than from ground to positive. A positive level from the driver will turn on the upper N-channel device, allowing current flow to the load. In contrast, a negative level will turn on the lower P-channel device, allowing load current flow via VSS.45 It is worth noting that the gate drive signal must swing higher and lower than the two power supplies. This is because when a device is on, VDS will be nearly zero, meaning that the source will be at the power rail. As VGS must be greater than VGS(th), this means that VG must be greater than the power supply.

An alternate connection scheme is shown in Figure 14.9. Here, the N- and P-channel devices have switched positions.

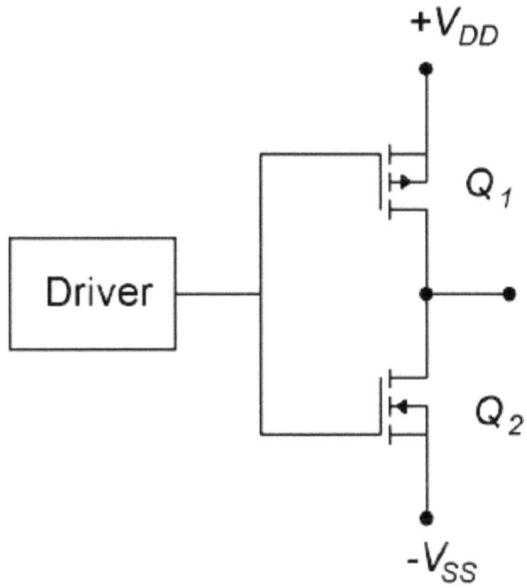

Figure 14.9 Alternate push-pull switching.

The logic here is reversed: the negative pulse turns on the upper device and the positive pulse turns on the lower device. The gate swing is lessened a little compared to the earlier circuit but it suffers from a flaw common to both configurations, namely, asymmetry between the N- and P-channel device characteristics. This includes variations between internal device capacitances and rDS(on) values. For the best possible matching, and thus the lowest distortion and highest performance, it would be better to configure the output using identical devices. This is shown in Figure 14.10.

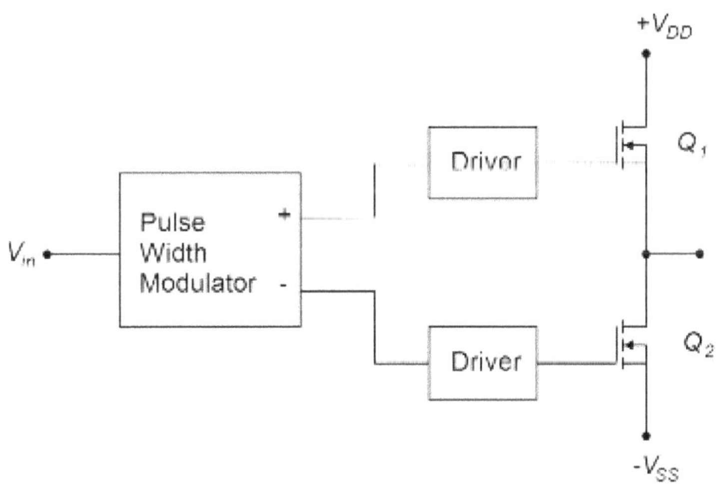

Figure 14.10 Output switching using identical devices.

This configuration complicates the drive signal in that we can no longer drive both gates with the same signal; instead, unique signals must be presented to each gate terminal. This circuit is known as a *half bridge*. Our final step will be to drive both ends of the load in differential fashion using a *full bridge*, or *H bridge* as it is sometimes known. This is shown in Figure 14.11.

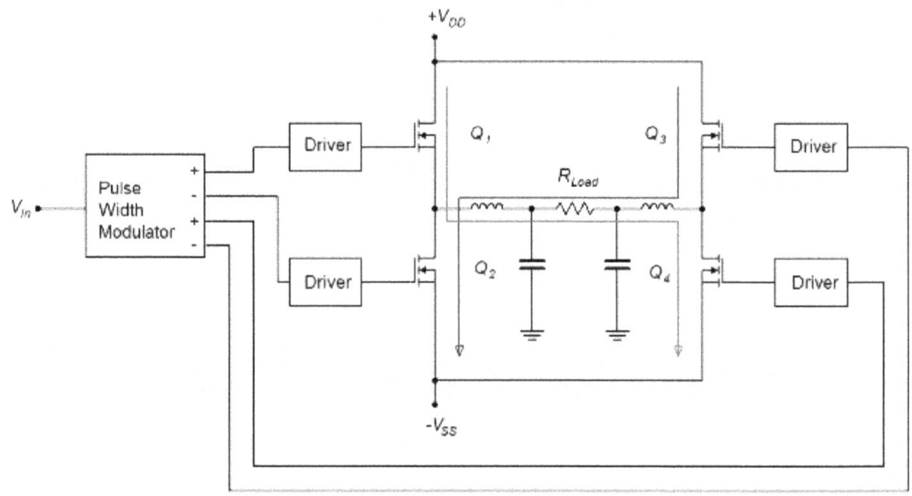

Figure 14.11 Full bridge output.

The output devices are controlled as diagonal pairs. When Q1 is on, Q4 is on, creating a path for load current from left to right (red trace). In contrast, when Q2 is on, Q3 will also be on, thus creating a load current path from right to left (blue trace). This effectively doubles the current amplitude which quadruples the load power (because power varies as the square of current). This is the same technique discussed in Chapter 9 with class B amplifiers. A dual LC filter is included in this diagram to remove unwanted frequency components.

Example 14.1

A pair of E-MOSFETs are configured to drive an 8 Ω load as in Figure 14.10. Assuming that ±50 volt sources are used and that each device has an *rDS(on)* of0.03 Ω, determine the peak load current and *VDS* for the MOSFETs.

At any given time, one MOSFET will be on, creating a path between one supply, itself, the load and ground. The total resistance to limit the current will be the load plus *rDS(on)*.

$$i_{load} = \frac{V_{DD}}{r_{load} + r_{DS(on)}}$$

$$i_{load} = \frac{50\,\text{V}}{8\,\Omega + 0.03\,\Omega}$$

$$i_{load} = 6.227\,\text{A}$$

The device voltage is found via Ohms law as load and drain current are identical.

$$v_{DS} = i_{load}\, r_{DS(on)}$$

$$v_{DS} = 6.227\,\text{A} \times 0.03\,\Omega$$

$$v_{DS} = 0.19\,\text{V}$$

There are a few details left that should not be overlooked. Two of them are related to the edge transition areas, another concerns the complexity of the drive circuits, and the final issue deals with the power supplies themselves

The first item of concern is precisely what happens during the transition. All of the output forms we have examined utilize two active devices configured in series between two power sources. There is nothing in that path to limit current. If both devices were to be simultaneously triggered to the on-state, a huge and possibly damaging current would flow. While it would be foolish to turn both devices on intentionally, the rise and fall times of the pulses effectively do this. As one device is turning on and the other is turning off, both devices are in a conducting state, even if it's not maximum conduction. Essentially, we have two low impedance devices in series between two sources. This results in a large current pulse known as shoot- through. This situation is depicted in Figure 14.12.

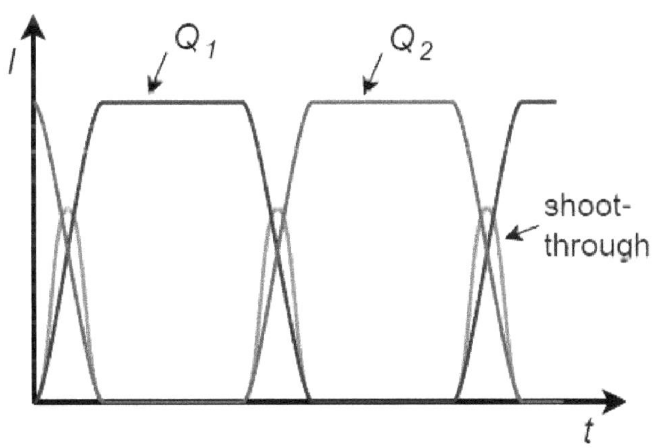

Figure 14.12 Shoot-through.

The current pulses for the two devices are shown in blue and violet. The maximal currents are directed to the load, but during the transition, a pulse of current, shown in red, "shoots through" the two devices, from one power supply directly to the other.

The solution to shoot-through is to adjust the turn-on and turn-off pulse timing so as to create a dead time, that is, a time span when neither device is directed to turn on. This is illustrated in Figure 14.13.

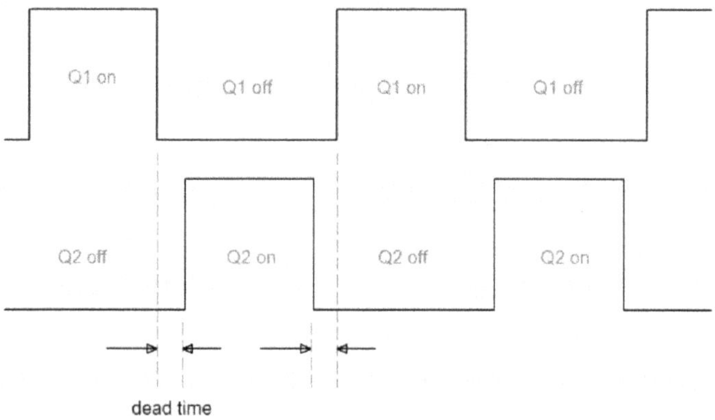

Figure 14.13 Dead time.

Dead time is adjusted to correspond with the rise and fall times of the output devices. Basically, a device is not allowed to turn on until the other device is, indeed, fully off. The inclusion of dead time alters the width of the pulse and consequently can introduce waveform distortion. A minimum amount of dead time should be used to avoid this. This is another reason to use very fast output devices as they will require shorter dead times.

The second issue regarding timing is one of device capacitance. Power MOSFETs exhibit relatively high device capacitances. For example, the FDMS86180 examined in Chapter 12 exhibits input and output capacitances of roughly 4.4 nF and 2.7 nF, respectively. Although the extremely high gate input resistance might seem to indicate that very little drive current is needed to turn these devices on, the capacitance tells a different story.

The rate of change of voltage ac itor is a function of the capacitance and the current driving it:

$$\frac{dv_C}{dt} = \frac{i_C}{C}$$

The larger the current, or the smaller the capacitor, the greater the rate of change of voltage. This can place a serious limit on how quickly a device may be controlled. For example, suppose the drive circuit can pump out up to 10 mA. At first glance that may seem like an enormous amount of current to drive a MOSFET. Now, consider what happens if the input capacitance is 2 nF:

$$\frac{dv_C}{dt} = \frac{i_C}{C}$$

$$\frac{dv_C}{dt} = \frac{10\,\text{mA}}{2\,\text{nF}}$$

$$\frac{dv_C}{dt} = 5\text{E}6\,\text{V/s}$$

While a 5 million volt-per-second slope might sound fast, it's only 5 volts per microsecond. Compared to the requirements of, say, a 200 kHz to 300 kHz switching frequency, that is horribly slow.

Computer Simulation

To see the effect of input capacitance, a two-stage amplifier is captured in a simulator, as shown in Figure 14.14. The circuit consists of a relatively standard small signal amplifier feeding a medium power E-MOSFET, the IRF7201. A 10 kHz square wave is used to drive the circuit. The input capacitance of the MOSFET is 550 pF, certainly larger than a small signal FET but not an extremely large value. A single capacitor is placed across the gate that will be used to simulate a much larger device and the associated increased input capacitance.

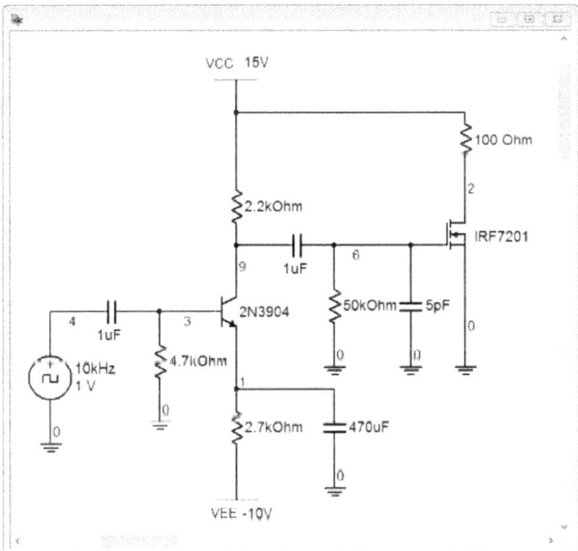

Figure 14.14 Circuit for input capacitance testing.

The initial transient analysis is run using a 5 pF gate capacitance which has no appreciable effect on the outcome. The result of the simulation is shown in Figure 14.15.

The red trace is the gate voltage at node 6 while the blue trace is the final output at node 2. The gate drive signal is suffering somewhat at the upper portion and the rise and fall times are evident. The output signal is swinging from the power supply of+15 volts down to ground, as expected. The rise time is somewhat quicker than the fall time but, in general, the output presents a decent square pulse at close to 50% duty cycle. The simulation is repeated but this time the gate capacitance is increased from 5 pF to 5 nF, a value more typical of a large power FET. The result is shown in Figure 14.16. The red gate drive signal has taken a serious hit and is no longer square in shape. The output pulse in blue still runs to and from the expected voltage levels, but the rising and falling edges are noticeably slowed. Also, the positive pulse width has been stretched due to the slowing of the gate signal which retards the turn on of the MOSFET. The result is a duty cycle that is greater than 50%.

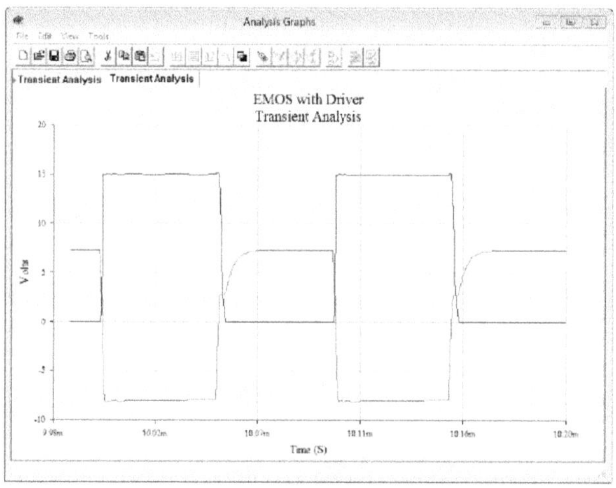

Figure 14.15 Waveforms for normal circuit.

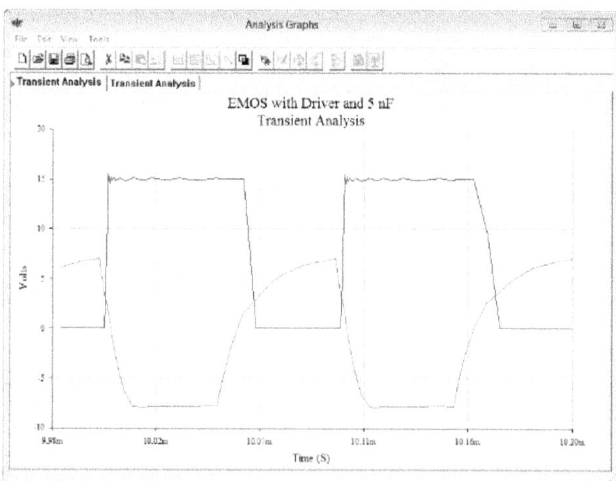

Figure 14.15 Waveforms for circuit with increased gate capacitance.

The bottom line is that, in order maximize speed, care must be taken to minimize capacitance, decrease the output impedance of the driver circuit and increase the drive current. Fortunately, this issue has been largely solved by IC manufacturers who offer FET driver integrated circuits designed specifically for these applications.

The final item of practical concern is the power supply itself, or more precisely, the quality of the supply voltage. Remember, the output device is being used as a switch. When the device is on, the power supply is directly connected to the output (with the exception of the small voltage drop across the output device). This means that any noise or ripple on the power supply will make its way to the output filter. Whatever noise components fall within the desired input signal range will not be filtered out, and thus are delivered to the load. In short, the output devices will "leak" the power supply noise into the output, so care must be taken to have as clean of a power supply voltage as possible.

CONCISE SUMMARY

- The class D amplifier boasts very high efficiency, theoretically approaching 100%. The amplifier operates its output devices as switches; they are either fully on or fully off. The power losses are mostly relegated to switching edge losses so it is important to not switch the output devices at too high of a frequency.
- The input signal is encoded as a series of pulses, typically via pulse width modulation. The pulse frequency is much higher than the highest input signal frequency, typically by an order of magnitude. The width of the pulses is a function of the amplitude of the input signal. That is, the higher the input signal, the greater the pulse width. The pulses are amplified by applying them to the output devices which then act as switches to alternately connect and disconnect the power supplies to the output terminal. The sequence of much larger amplitude pulses are then fed to a low-pass filter, typically, an LC filter, to remove the high frequency components of the pulse train. The reconstituted input signal is what remains, but at a much higher amplitude.
- The output can be configured using either two-device half-bridge or four-device full- bridge arrangements. The full-bridge is preferred for higher performance. The input capacitance of the output devices can be relatively high, so care must be taken to ensure that sufficient capacity is available from the driver circuit.

CHAPTER-15

OPERATION OF INSULATED GATE BIPOLAR TRANSISTORS (IGBTS)

15.1 INTRODUCTION: AN OVERVIEW

Insulated Gate Bipolar Transistors (IGBTs) stand as pivotal components in power electronics, facilitating the efficient control of high-power applications. Combining the advantages of both MOSFETs and bipolar transistors, IGBTs operate as voltage-controlled devices that exhibit the high input impedance of MOSFETs and the robust current-carrying capability of bipolar transistors. This unique amalgamation positions IGBTs as key elements in various applications, including motor drives, power inverters, and amplifiers, where precise control of high currents and voltages is paramount. Understanding the operation of IGBTs involves delving into their voltage-controlled switching behavior, on-state conduction characteristics, and the crucial role they play in enhancing the efficiency of power conversion systems. As power electronics continues to evolve, the operational principles of IGBTs remain central to engineers and designers aiming to optimize performance and efficiency in diverse electronic applications.

The *Insulated Gate Bipolar Transistor*, or *IGBT*, is a power semiconductor that first became available to the commercial market during the 1980s. Initial devices had certain performance issues but these problems largely have been taken care of with subsequent generations. Today, the IGBT is in wide use, competing with power BJTs and power E-MOSFETs across a range of applications. The IGBT is designed to be used as a high voltage/high current switch and typically is not used for linear applications such as an audio class B power amplifier. The IGBT has also overtaken the older thyristor devices (e.g., SCR) in many areas due to its speed and the relative simplicity of the driving circuits used to control it.

The IGBT offers a mix of performance characteristics of both the power BJT and the power E-MOSFET. Like the BJT, the IGBT offers low on-state power loss and the ability to handle large currents and voltages. Like the power E-MOSFET, it is relatively easy to drive, being a voltage-controlled device rather than a current-controlled device. On the down-side, it is not as fast as the current generation of power E-MOSFETs and tends to be more costly than both the power BJT and power E-MOSFET. Consequently, the choice of which of these three devices should be used for a given power switching application will depend on the specifics of the design. For example, a medium to high power design that focuses on lowest cost may favor the BJT, a low to medium power application that requires very high switching speeds may be best solved with a power E-MOSFET, while an IGBT might be ideal for a very high power application utilizing low to medium speed clocking. This chapter will examine a number of power switching applications, and while they will all revolve around using the IGBT, please bear in mind that, depending on the specifics, power BJTs and E-MOSFETs might also be used.

The IGBT is available in two variants: the punch through, or PT; and the non-punch through, or NPT, versions. We shall look at both.

15.2 IGBT INTERNALS

The IGBT is a multi-layer device. The cutaway shown in Figure 15.1 uses an N-channel, although P-channel is possible. This device has many features in common with the power E-MOSFET discussed in Chapter 12.

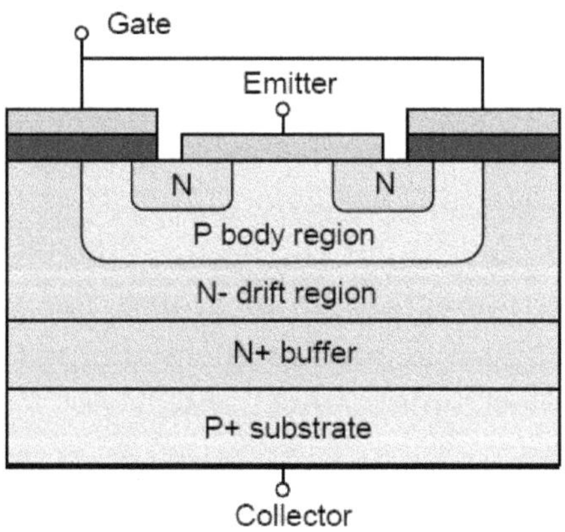

Figure 15.1 Cutaway view of IGBT.

To begin, note the labeling of the three terminals: gate, emitter and collector. The gate is isolated just as it is in MOSFETs. When a positive gate-emitter potential is applied, an N-type inversion layer will develop in the the P body region, allowing current to flow. The middle N region is split into two sections, the main N- drift region and the N+ buffer layer (the +/- signs indicate heavy and light doping, respectively). The device shown is the punch through (PT) variety. The non-punch through (NPT) omits the buffer layer. The functional differences between the two are that splitting the N material to include the buffer layer improves speed and lowers on-state voltage. Consequently, the PT version exhibits lower switching and conduction losses.

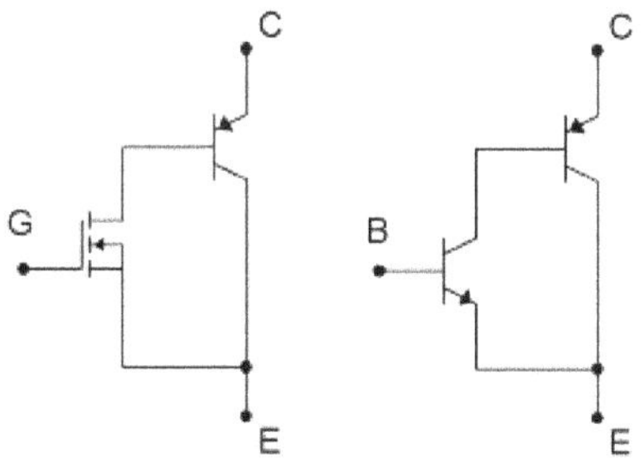

Figure 15.2 Simple model of an IGBT (left)compared to a Sziklai pair(right).

The upper three layers of Figure 15.1 form an E-MOSFET while the lower section (P body, N drift/buffer and P substrate) form a PNP transistor. Thus, we can make a simplified model of the IGBT using these other devices, as shown in Figure 15.2.

This simplified model is reminiscent of the NPN version of the Sziklai pair examined in Chapter 9. The input device has been replaced by an E-MOSFET. Therefore we expect the very small gate current and relatively simple drive requirements of the E- MOSFET with the power handling of a BJT.

The operational device curves are, unsurprisingly, also reminiscent of these two components. A set of collector curves is presented in Figure 15.3 using representative values for voltage and current.

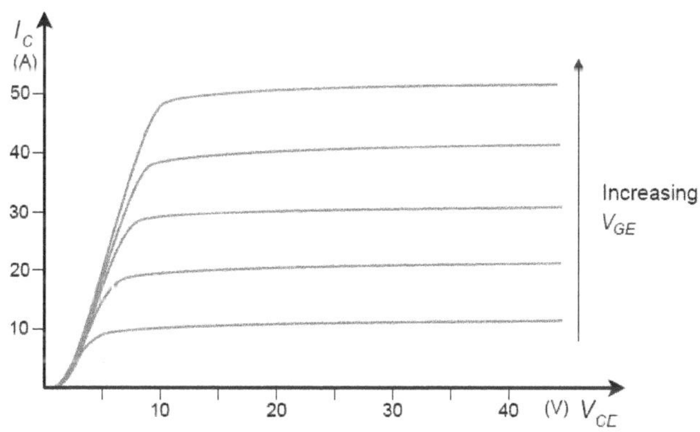

Figure 15.3 IGBT collector curve family.

This set of curves appears as a cross between the collector curve family seen in Chapter 4 and the drain curve family seen in Chapter 10, with one slight twist: the entire set of curves is displaced positively from the origin by about one volt. As with any E-MOSFET, channel current does not begin to flow until the gate threshold voltage is reached, here referred to as VGE(th). Increasing values of VGE cause increased conduction and current flow. Finally, note that the curves do not begin to "flatten" until VCE has reached several volts, unlike the saturation voltage of a single BJT which might be only tenths of a volt.

The forward current-voltage characteristic curve reflects the E-MOSFET portion of the model. This is shown in Figure 15.4.

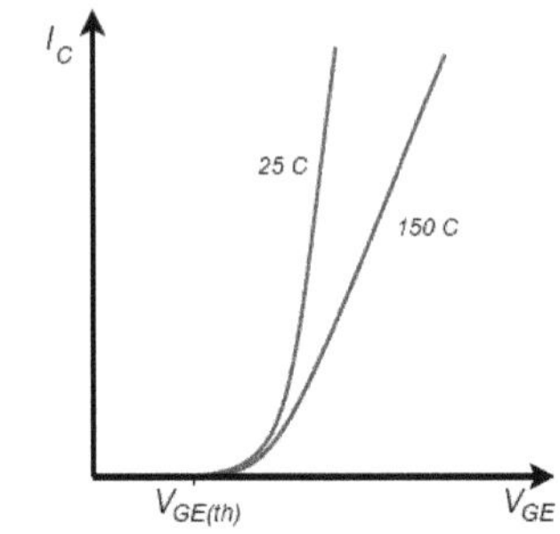

Figure 15.4 IGBT characteristic curve and variation with temperature.

Two traces are plotted for two different temperatures. The curves are essentially the same as the characteristic curve of the E-MOSFET, as seen in the introductory MOSFET chapter, Figure 12.11 (here the current axis has been expanded, making the traces appear steeper, in order to more clearly show the variation with temperature). Conduction begins at the threshold voltage, *VGE(th)*, and then rises rapidly, following a square law trajectory. Once a sufficient current level is reached, the curve can be approximated as a straight line.

Of particular interest here is how the curve varies with temperature. As temperature increases (red trace), the slope decreases. Recalling that the slope of the current- voltage characteristic curve is the device trans-conductance, this means that the trans-conductance decreases with temperature. In other words, the IGBT exhibits a negative temperature coefficient of trans-conductance, just like power E-MOSFETs, and thus are also less inclined to suffer from current hogging and thermal runaway problems than BJTs.

Unfortunately, there isn't a single, standardized schematic symbol for the IGBT. Two versions are shown in Figure 15.5.

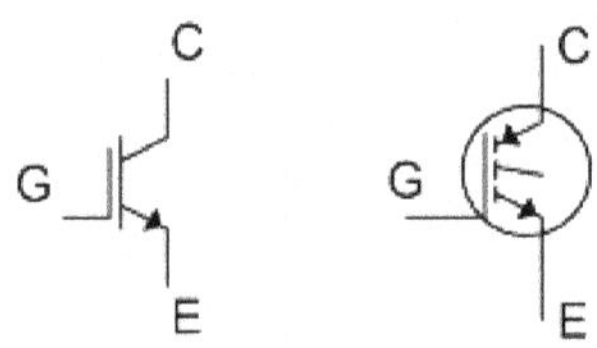

Figure 15.5 IGBT schematic symbols.

Both symbols attempt to reflect the dual E-MOSFET/BJT character of the device. The symbol on the left appears to have wider use currently, although there is also a third variation based on it where the gate connection is drawn toward the center rather than closer to the emitter.

15.3 IGBT DATA SHEET INTERPRETATION

A portion of the data sheet for the Fairchild/ON Semiconductor FGH50T65SQD IGBT is shown in Figure 15.6. This is a fourth generation IGBT featuring trench construction. It is rated for 650 volts and 50 amps. The device includes an anti- parallel diode. This is useful for bridge applications that drive inductive loads (recall that the current through an inductor cannot change instantaneously, thus, when devices are switched on/off in a bridge, the diode serves to create a path around the IGBT for this current).

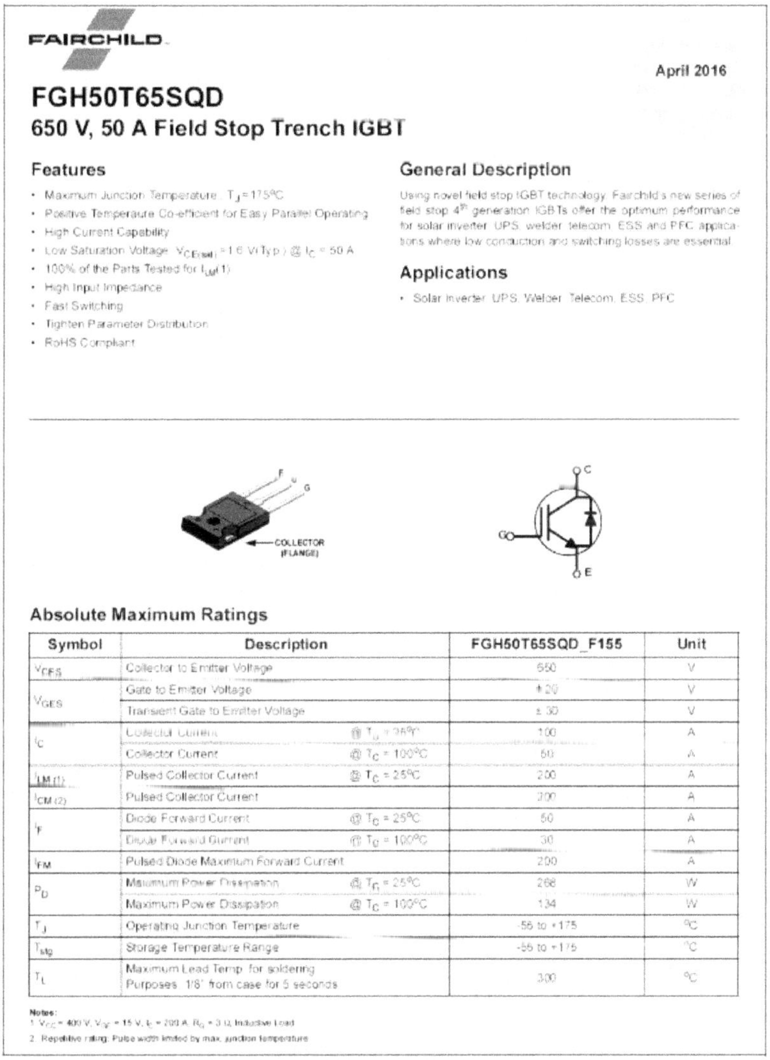

FAIRCHILD

April 2016

FGH50T65SQD

650 V, 50 A Field Stop Trench IGBT

Features

- Maximum Junction Temperature : T_J = 175°C
- Positive Temperaure Co-efficient for Easy Parallel Operating
- High Current Capability
- Low Saturation Voltage: $V_{CE(sat)}$ = 1.6 V(Typ.) @ I_C = 50 A
- 100% of the Parts Tested for I_{LM}(1)
- High Input Impedance
- Fast Switching
- Tighten Parameter Distribution
- RoHS Compliant

General Description

Using novel field stop IGBT technology, Fairchild's new series of field stop 4th generation IGBTs offer the optimum performance for solar inverter, UPS, welder, telecom, ESS and PFC applications where low conduction and switching losses are essential.

Applications

- Solar Inverter, UPS, Welder, Telecom, ESS, PFC

Absolute Maximum Ratings

Symbol	Description		FGH50T65SQD_F155	Unit
V_{CES}	Collector to Emitter Voltage		650	V
V_{GES}	Gate to Emitter Voltage		± 20	V
	Transient Gate to Emitter Voltage		± 30	V
I_C	Collector Current	@ T_C = 25°C	100	A
	Collector Current	@ T_C = 100°C	50	A
$I_{LM (1)}$	Pulsed Collector Current	@ T_C = 25°C	200	A
$I_{CM (2)}$	Pulsed Collector Current		200	A
I_F	Diode Forward Current	@ T_C = 25°C	50	A
	Diode Forward Current	@ T_C = 100°C	30	A
I_{FM}	Pulsed Diode Maximum Forward Current		200	A
P_D	Maximum Power Dissipation	@ T_C = 25°C	268	W
	Maximum Power Dissipation	@ T_C = 100°C	134	W
T_J	Operating Junction Temperature		-55 to +175	°C
T_{stg}	Storage Temperature Range		-55 to +175	°C
T_L	Maximum Lead Temp. for soldering Purposes, 1/8" from case for 5 seconds		300	°C

Notes:
1. V_{CC} = 400 V, V_{GE} = 15 V, I_C = 200 A, R_G = 3 Ω, Inductive Load
2. Repetitive rating: Pulse width limited by max. junction temperature

Figure 15.6a FGH50T65SQD data sheet. Used with permission from SCILLCdba ON Semiconductor.

Thermal Characteristics

Symbol	Parameter	FGH50T65SQD_F155	Unit
$R_{\theta JC}$(IGBT)	Thermal Resistance, Junction to Case, Max	0.56	°C/W
$R_{\theta JC}$(Diode)	Thermal Resistance, Junction to Case, Max	1.25	°C/W
$R_{\theta JA}$	Thermal Resistance, Junction to Ambient, Max	40	°C/W

Package Marking and Ordering Information

Part Number	Top Mark	Package	Packing Method	Reel Size	Tape Width	Qty per Tube
FGH50T65SQD_F155	FGH50T65SQD	TO-247 G03	Tube	-	-	30

Electrical Characteristics of the IGBT T_C = 25°C unless otherwise noted

Symbol	Parameter	Test Conditions	Min.	Typ.	Max.	Unit
Off Characteristics						
BV_{CES}	Collector to Emitter Breakdown Voltage	V_{GE} = 0V, I_C = 1 mA	650	-	-	V
ΔBV_{CES} / ΔT_J	Temperature Coefficient of Breakdown Voltage	I_C = 1 mA, Reference to 25°C	-	0.6	-	V/°C
I_{CES}	Collector Cut-Off Current	V_{CE} = V_{CES}, V_{GE} = 0 V	-	-	250	µA
I_{GES}	G-E Leakage Current	V_{GE} = V_{GES}, V_{CE} = 0 V	-	-	±400	nA
On Characteristics						
$V_{GE(th)}$	G-E Threshold Voltage	I_C = 50 mA, V_{CE} = V_{GE}	2.6	4.5	6.4	V
$V_{CE(sat)}$	Collector to Emitter Saturation Voltage	I_C = 50 A, V_{GE} = 15 V, T_C = 25°C	-	1.6	2.1	V
		I_C = 50 A, V_{GE} = 15 V, T_C = 175°C	-	1.92	-	V
Dynamic Characteristics						
C_{ies}	Input Capacitance	V_{CE} = 30 V, V_{GE} = 0 V, f = 1MHz	-	3275	-	pF
C_{oes}	Output Capacitance		-	84	-	pF
C_{res}	Reverse Transfer Capacitance		-	12	-	pF
Switching Characteristics						
$t_{d(on)}$	Turn-On Delay Time	V_{CC} = 400 V, I_C = 12.5 A, R_G = 4.7 Ω, V_{GE} = 15 V, Inductive Load, T_C = 25°C	-	22	-	ns
t_r	Rise Time		-	8.7	-	ns
$t_{d(off)}$	Turn-Off Delay Time		-	105	-	ns
t_f	Fall Time		-	2.5	-	ns
E_{on}	Turn-On Switching Loss		-	180	-	uJ
E_{off}	Turn-Off Switching Loss		-	45	-	uJ
E_{ts}	Total Switching Loss		-	225	-	uJ
$t_{d(on)}$	Turn-On Delay Time	V_{CC} = 400 V, I_C = 25 A, R_G = 4.7 Ω, V_{GE} = 15 V, Inductive Load, T_C = 25°C	-	19	-	ns
t_r	Rise Time		-	13	-	ns
$t_{d(off)}$	Turn-Off Delay Time		-	93	-	ns
t_f	Fall Time		-	6.4	-	ns
E_{on}	Turn-On Switching Loss		-	410	-	uJ
E_{off}	Turn-Off Switching Loss		-	88	-	uJ
E_{ts}	Total Switching Loss		-	498	-	uJ

Figure 15.6b FGH50T65SQD data sheet(cont).

The main features are the 650 volt rating for *VCE*, 100 amp continuous collector current and 268 watt dissipation at 25°C. The current and power ratings are essentially halved at the more practical temperature of 100°C. The threshold voltage, *VGE(th)*, is specified as 4.5 volts with a ±1.9 volt spread. The saturation voltage typically is 1.6 volts at room temperature with 50 amps of collector current. This compares favorably to basic power BJTs. Like power MOSFETs, the input capacitance is relatively high at 3275 pF, so the same gate drive precautions must be followed. Finally, note the asymmetry in switching times. At room temperature and 12.5 amps of collector current, the turn-on delay plus rise time is specified as approximately 31 nanoseconds while the turn-off delay and fall time is nearly 110 nanoseconds. This relative slowing of the off-state transition is typical of IGBTs. Further, as both current and temperature increase (Figure 15.6c), these times increase by a few percent. By comparison, the FDMS86180 power E-MOSFET examined in Chapter 12 exhibited symmetrical values in the mid-30 nanosecond region at a drain current of 67 amps.

Electrical Characteristics of the IGBT (Continued)

Symbol	Parameter	Test Conditions	Min.	Typ.	Max	Unit
$t_{d(on)}$	Turn-On Delay Time	V_{CC} = 400 V, I_C = 12.5 A, R_G = 4.7 Ω, V_{GE} = 15 V, Inductive Load, T_C = 175°C	-	20	-	ns
t_r	Rise Time		-	9.8	-	ns
$t_{d(off)}$	Turn-Off Delay Time		-	116	-	ns
t_f	Fall Time		-	3.5	-	ns
E_{on}	Turn-On Switching Loss		-	402	-	uJ
E_{off}	Turn-Off Switching Loss		-	110	-	uJ
E_{ts}	Total Switching Loss		-	512	-	uJ
$t_{d(on)}$	Turn-On Delay Time	V_{CC} = 400 V, I_C = 25 A, R_G = 4.7 Ω, V_{GE} = 15 V, Inductive Load, T_C = 175°C	-	18	-	ns
t_r	Rise Time		-	15	-	ns
$t_{d(off)}$	Turn-Off Delay Time		-	102	-	ns
t_f	Fall Time		-	8	-	ns
E_{on}	Turn-On Switching Loss		-	[illegible]	-	uJ
E_{off}	Turn-Off Switching Loss		-	203	-	uJ
E_{ts}	Total Switching Loss		-	844	-	uJ
Q_g	Total Gate Charge	V_{CE} = 400 V, I_C = 50 A, V_{GE} = 15 V	-	99	-	nC
Q_{ge}	Gate to Emitter Charge		-	17	-	nC
Q_{gc}	Gate to Collector Charge		-	23	-	nC

Electrical Characteristics of the Diode T_C = 25°C unless otherwise noted

Symbol	Parameter	Test Conditions		Min.	Typ.	Max	Unit
V_{FM}	Diode Forward Voltage	I_F = 30 A	T_C = 25°C	-	2.2	2.6	V
			T_C = 175°C		1.9	-	
E_{rec}	Reverse Recovery Energy	I_F = 30 A, dI_F/dt = 200 A/μs	T_C = 175°C	-	40	-	uJ
t_{rr}	Diode Reverse Recovery Time		T_C = 25°C	-	31	-	ns
			T_C = 175°C	-	207	-	
Q_{rr}	Diode Reverse Recovery Charge		T_C = 25°C	-	48	-	nC
			T_C = 175°C	-	820	-	

Figure 15.6c FGH50T65SQD data sheet(cont).

Finally, consider the graphical data presented in Figure 15.6d.

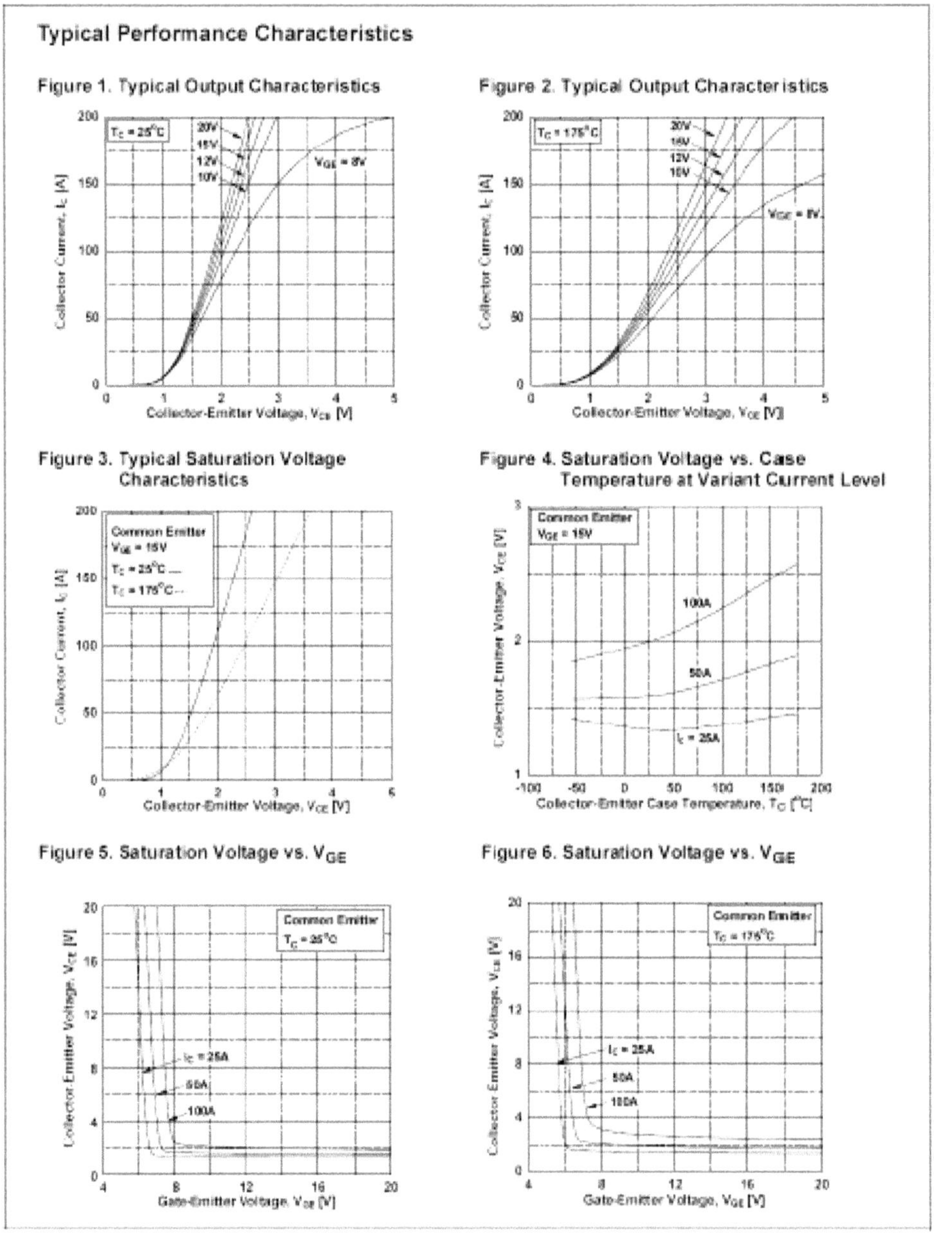

Figure 15.6d FGH50T65SQD data sheet(cont).

The top two graphs show the collector curves in the low voltage region. Compare these with Figure 15.3. Similarly, the middle-left graph shows the current-voltage characteristic with temperature variation. This reflects Figure 15.4. Finally, the bottom two graphs show the collector-emitter saturation voltage with respect to gate- emitter drive voltage for three different collector currents. Note that for the highest current at room temperature, saturation voltage is around 2 volts for a gate drive of at least 8 volts. This rises to about 3 volts at 175°C.

Computer Simulation

To highlight the performance of the IGBT, a simple series load switch is simulated. The circuit is shown in Figure 15.7.

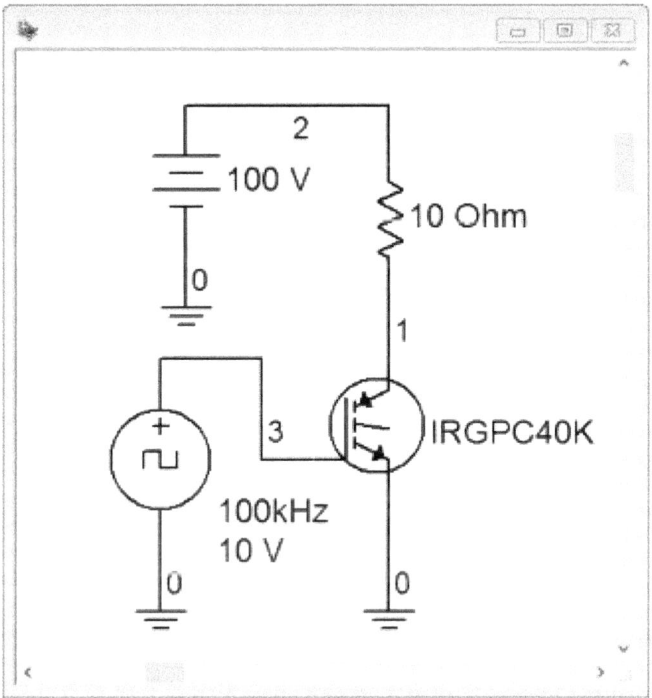

Figure 15.7 IGBT simulation schematic.

A 10 ohm load is switched from a 100 volt DC power supply via an International Rectifier IRGPC40K IGBT. The gate is driven from a 10 volt peak square wave running at 100 kHz.

Ideally, if the IGBT produced no losses, the full 100 volt source would drop across the load, producing 10 amp current pulses. According to the device data sheet, collector-emitter saturation voltage typically is 2.1 volts (3.2 volts maximum) with a 25 amp collector current. The total turn-on/rise time and turn-off/fall time values are 62 nanoseconds and 290 nanoseconds typically, at 25 amps and room temperature.

A transient analysis is run, with the results shown in Figure 15.8.

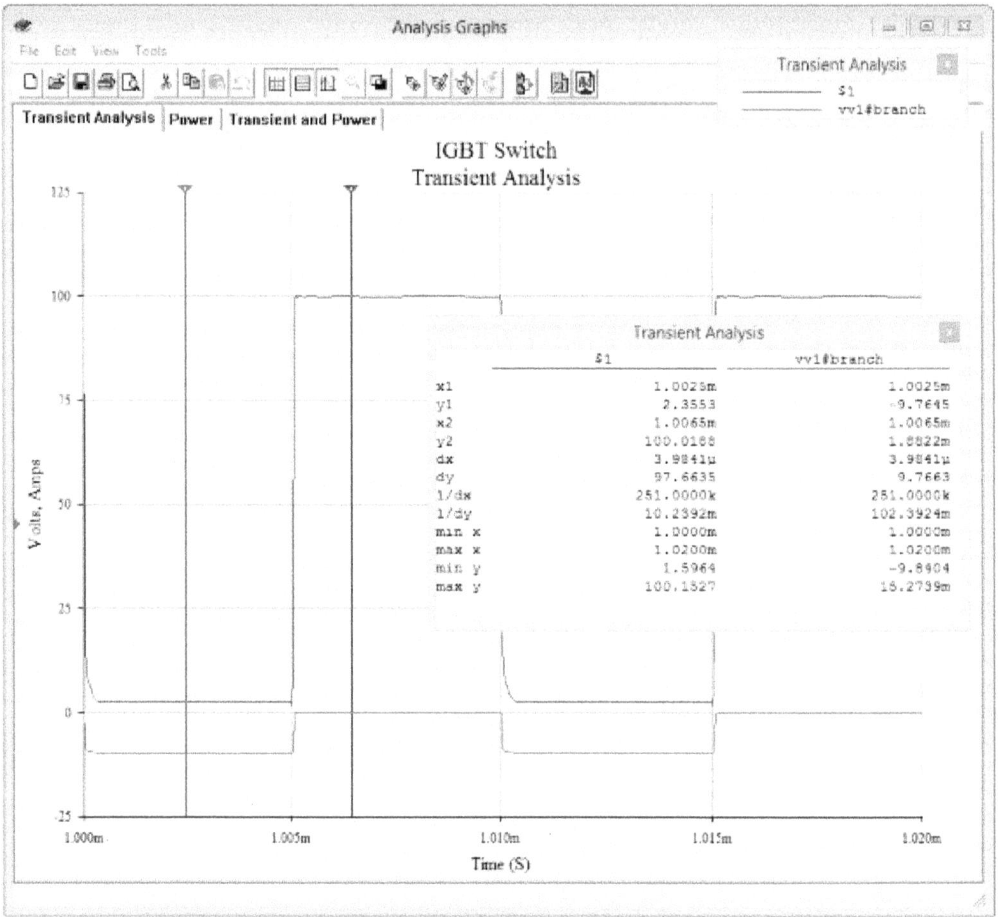

Figure 15.8 Transient analysis of simple IGBT switch.

The drain voltage (node 1) is shown as the purple trace and the drain current is shown in green (reversed in polarity to see it more clearly). The two cursors (red is#1, blue is #2) show the various levels. As expected, the drain voltage peaks at 100 volts in the off-state, at which time the current is nil. During the on-state, the drain voltage drops to about 2.36 volts, very close to the data sheet's typical saturation potential. This small drop reduces the voltage across the load to about 97.64 volts. This is verified by cursor #1 showing a current of 9.765 amps. Also, although it is not possible to determine the edge timings with great precision from this plot, the asymmetry between rise and fall is apparent, and the edges are generally consistent with the numeric values from the data sheet.

Figure 15.9 shows the result of multiplying the current and voltage waveforms. This new waveform represents the power dissipation of the IGBT.

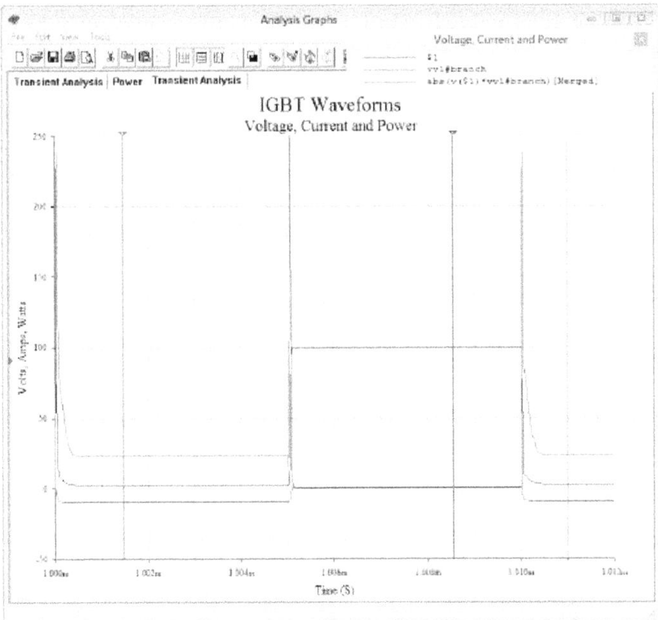

Figure 15.9 Power waveform of simple IGBT switch.

The power trace is shown in fuchsia, or magenta, or screaming purple-pink, or whatever-you-want-to-call-it. The edge spikes dominate but the on-state power is apparent as well. Remember, during the on-state, the load is dissipating close to 1000 watts. The cursor output window is shown separately in Figure 15.10. The on- state power dissipation is approximately 23 watts which represents less than 2.5 percent of the load power. In contrast, the edge spikes are maxing out at around 250 watts. Of course, the time duration is very short, being only a few percent of the cycle period, but it cannot be ignored.

Voltage, Current and Power

	$1	vv1#branch	abs(v($1)*vv1#branch) (Mer
x1	1.0015m	1.0015m	1.0015m
y1	2.3509	9.7649	22.9553
x2	1.0086m	1.0086m	1.0086m
y2	100.0292	2.9179m	417.2293m
dx	7.0916u	7.0916u	7.0916u
dy	97.6783	9.7678	-22.5387
1/dx	141.0112k	141.0112k	141.0112k
1/dy	10.2377m	102.3769m	-44.3681m
min x	1.0000m	1.0000m	1.0000m
max x	1.0200m	1.0200m	1.0200m
min y	1.5061	-9.8404	28.1911m
max y	100.1527	15.2739m	249.7386

Figure 15.10 Numeric values at waveform cursors for the IGBT switch.

15.4 IGBT APPLICATIONS

IGBTs lend themselves to a variety of high power switching applications. In this section, we shall look at four of them. Bear in mind that power BJTs and power E-MOSFETs might also be used for these applications, depending on the specifics of the design. In general, power E-MOSFETs will be preferred when using high switching frequencies at medium to low powers and voltages, while IGBTs are favored at higher voltages, currents and powers.

Induction Heating

Although thermal conduction is the first method commonly thought of when it comes to heating something, magnetic induction may also be used. Magnetic induction creates heat through the Joule effect and can be used for large scale industrial processes, such as creating metal alloys via an induction furnace, too much smaller scale consumer applications, such as an inductive cook top. Induction heating is efficient because the vessel itself is heated directly and less heat is lost to the immediate environment. Also, control of heating can be very precise. The basic idea is to create a rapidly changing magnetic field placed next to the container to be heated. If this vessel is ferromagnetic, eddy currents will be induced in the vessel, creating heat. Thus, if we were to place a cast iron pot within the field, the pot itself would heat up as eddy currents are induced within it, thus heating the pot's contents. There are no open flames or surface heating elements involved. The only downside to this process is that the vessel must be made of ferromagnetic material. For a cook top, that means that pots and pans must be made of iron or certain steel alloys. An aluminum sauté pan or ceramic container will not work with this system.

As an example, let's consider an inductive cook top. A sophisticated design could feature a full- or half-bridge arrangement of IGBTs, but for illustrative purposes we'll focus on a simple single-ended system using just one IGBT.

The system features four main components: the rectifier/EMI filter, the control/drive circuit, the IGBT switch and the LC resonant tank sub-circuit that generates the field. This is illustrated in Figure 15.11.

The rectifier produces full-wave pulsating DC, and the associated capacitor, C2, is used to help minimize EMI (electromagnetic interference) and also provide a return path for the tank. The control circuit produces a variable duty cycle pulse train to drive the gate of the IGBT. The greater the duty cycle, the longer the on-state of the IGBT, and ultimately, the greater the heating. Between the IGBT and the rectified power signal is a parallel resonant tank circuit comprised of C1 and L1. The inductor is comprised of a series of loops of large gauge wiring or copper tubing embedded in the cook surface, typically under a glass or ceramic top. The resonant frequency of the tank is tuned to the frequency of the controlling pulse train. This will maximize the tank current and thus produce a more powerful magnetic field. The switching frequency is usually placed just above the range of human hearing to avoid audible microphonics.46 Values in the range of 20 kHz to 30 kHz are typical, and the base frequency may change as the heat demand changes. For example, to minimize switching losses, the controlling frequency might start at 30 kHz for modest heating and decrease to 20 kHz for maximum heating.

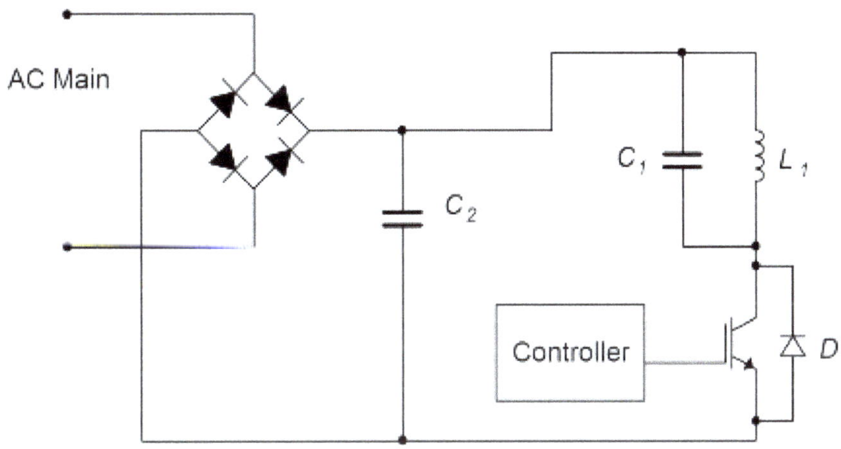

Figure 15.11 Simple inductive heater.

From the cook's perspective there is no change between using the inductive cook top and an ordinary electric cook top using resistive heating elements: The cook places the pot or pan on the surface, under which lies the coil. A heat level control knob is provided for them to adjust the heat intensity. To their advantage, when they remove the pot or pan, the cooking surface itself will not be as hot as an ordinary cook top.

From the designer's perspective, the heat control knob simply changes the duty cycle of the controlling pulse train (and optionally, its frequency, as mentioned previously). Other refinements might include sensing whether or not a vessel is on the cook top and throttling back control if nothing is detected. Finally, an even simpler system could switch the IGBT on and off at a much slower rate (think in terms of seconds) to greatly reduce switching losses but this runs the risk of heat cycling if the pots and pans used are of very light gauge construction (i.e., their thermal time constant will be faster).

DC-to-AC Inversion

There are many instances where we wish to derive an AC voltage from an existing DC voltage. Examples include an uninterruptible power supply (UPS) that would draw current from a battery and deliver standard AC power when there is a disruption in the power grid, and the need to operate electronic devices designed for the home in a remote location. This process is known as DC-to-AC inversion.

The simplest method to create AC from DC is to just "chop" the DC at the desired line frequency and then scale it, that is, feed the DC into a simple IGBT switch which will produce a square wave and then feed the square wave into a transformer to arrive at the desired voltage. The obvious problem with this technique is that the AC signal will not be a nice, smooth sine wave, but rather, a distorted square wave.

Unfortunately, for many electronic components this will present a challenge to their power supply circuits. A possible refinement involves making a step-wise approximation of a sine wave but this is still not ideal.

A more accurate scheme involves synthesizing a sine wave via PWM. We still chop the DC, but now the action is performed at a higher frequency and with a varying duty cycle such that, when the output is averaged, we arrive at a sine wave. A block diagram of this scheme is shown in Figure 15.12.

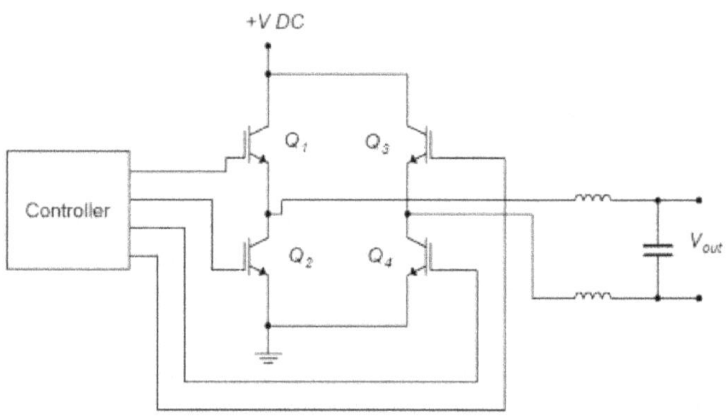

Figure 15.12 DC-to-AC inverter (anti-parallel diodes not shown).

The controller generates a set of PWM signals to drive the four IGBTs configured as a full-bridge. The bridge output is then fed to a balanced *LC* filter that removes the high frequency PWM components, leaving a smoothed sine wave. This signal can be used as is, or fed into a step-up transformer if the desired AC voltage is greater than the starting DC voltage. An example of this would be the need to supply a device designed to run on 120 VAC from the nominal 12 VDC system found in a car.

There is a side item worthy of mention here. Some devices derive timing signals from the line frequency (a classic example is an electronic alarm clock/radio). This is possible because the power generation utility monitors this frequency with great accuracy. If the controller shown in Figure 15.12 does not produce an accurate base frequency, then that clock/radio will not tell time accurately.

Motor Control

IGBTs can be used to control the speed of electric motors. The configuration of the control circuit will depend on the kind of motor being controlled. In simplest terms, the speed of a DC motor is controlled by the voltage applied to it: the higher the voltage, the higher the speed. In contrast, the speed of an AC motor depends on the frequency of the applied source (they are proportional).

Controlling a DC motor is a straightforward situation. If all we need to do is start and stop the motor, the IGBT can be inserted in series with the motor and used as a switch to open and close the circuit. Being solid-state, the IGBT has numerous advantages over a mechanical switch or relay including long-term reliability and simplicity of the control circuit. To alter the speed, the IGBT can be controlled via PWM. This is illustrated in Figure 15.13.

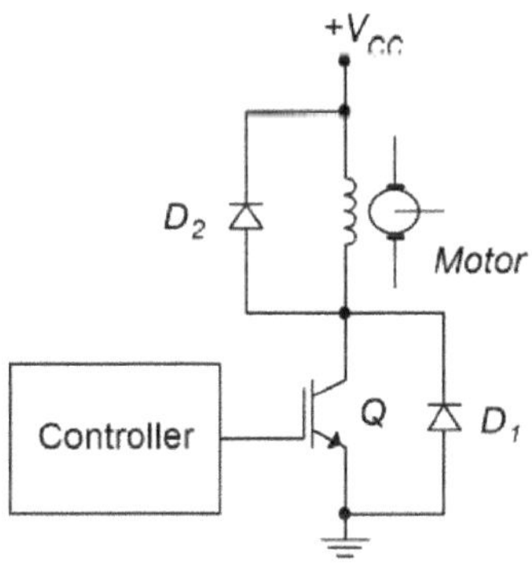

Figure 15.13 DC motor control via IGBT.

D1 and D2 are flyback or snubber diodes used for protection from inductive current spikes caused by the motor's current being switched on and off. As noted earlier, some IGBTs are co-packaged with an anti-parallel diode (D1).

To vary the motor's speed, the controller produces a PWM drive signal. The smaller the duty cycle of this pulse train, the lower the average applied voltage to the motor, and therefore the slower its speed. The base frequency of the PWM signal does not have to be particularly high in this scenario; a few hundred hertz may prove sufficient. This will help minimize switching losses.

For an AC motor the situation is a little more complex. One approach is to use the PWM technique explained under the DC-to-AC converter section. The difference is that the power source would not be DC, but rather, AC. Consequently, we would need to transform the AC power source into a more usable signal and then apply the circuit depicted in Figure 15.12 to power the motor. The controller itself will need to be considerably more sophisticated. In Figure 15.12, the duty cycle is continuously changed such that the "area under the curve" approximates a sine wave. Eventually, the pattern will repeat itself for subsequent cycles of the sine wave. In other words, the rate at which the pattern repeats is the sine wave's period. In the DC-to-AC inverter application, this rate never changes because we need a constant output frequency (e.g., 60 Hz). In the AC motor control application, such is not the case.

This repetition rate needs to be adjustable because that's what controls the motor's speed. One way to do this is to simply increase the base frequency of the PWM pulse train. This method is simple and direct but has the disadvantage of creating more transient edges per unit time and therefore tends to increase switching losses. An alternate approach is to keep the base frequency constant and instead alter the duty cycle pattern. This helps minimize the switching loss issues but has the disadvantage of requiring a more complex control circuit and possibly producing a lower quality sine wave at higher output frequencies.

DC-to-DC Conversion

Our final application is DC-to-DC up-conversion, that is, producing a new DC voltage that is higher than the original source and also capable of high output current (voltage doublers and triplers can be made from diode/capacitor lattices but they are not designed to deliver high, continuous currents). Applications requiring up- conversion include photovoltaic systems (i.e, combining the outputs of several solar panels and tying them into the power system) and high output car audio systems. Specifically, the nominal 12 volt automotive power system is insufficient to supply an amplifier intended to deliver hundreds or even thousands of watts to a subwoofer. The 12 volt source will need to be increased, perhaps by a factor of ten, to achieve the desired output levels.[47]

As we have already seen how DC can be translated into AC via an inverter, it is possible to simply rectify and filter the new AC, producing a higher DC level. This scheme is illustrated in Figure 15.14.

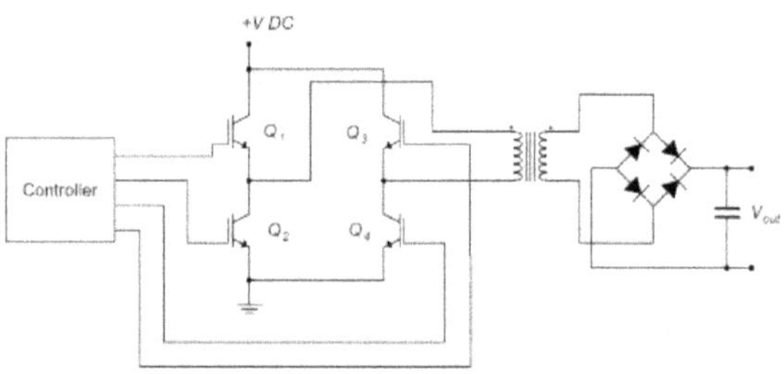

Figure 15.14 DC-to-DC conversion via transformer (anti-parallel diodes not shown).

The transformer pictured here will need to be a step-up variety in order to achieve the desired output voltage level. Note that the AC generation side does not have to produce a particularly nice sine wave nor does it have to be at the usual line frequency. In fact, increasing the frequency will likely result in reduced sizes for the transformer and filter capacitors.

A completely different approach is to use a switching regulator. Switching regulators use a feedback control system to generate a very stable output voltage by comparing it to a reference voltage. They can be configured in step-down, step-up or polarity inversion forms.[48] In this case, we can use the step-up, or *boost*, form. An example is shown in Figure 15.15.

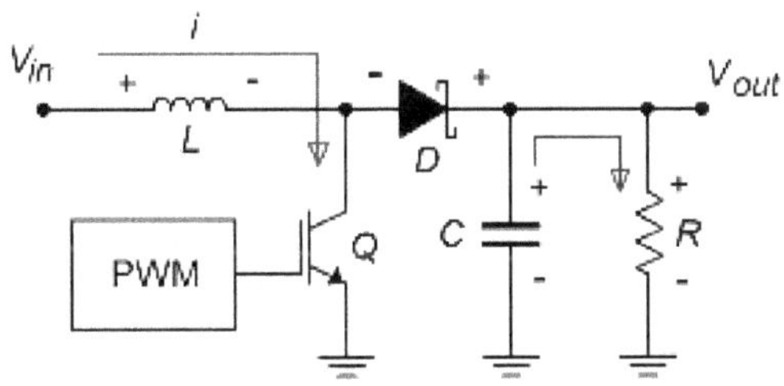

Figure 15.15a Step-up switching regulator, on-state.

As with the other applications presented so far, the IGBT is being used as a switch. Figure 15.15a illustrates the on-state of the IGBT. During this phase, current is drawn through the inductor, L, storing energy in the associated magnetic field. The reverse- biased Schottky diode, D, isolates this section from the output section, where C is delivering the load voltage and current.

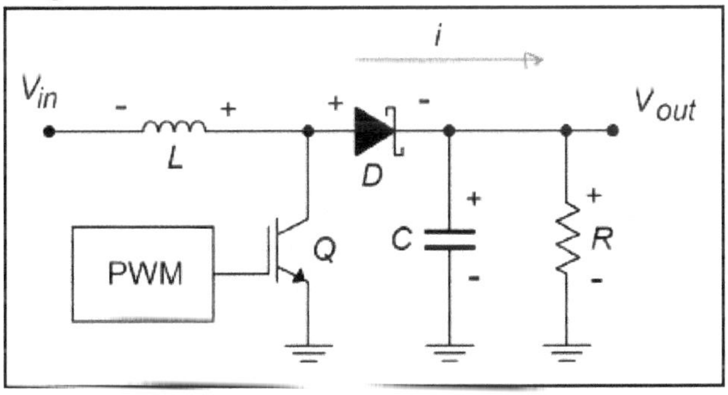

Figure 15.15b Step-up switching regulator, off-state.

The off-state is depicted in Figure 15.15b. During this portion, the inductor discharges and appears as a source. As it is in series with the input voltage, the voltage at the output will equal the inductor's voltage plus the input voltage. Also, during this phase the capacitor is recharged, ready for the next on-state of the IGBT. Note that a Schottky diode is used here because it exhibits fast switching times and a low forward voltage drop. Switching frequencies for these circuits tend to be high (100 kHz and up are common) as that minimizes the sizes of the inductor and capacitor.

The Insulated Gate Bipolar Transistor, or IGBT, can be thought of as a combination of a power BJT and a power E-MOSFET. As such, it combines the low on-state conduction losses of the BJT with the relatively easy drive requirements of the E- MOSFET. The IGBT is available in two variants; the PT, or punch through, and the NPT, or non-punch through types. The PT type includes an N+ buffer layer in its construction and this endows the device with faster switching speed and lower on- state losses.

The IGBT's characteristic curves tend to echo that of the E-MOSFET. Conduction does not begin until the gate-emitter voltage exceeds a threshold voltage, $V_{GE(th)}$. From there, the current-voltage characteristic follows a square-law trajectory, and at sufficiently high current levels it can be approximated as a straight line. The IGBT exhibits a negative temperature coefficient of trans-conductance, like the MOSFET, making it less prone to thermal runaway and current hogging issues. A family of collector curves (i.e., V_{CE} vs. I_C) shares attributes with BJT collector curves and MOSFET drain curves. The curves echo the same overall shape, starting with a section where current rises rapidly compared to voltage, and then leveling out into a constant current region. The initial region of rapid change is somewhat drawn out as it is in the MOSFET. Also, the entire set of curves is displaced positively by about a volt, rather than current increasing immediately from the origin.

In general, the IGBT offers higher voltage, current and power capability than the power E-MOSFET although it lags behind in switching speed. Further, switching times for the on- and off-state are asymmetrical. Compared to the power BJT, the IGBT tends to be more expensive. Consequently, power E-MOSFETs tend to be favored at low and moderate power levels when high switching speeds are needed and BJTs tend to be preferred when cost is a major component in more modest designs. As such, IGBTs find use across a range of applications including power inverters, uninterruptible power supplies, induction heaters, solar power systems, motor controllers and so forth.

CONCISE SUMMARY

- The Insulated Gate Bipolar Transistor, or IGBT, can be thought of as a combination of a power BJT and a power E-MOSFET. As such, it combines the low on-state conduction losses of the BJT with the relatively easy drive requirements of the EMOSFET. The IGBT is available in two variants; the PT, or punch through, and the NPT, or non-punch through types. The PT type includes an N+ buffer layer in its construction and this endows the device with faster switching speed and lower onstate losses.
- The IGBT's characteristic curves tend to echo that of the E-MOSFET. Conduction does not begin until the gate-emitter voltage exceeds a threshold voltage, VGE(th). From there, the current-voltage characteristic

follows a square-law trajectory, and at sufficiently high current levels it can be approximated as a straight line. The IGBT exhibits a negative temperature coefficient of trans-conductance, like the MOSFET, making it less prone to thermal runaway and current hogging issues. A family of collector curves (i.e., VCE vs. IC) shares attributes with BJT collector curves and MOSFET drain curves. The curves echo the same overall shape, starting with a section where current rises rapidly compared to voltage, and then leveling out into a constant current region. The initial region of rapid change is somewhat drawn out as it is in the MOSFET. Also, the entire set of curves is displaced positively by about a volt, rather than current increasing immediately from the origin.

- In general, the IGBT offers higher voltage, current and power capability than the power E-MOSFET although it lags behind in switching speed. Further, switching times for the on- and off-state are asymmetrical. Compared to the power BJT, the 395 Figure 15.15b Step-up switching regulator, off-state. IGBT tends to be more expensive. Consequently, power E-MOSFETs tend to be favored at low and moderate power levels when high switching speeds are needed and BJTs tend to be preferred when cost is a major component in more modest designs. As such, IGBTs find use across a range of applications including power inverters, uninterruptible power supplies, induction heaters, solar power systems, motor controllers and so forth.

FOOTNOTE

CHAPTER-1

1. It is worth noting that the construction of an integrated circuit does not involve the creation and interconnection of millions or billions of single discrete transistors. Instead, the manufacturing process builds all of the transistors simultaneously, rather like a layer cake.
2. An ion is an atom or molecule that does not have a neutral net charge, i.e., the numbers of protons and electrons are not equal. If it loses electrons, resulting in a net positive charge, it is called a cation. If it gains electrons resulting in a net negative charge it is called an anion.

CHAPTER-2

3. They say it has 1001 uses but this ain't one of them
4. Cathode is often denoted by a k. This is likely due to the word's Greek root, kathodos.
5. The I-V plot is not a straight line (linear) and the forward and reverse quadrants are not identical (bilateral).
6. The dynamic resistance of a PN junction may be approximated as 26 mV/Ijunction. This will be shown in an upcoming chapter.
7. Although they are called Zener diodes, they rely on either the Zener or avalanche effects, depending on the magnitude of the voltage.
8. There are advantages to using the infrared over the visible spectrum for this application. It tends to be less sensitive to room lighting conditions and there are no potentially annoying visible flashes of light coming from the remote.
9. It is interesting to observe that the human visual system operates over a frequency range of less than 2:1 while the human auditory system operates over a frequency range of about 1000:1 (20 hertz to 20,000 hertz). If human hearing had a range equivalent to that of our sight, we'd hear less than a full

octave of pitches in total. In other words, do-re-mifa-sol-la-ti-do would end at ti and anything beyond would be inaudible. In such a case one thing is certain: piano keyboards would be much shorter.

10. As a side note, depending on their construction some LEDs can be used as crude photodiodes. Although they are not optimized for this use it can be entertaining to shine a light on an LED and watch it produce a voltage.
11. The mechanical version would require a rotary-style adjustable capacitor connected to some form of small motor or solenoid to move the capacitor plates. While this can work at lower frequencies, if rapid changes are needed the resulting friction-generated heat may cause this contraption to burst into flames. Generally speaking, this is not something we want our circuits to do.

CHAPTER-3

12. If you're wondering why we don't just use DC distribution instead in order to "cut out the middle man", the reasons are manifold. First, it is generally more efficient to distribute power via AC rather than DC. Second, even if DC is available, it may not be at the amplitude the circuitry requires. Therefore some form of DC-to-DC conversion would be needed. Depending on the application, this can turn out to be more expensive than AC-to-DC conversion.
13. As you might guess, this is not particularly efficient because even when the load demand is nil, the Zener diode is still drawing current from the transformer. An improved circuit may include a bipolar transistor, as examined in Chapter 4. For details on more sophisticated techniques to regulate voltage, see Fiore, J, Operational Amplifiers and Linear Integrated Circuits: Theory and Application, another free OER text.
14. We will take a closer look at amplifier clipping in the chapters that cover power amplifiers.
15. Or eleven, if you have a custom Spinal Tap amplifier.

CHAPTER-4

16. Homer says, "Mmm, NPN layer cake sandwich..."
17. Both being equally effective.
18. It is also known as a commutating diode, clamp diode, flyback diode and by a host of other names. But as Shakespeare said, "A snubbing diode by any other name would clamp a flyback voltage as well". Or something like that.
19. This is in reference to the old phrase "pulling yourself up by your bootstraps". To be honest, that saying never made sense to this author and all that ever happened when I tried to do it was that my arms got tired.

CHAPTER-6

20. To be strictly accurate, the residual consists of the harmonics plus any noise produced by the amplifier. Therefore it is more accurate to refer to this as a THD+noise spec.
21. 20 Hz – 20 kHz is the range of frequencies heard by a typical healthy young human.
22. Decidedly not hi-fi, but do we really need high fidelity to call-in a take-out order?

CHAPTER-7

23. The values plotted along the current axis are typical of a generic device and do not represent the current values for all BJTs.
24. Technically, this value is called the device's trans-conductance and is denoted as gm. We shall be seeing this again in upcoming work.
25. The obvious question is, "How do we get both high gain and low distortion?" One answer is to use multiple low gain stages in cascade.
26. Is "funsies" a real word? It is if we all agree that it is. Besides, if it was an imaginary word, we'd spell it "j funsies".
27. Don't believe it? Just try screaming into one and listen to what comes out of the guitar amp.
28. For details on alternate methods, see Fiore, J, Operational Amplifiers and Linear Integrated Circuits: Theory and Application, another free OER text.
29. This circuit does use emitter bypass capacitors so the DC gain will be less than the AC gain. In that sense we might say that this amplifier is not fully DC coupled.

CHAPTER-8

30. This is the case if the AC and DC load lines are identical. This is atypical. Consequently, the power supply will tend to be larger than twice VCEQ which makes the situation even worse.
31. It's a rather odd name given that all loudspeakers are dynamic in some respect. If they weren't they wouldn't produce sound.
32. Adapted from R. H. Small, "Direct Radiator Loudspeaker System Analysis", Journal of the Audio Engineering Society, June, 1972.

CHAPTER-9

33. Think of a car engine: How much sense would it make to have an engine with a 6000RPM maximum run at 3000 RPM when you're sitting motionless at a red light?

34. Of course, the dynamic resistance is a function of the current flowing through the diodeso it will fluctuate as the signal changes. This is best thought of as a distortion generating mechanism because it will slightly alter the signal that reaches the base. This distortion is most likely orders of magnitude smaller than the notch distortion the diode mitigates, so it's a good trade.
35. Such as bits of wire, aluminum foil, small bolts or screws, etc. Just ask any seasoned repair technician.
36. Not to be confused with the quasi-complimentary output of which we have only partially nice things to say.
37. OK, not huge, but probably still loud enough to obscure the sirens of emergency vehicles until they're right behind you.
38. Oopsy!

CHAPTER-10

39. As evidenced in the Shockley equation, Equation 2.1.
40. We could add a third axis for k and plot a surface, and while it might be pretty, a 3D plot like this rendered onto a 2D surface, such as a page in a textbook, is of marginal utility.

CHAPTER-11

41. In fact, we can say that 1/r'e is gm for a BJT.
42. rDS(on) can be as little as a few milliohms for specialized high power MOSFETs (Chapters 12 and 13).

CHAPTER-12

43. Which brings to mind the old question of what to store a universal solvent in.

CHAPTER-14

44. For an audio amplifier, it is important that these components do not reach the loudspeaker. Even though they are beyond the range of human hearing, they can damage loudspeaker sub-components and, at the very least, present an extra power dissipation burden to them. Other kinds of loads may not be effected by the harmonics and filtering may not be needed.
45. Yes, it is labeled VSS in spite of the fact that it's connected to the drain of the P-channel device. It's a matter of consistency with other circuits. "A rose by any other name" and all that...

CHAPTER-15

46. Granted, it may still fall within the hearing range of your dog, so don't be surprised if your border collie prefers a gas cook top to make a balsamic reduction.
47. In the process, the current demand will be increased greatly as well, perhaps beyond the capabilities of the vehicle's alternator (which will also require upgrading), but these are the prices one must pay if one desires very high sound pressure levels in what is arguably the worst acoustical environment in which to listen to music. Of course, we should also admit that the act of critically listening to and enjoying music may not be the point of such an exercise.
48. For details on switching regulators, see Fiore, J, Operational Amplifiers and Linear Integrated Circuits: Theory and Application, another free OER text.

APPENDIX-A

MANUFACTURER'S DATA SHEET LINKS

These links are current as of Winter, 2019 and are listed here for convenience. Always check the manufacturer's site for the latest information before beginning any new design. Note that some devices are available from several different manufacturers.

DIODES

1. 1N4002rectifier: https://www.onsemi.com/pub/Collateral/1N4001-D.PDF 1N4148/1N914 switching: https://www.onsemi.com/pdf/datasheet/1n914-d.pdf NZX5V1B Zener: https://assets.nexperia.com/documents/data-sheet/NZX_SER.pdf

LED

2. LED: https://www.cree.com/led-components/media/documents/ds-C566D-1275.pdf

TRANSISTORS

3. 2N3904 NPN BJT: https://www.onsemi.com/pub/Collateral/2N3903-D.PDF
4. 2N306 PNP BJT: https://www.onsemi.com/pub/Collateral/2N3906-D.PDF
5. J112 N JFET: https://www.onsemi.com/pub/Collateral/J111-D.PDF
6. 2N3055 Power NPN BJT: http://www.onsemi.com/pub_link/Collateral/2N3055-D.PDF
7. FDMS86180 Power E-MOSFET: https://www.onsemi.com/pub/Collateral/FDMS86180-D.pdf
8. IRF7201 Power E-MOSFET: http://www.infineon.com/dgdl/irf7201pbf.pdf?
9. fileId=5546d462533600a4015355f13efb1ada
10. FGH50T65SQD IGBT: https://www.onsemi.com/pub/Collateral/FGH50T65SQD-D.PDF
11. IRGPC40K IGBT: http://www.irf.com/product-info/datasheets/data/irgpc40k.pdf

APPENDIX-B

STANDARD COMPONENT SIZES

Passive components (resistors, capacitors and inductors) are available in standard sizes. The tables below are for resistors. The same digits are used in subsequent decades up to at least 1 Meg ohm (higher decades are not shown). Capacitors and inductors are generally not available in as many standard values as are resistors.

Capacitors below 10 nF (.01 μF) are usually available at the 5% standard digits while larger capacitances tend to be available at the 20% standards.

5% and 10% standard values,, EIIA E24 and EIIA E12

10% values (EIA E12) are bold

20% values (seldom used) are every fourth value starting from 10

(i.e., every other 10% value)

10	11	**12**	13	**15**	16	**18**	20	**22**	24	**27**	30
33	36	**39**	43	**47**	51	**56**	62	**68**	75	**82**	91

5% and 10% standard values,, EIIA E24 and EIIA E12

2% values (EIA E48) are bold

10.0	10.2	10.5	10.7	11.0	11.3	11.5	11.8	12.1	12.4	12.7	13.0
13.3	13.7	14.0	14.3	14.7	15.0	15.4	15.8	16.2	16.5	16.9	17.4
17.8	18.2	18.7	19.1	19.6	20.0	20.5	21.0	21.5	22.1	22.6	23.2
23.7	24.3	24.9	25.5	26.1	26.7	27.4	28.0	28.7	29.4	30.1	30.9
31.6	32.4	33.2	34.0	34.8	35.7	36.5	37.4	38.3	39.2	40.2	41.2
42.2	43.2	44.2	45.3	46.4	47.5	48.7	49.9	51.1	52.3	53.6	54.9
56.2	57.6	59.0	60.4	61.9	63.4	64.9	66.5	68.1	69.8	71.5	73.2
75.0	76.8	78.7	80.6	82.5	84.5	86.6	88.7	90.9	93.1	95.3	97.6

APPENDIX-C

"""There iiss life beyond the cymbals..""

-BillBruford

After many years of observing the food ingestion habits of college students and faculty, it can be said that many do not cook, bake and eat food so much as they consume prepackaged *food-like substances* of dubious nutritional quality and limited flavor. There is no need for this as basic cooking is not difficult, and besides, it gives **you** control of what goes inside your body. Here is an excellent bit of snack-food: tasty, nutritious and relatively easy to make. And when you're done, you can feel good that you didn't drop a bunch of cash on some assembly-line manufactured excuse for sustenance that's loaded with saturated fat, refined sugar, salt and who-knows-what-else. Also, your cooking challenged friends will think you're a genius.

AUTUMN BREAK

Autumn bread is a variation on banana bread, but is very low in fat. If you add the optional nuts, the fat content goes up but it's not saturated fat or trans fat as in butter, margarine, or shortening. Walnuts in particular are a decent source of alpha-linolenic acid, a "good fat". Autumn bread is fairly dense and moist. It definitely *sticks with you.* For variations, you could swap one of the bananas for another apple if desired. In this case you might consider adding a little nutmeg. Also, if you like it less sweet, you can drop the honey back to 1/3 cup. For many people, there are sufficient raisins in the raisin bran but an extra half cup has been added "just because". In fact, extra raisins will help sweeten the flavor and that's another reason to cut back on the honey. Actually, don't bother to measure out the raisins and walnuts, just eyeball them and add what looks good at the moment. Cooking should be fun and a little bit experimental. "t" = teaspoon, "T" = tablespoon, "C" = 8 ounce cup.

3 egg whites
3 overripe mashed bananas
1 medium apple, diced
3/4 C plain non-fat yogurt

2 T molasses
1/2 C honey
1 t cinnamon

1 t vanilla
1/2 C walnuts, almonds, or pecans; or combination thereof (optional)
1/2 C raisins
1/2 C raisin bran

1/2 C flour (try using 50/50 whole wheat/all-purpose)

1/2 t baking soda

2 1/2 t baking powder

Preheat oven to 350° and coat a 9x5 loaf pan with non-stick spray. Make sure that was 350°F and not 350°C; otherwise, prepare to eat carbon chunks when this comes out of the oven.

In a large bowl beat egg whites for about a minute by hand, then add next nine ingredients (through raisins), mixing well. Make sure you do this in a large bowl as doing it directly on the kitchen counter probably will be messy. Add raisin bran, mixing to make sure it's all wet. Let stand 10 minutes or so. While you're waiting, combine flours, baking soda and baking powder in a second bowl. Go back to first bowl and mix it around to make sure that the raisin bran is broken up. If it's not broken up, tell it a funny story and see if that breaks it up. Add the flour mixture to the wet bowl in batches, mixing thoroughly. If you don't know how to spell "thoroughly", just do the best you can. Now, grab the loaf pan. If you sprayed the *outside* of the pan instead of the inside, discard it, get another pan, and only spray the *inside* this time. Dump the contents of the bowl (not the empty one, because that would just be confusing) into the loaf pan and pop it in the oven (the loaf pan, not the bowl). Bake 60 – 65 minutes or until toothpick comes out clean when inserted in center of pan. Cool on rack for 15 minutes. You might wish to let the loaf cool and then cover it in plastic wrap for a couple of hours as this softens the crust.

This makes one big "gigundo" size loaf, enough to feed a family of four for a week, or one hungry distance runner for the afternoon.

INDEX

A

Accuracy 30, 332

B

C

www.ingramcontent.com/pod-product-compliance
Ingram Content Group UK Ltd.
Pitfield, Milton Keynes, MK11 3LW, UK
UKHW050948290926
15470UKWH00005B/1705

9 781787 152717